Free Energy Calculations in Rational Drug Design

Free Energy Calculations in Rational Drug Design

Edited by

M. Rami Reddy

and

Mark D. Erion

Metabasis Therapeutic, Inc.
San Diego, California

Kluwer Academic / Plenum Publishers
New York, Boston, Dordrecht, London, Moscow

Library of Congress Cataloging-in-Publication Data

ISBN: 0-306-46676-7

233 Spring Street, New York, N.Y. 10013

http://www.wkap.nl/

10 9 8 7 6 5 4 3 2 1

A C.I.P. record for this book is available from the Library of Congress

Printed in the United States of America

In Memory of
Peter Andrew Kollman, Ph.D.
(7/26/44 - 5/25/01)

Few scientists have had an impact on their chosen research field as much as Peter Kollman on computer-aided molecular modeling and its application to problems in both chemistry and biology. Sadly, Peter died in May leaving behind many friends and scientific collaborators from around the world.

Peter was born in Iowa City, Iowa in 1944. He received his BS in Chemistry from Grinnell College in 1966 and his doctorate from Princeton University in 1970. He spent a year at Cambridge University as a postdoctoral fellow before joining the faculty of the University of California at San Francisco in 1971. At UCSF, Peter rapidly became a well-recognized leader in computational chemistry through his publications and lectures about his numerous insightful discoveries. In 1980 he was awarded a full professorship and later he became the Associate Dean for Academic Affairs in the School of Pharmacy. During his tenure at the University, Peter trained 17 Ph.D. candidates and over 60 post-doctoral fellows as well as a multitude of visiting scholars.

Peter was instrumental in the design and development of AMBER, a suite of molecular mechanics programs currently used in over 300 laboratories around the world conducting research in the fields of biophysics and pharmaceutical chemistry. Using AMBER, Peter and his team showed that free energy calculations provide accurate predictions and valuable insight into reaction mechanisms, protein structure-function and drug design.

Peter was a unique, vibrant and friendly individual who exhibited enormous enthusiasm for both science and life.

He will be missed.

CONTRIBUTORS

Atul Agarwal
Wyeth-Ayerst Research/AHP
401 N. Middletown Rd.
Pearl River, NY 10965

Krzysztof Appelt
Agouron Pharmaceuticals, Inc.
3565 General Atomics Ct.
San Diego, CA 92121

Johan Åqvist
Dept. of Cell & Molecular Biology
Uppsala University
Biomedical Center
Box 596
SE-751 24 Uppsala, Sweden

Shinichi Banba
Mitsui Chemical, Inc.
580-32 Nagura
Sodegaura-City
Chiba, Japan 299-0265

J. Phillip Bowen
Department of Chemistry
Computational Center for Molecular Structure and Design
University of Georgia
Athens, GA 30602

Charles L. Brooks, III
The Scripps Research Institute
Department of Mol. Biology
10550 North Torrey Pines Road
La Jolla, CA 92037

Frank K. Brown
R.W. Johnson, PRI
1000 Route 202
P.O. BOX 300
Raritan, NJ 08869

Christopher J. Cramer
Department of Chemistry &
Supercomputer Institute
University of Minnesota
Kolthoff and Smith Halls
207 Pleasant Street SE
Minneapolis, MN 55455

Peter L. Cummins
Department of Biochemistry and
Molecular Biology
John Curtin School of Med. Research
Australian National University
P.O. Box 334
Canberra, ACT 2601 Australia

Oreola Donini
Kinetek Pharmaceuticals, Inc.
1779 75th Ave. W.
Vancouver, BC V6P 6P2 Canada

Mark D. Erion
Metabasis Therapeutics, Inc.
9390 Towne Centre Drive
San Diego, CA 92121

Jill E. Gready
Department of Biochemistry and
Molecular Biology
John Curtin School of Med. Research
Australian National University
P.O. Box 334
Canberra, ACT 2601 Australia

Zhuyan Guo
Schering-Plough Research Institute
2000 Galloping Hill Road
Kenilworth, NJ 07033

Frederick H. Hausheer
BioNumerik Pharmaceuticals, Inc.
8122 Datapoint #1250
San Antonio, TX 78229

Shuanghong Huo
Dept. of Pharmaceutical Chemistry
Univ. of California at San Francisco
Box 0446, S-924
San Francisco, CA 94143

William L. Jorgensen
Department of Chemistry
Yale University
P.O. Box 208107
New Haven, CT 06520-8107

Paul M. King
School of Biological and
Chemical Sciences
Birkbeck College
29 Gordon Square
London WC1H 0PP U.K.

Peter A. Kollman
Dept. of Pharmaceutical Chemistry
Univ. of California at San Francisco
Box 0446, S-924
San Francisco, CA 94143

Bernd Kuhn
Dept. of Pharmaceutical Chemistry
Univ. of California at San Francisco
Box 0446, S-924
San Francisco, CA 94143

Tai-Sung Lee
Molecular Simulations, Inc.
9685 Scranton Road
San Diego, CA 92121

John Marelius
Dept. of Cell & Molecular Biology
Uppsala University
Biomedical Center
Box 596
SE-751 24 Uppsala, Sweden

J. Andrew McCammon
Howard Hughes Med. Institute
Dept. of Chemistry & Biochemistry
Department of Pharmacology
University of California at San Diego
9500 Gilman Drive 0365
La Jolla, CA 92093-0365

David Pearlman
Vertex Pharmaceuticals, Inc.
130 Waverly St.
Cambridge, MA 02139

Albert C. Pierce
Department of Chemistry
Yale University
P.O. Box 208107
New Haven, CT 06520-8107

Daniel J. Price
Department of Chemistry
Yale University
P.O. Box 208107
New Haven, CT 06520-8107

Melissa L. P. Price
Department of Chemistry
Yale University
P.O. Box 208107
New Haven, CT 06520-8107

K. Ramnarayan
ImmunoPharmaceutics, Inc.
11011 Via Frontera
San Diego, CA 92127

B. Govinda Rao
Vertex Pharmaceuticals, Inc.
130 Waverly Street
Cambridge, MA 02139

M. Rami Reddy
Metabasis Therapeutics, Inc.
9390 Towne Center Drive
San Diego, CA 92121

Robert C. Rizzo
Department of Chemistry
Yale University
P.O. Box 208107
New Haven, CT 06520-8107

Suresh B. Singh
Merck Research Laboratories
P.O. Box 2000
Rahway, NJ 07065

U. Chandra Singh
AM Technologies, Inc.
4 Thornhurst Road
San Antonio, TX 78218

Julian Tirado-Rives
Department of Chemistry
Yale University
P.O. Box 208107
New Haven, CT 06520-8107

Pat Metthe Todebush
Department of Chemistry
Computational Center for Molecular
Structure and Design
University of Georgia
Athens, GA 30602

Donald G. Truhlar
Department of Chemistry &
Supercomputer Institute
University of Minnesota
Kolthoff and Smith Halls
207 Pleasant Street SE
Minneapolis, MN 55455

Deping Wang
Department of Chemistry
Yale University
P.O. Box 208107
New Haven, CT 06520-8107

Junmei Wang
Dept. of Pharmaceutical Chemistry
Univ. of California at San Francisco
Box 0446, S-924
San Francisco, CA 94143

Graham A. Worth
Theoretical Chemistry
Dept. Chemistry
King's College London
Strand WC2R 2LS U.K.

Preface

The holy grail of structure-based drug design is the design and rapid identification of highly potent and specific enzyme inhibitors using only computational methods and protein structural information to determine ligand binding affinities. Success depends upon the accuracy of the calculated binding affinities, which until recently was severely compromised by limitations in computer power and the approximations associated with the force field potential energy equations used to describe the ligand binding energetics. Advances in computer speed in the 1990s led to the inclusion of additional terms in the energy equations and ultimately to an increase in calculation accuracy. The aim of this book is to provide computational chemists and medicinal chemists with a comprehensive review of the methods used to calculate free energies and of the studies applying these methods to drug design.

The potential of free energy calculations for predicting inhibitor binding affinities was first realized in 1986 following calculations conducted by Wong and McCammon on two benzamidine inhibitors of trypsin. Since then, numerous studies have appeared in the literature demonstrating the value of FEP calculations in predicting ligand binding affinities and identifying molecular factors that influence substrate binding and catalysis. J. Andrew McCammon's overview of the free energy perturbation (FEP) approach in Chapter one provides a historical perspective for these studies as well as the challenges that lie ahead. While the FEP approach remains the method that consistently generates the most accurate free energies, its high CPU requirements and inability to evaluate compounds that differ significantly in structure, clearly limit the impact and value of FEP calculations on drug design. Accordingly, efforts are on-going to develop faster methods that have the potential to evaluate large compound libraries semi-quantitatively. These methods include the ligand interaction energy approach, λ-dynamics or Chemical Monte Carlo/molecular dynamics (CMC/MD), Molecular Mechanics-Poisson Boltzmann Surface Area (MM-PBSA) and ligand interaction scanning. With these advances, free energy calculations are becoming more common in the design and analysis of potential drug candidates as evidenced by the exponential increase in the number of studies appearing in the literature over the past 10 years.

The background theory that underlies the FEP method as well as the molecular mechanics force fields that relate molecular structure to energy are reviewed in section one of the book. Section two describes the use of free energy calculations for determining molecular properties of ligands, including solvation, as calculated using both implicit and explicit water models, ionization and tautomerization. Section three reviews some of the original calculations that showed that the FEP method produced accurate relative binding free energies for small molecules interacting with macromolecules such as enzymes, as represented by the proteases Thermolysin and Rhizopus pepsin, as well as DNA.

Section four reviews several alternative methods for estimating ligand binding affinities and how these methods are used in ligand design and analysis. The scope and limitations of each method are discussed as well as the advantage of the method relative to FEP. Promising results are reported for the linear interaction energy method as well as the MM-PBSA method. These methods, as well as methods designed to screen multiple ligands simultaneously or to scan binding site interactions, are expected to enable rapid analysis of the ligand and binding site SAR and therefore to be useful in drug design. The final chapter of this section describes the combining of quantum mechanical calculations with molecular mechanics for predicting reaction free energy profiles, which are often useful in drug design since they can provide valuable insight into the enzyme catalytic mechanism, the transition state structure stabilized by the enzyme and possible compounds that could act as high affinity transition state mimetics.

Studies using free energy calculations for the design and analysis of potential drug candidates are reviewed in section five. The chapters in this section cover drug discovery programs targeting fructose 1,6-bisphosphatase (diabetes), COX-2 (inflammation), SRC SH2 domain (osteoporosis and cancer), HIV reverse transcriptase (AIDS), HIV-1 protease (AIDS), thymidylate synthase (cancer), dihydrofolate reductase (cancer) and adenosine deaminase (immunosuppression, myocardial ischemia).

Overall, this book provides for the first time an extensive overview of the scope and limitations of free energy calculations and their application to rational drug design. The authors contributing to the book are well-recognized leaders of this field of research representing academic institutions and pharmaceutical companies located in the U.S., Europe, Australia, and Asia. The editors would like to thank the authors for their chapters and their input. The editors would also like to thank Ms. Lisa Weston and Ms. Juliette Jomini for their efforts formatting the chapters and assembling the book. Last, we are grateful to Mr. Kenneth Howell and Kluwer Academic/Plenum Publishers for their enthusiasm and support for this project.

Mark D. Erion
M. Rami Reddy

June 2001

CONTENTS

5. Relative Solvation Free Energies Calculated Using Explicit Solvent

Atul Agarwal, Frank K. Brown, and M. Rami Reddy

6. Tautomerism and Ionisation Studies Using Free Energy Methods

Graham A. Worth and Paul M. King

SECTION THREE: LIGAND BINDING

SECTION FOUR: LIGAND DESIGN AND ANALYSIS

SECTION FIVE: DRUG DESIGN CASE STUDIES

Chapter 1

Historical Overview and Future Challenges

J. Andrew McCammon
Howard Hughes Medical Institute, Department of Chemistry and Biochemistry, and Department of Pharmacology, University of California, San Diego, La Jolla, CA 92093

1. INTRODUCTION

The selective binding of molecules to form productive complexes is of central importance to pharmacology and medicinal chemistry. Although kinetic factors can influence the yields of different molecular complexes in cellular and other non-equilibrium environments,[1] the primary factors that one must consider in the analysis of molecular recognition are thermodynamic. In particular, the equilibrium constant for the binding of molecules A and B to form the complex AB depends exponentially on the standard free energy change associated with complexation.

It has long been recognized that if one could compute the standard free energy change of complexation of biologically active molecules, it would be possible both to gain a deeper understanding of the origins of molecular recognition in biology, and to contemplate the "first principles" design of pharmaceuticals and other compounds. Such calculations were attempted, for example, by the Scheraga group as early as 1972,[2] although limitations in computer power did not allow inclusion of solvation or entropic effects in this work. In 1986, Wong and McCammon[3] combined the statistical mechanical theory of free energy with atomistic simulations of solvent and solutes to calculate the relative standard free energy of binding of different small inhibitor molecules to an enzyme. The necessary statistical mechanical theory had been available for many years. Two new elements were required to make the calculation possible. One was the increased power of computers, which allowed molecular dynamics simulation of the enzyme trypsin in a bath of explicitly represented water molecules. The other was the concept of using thermodynamic cycles to relate the desired relative free energy to that of two nonphysical processes: computational "alchemical" transformations of one inhibitor into another one, in solution and in the binding site.[4]

Subsequent work has shown that free energy calculations that involve systems as large as proteins or other macromolecules can provide usefully accurate results in favorable cases. But, in general, there are difficulties in achieving precise and accurate results with reasonable amounts of computer time, even using current state-of-the-art machines. These difficulties arise primarily from the incomplete sampling of the rough, many-dimensional potential energy surfaces of such systems. Below, I mention several lines of work that hold promise for making free energy calculations faster and more accurate for biomolecular systems. The subsequent chapters in this volume describe some of these lines of work in more detail. Excellent reviews of this work can also be found elsewhere.[5-9]

2. THEORY AND METHODS

For calculations of relative free energies of binding, the theoretical framework outlined by Tembe and McCammon[4] has been used essentially without change. This framework recognizes that brute force calculations of standard free energies of binding will encounter convergence problems related to the dramatic changes in solvation of the binding partners, conformational changes that require physical times longer that those that can be explored by simulation, etc. Tembe and McCammon[4] introduced the use of thermodynamic cycle analyses that allow the desired relative free energies to be computed in terms of "alchemical" transformations, as described above. The advantage is that only relatively localized changes occur in the simulated system, at least in favorable cases.

Calculation of the standard free energy of binding itself can be viewed as a special case of the above, in which one of the pair of ligands contains no atoms.[10] Some care is required to be sure that such calculations yield answers that actually correspond to the desired standard state.[11, 12] Unfortunately, many calculations of free energies of binding have not made appropriate contact with a standard state, so that results in the literature must be interpreted with caution.

It has been mentioned that perhaps the greatest limitation to the precision of free energy calculations to date has been the often-inadequate sampling of a representative set of configurations of the system. Increases in computer power of course increase the "radius of convergence" of such calculations. Such increases come not only from the "Moore's Law" improvements in hardware, but also from algorithmic advances for parallelization and for increasing time steps in molecular dynamics.[13] New methods on the physical/theoretical side have also been developed to speed convergence. One such method is the use of soft-core solute models, so that one simulation can generate an adequate reference ensemble for a family of alchemical changes.[14, 15] The "lambda dynamics" method of Kong and Brooks[16] increases the efficiency of free energy calculations by treating the coupling parameter as a dynamic variable.

More rapid convergence of free energy calculations can also be obtained by replacing part of the system with a simpler model, such as a continuum model for the solvent. This has the advantage of obviating the need for sampling the configurations of this part of the system, and it also reduces the computation time so that longer simulations are possible for the rest of the system. Reasonable agreement has sometimes been obtained with fully atomistic simulations when solvent regions near binding sites have been replaced by continuum.[17, 18] But in view of the important role that specific hydrogen bonds may play, the combination of fully atomistic simulations with subsequent continuum analyses is probably a more reliable procedure.[19] The Kollman group has demonstrated impressive success with this approach to calculations of free energies of binding.[20]

Calculations of relative free energies of binding often involve the alteration of bond lengths in the course of an alchemical simulation. When the bond lengths are subject to constraints, a correction is needed for variation of the Jacobian factor in the expression for the free energy. Although a number of expressions for the correction formula have been described in the literature, the correct expressions are those presented by Boresch and Karplus.[21]

3. OUTSTANDING PROBLEMS

It was noted above that a continuum treatment of the solvent can be helpful, although representing certain solvent molecules explicitly may be necessary. The expressions for handling the free energy contributions in such hybrid models have been derived by Gilson et al.[11]

Two remaining problems relating to the treatment of solvation include the slowness of Poisson-Boltzmann calculations, when these are used to treat electrostatic effects, and the difficulty of keeping buried, explicit solvent in equilibrium with the external solvent when, e.g., there are changes in nearby solute groups in an alchemical simulation. Faster methods for solving the Poisson-Boltzmann equation by means of parallel finite element techniques are becoming available, however.[22-24]

For buried solvent molecules, open ensemble methods should be helpful, although extension of the existing methods to allow for solute flexibility is needed.[18]

It is not uncommon for protons to be taken up or released upon formation of a biomolecular complex. Experimental data on such processes can be compared to computational results based on, for example, Poisson-Boltzmann calculations.[25] There is a need for methods that automatically probe for the correct protonation state in free energy calculations. This problem is complicated by the fact that proteins adapt to and stabilize whatever protonation state is assigned to them during the course of a molecular dynamics simulation.[19] When the change in protonation state is known, equations are available to account for the addition or removal of protons from the solvent in the overall calculation of the free energy change.[11]

4. PROSPECTS

Although challenges remain, and provide fruitful grounds for basic research, it is clear that computational methods for free energy calculations are becoming increasingly useful. Computations are already of sufficient reliability for medium sized molecules such as synthetic host-guest systems, that they are an important tool for interpreting and even correcting experimental data in this area.[7] Recent years have seen growing interest in these methods for protein-small molecule systems, as shown in the following chapters.

Acknowledgments

Work in the author's laboratory is supported in parts by the NSF, NIH, the W. M. Keck Foundation, and the Howard Hughes Medical Institute.

5. REFERENCES

1. J. A. McCammon, Theory of biomolecular recognition, Curr. Op. Struct. Biol. **8**:245 (1998).
2. K. E. B. Platzer, F. A. Momany, and H. A. Scheraga, Conformational energy calculations of enzyme-substrate interactions. II. Computation of the binding energy for substrates in the active site of alpha-chymotrypsin, Int. J. Peptide Protein Res. **4**:201 (1972).
3. C. F. Wong and J. A. McCammon, Dynamics and design of enzymes and inhibitors, J. Am. Chem. Soc. **108**:3830 (1986).
4. B. L. Tembe and J. A McCammon, Ligand–receptor interactions, Comput. Chem. **8**:281 (1984).

5. T. P. Straatsma, Free energy by molecular simulation, in: Reviews in Computational Chemistry, vol. 9, K. B. Lipkowitz and D. B. Boyd, eds., VCH Publishers Inc., New York (1996), pp. 217-309.
6. P. A. Kollman, Advances and continuing challenges in achieving realistic and predictive simulations of the properties of organic and biological molecules, Acc. Chem. Res. **29**:461 (1996).
7. M. L. Lamb and W. L. Jorgensen, Computational approaches to molecular recognition, Curr. Opin. Chem. Biol. **1**:449 (1997).
8. D. A. Pearlman and B. G. Rao, Free energy calculations: Methods and applications, in: Encyclopedia of Computational Chemistry, P. v. R. Schleyer, ed., Wiley, New York (1999), pp. 1036-1061.
9. M. R. Reddy, M. D. Erion, and A. Agarwal, Free energy calculations: use and limitations in predicting ligand binding affinities, in: Reviews in Computational Chemistry, vol. 16, K. B. Lipkowitz and D. B. Boyd, eds., Wiley-VCH Inc., New York (2000), pp. 217-304.
10. W. L. Jorgensen, J. K. Buckner, S. Boudon, and J. Tirado-Rives, Efficient computation of absolute free energies of binding by computer simulations. Application to methane dimer in water, J. Chem. Phys. **89**:3742 (1988).
11. M. K. Gilson, J. A. Given, B. L. Bush, and J. A. McCammon, The statistical–thermodynamic basis for computation of binding affinities: A critical review, Biophys. J. **72**:1047 (1997).
12. J. Hermans and L. Wang, Inclusion of loss of translational and rotational freedom in theoretical estimates of free energies of binding. Application to a complex of benzene and mutant T4-lysozyme, J. Am. Chem. Soc. **119**:2707 (1997).
13. T. Schlick, R. D. Skeel, A. T. Brunger, L. V. Kale, J. A. Board, J. Hermans, and K. Schulten, Algorithmic challenges in computational molecular biophysics, J. Comp. Phys. **151**:9 (1999).
14. H. Liu, A. E. Mark, and W. F. van Gunsteren, On estimating the relative free energy of different molecular states with respect to a single reference state, J. Phys. Chem. **100**:9485 (1996).
15. T. Z. Mordasini and J. A. McCammon, Calculations of relative hydration free energies: a comparative study using thermodynamic integration and an extrapolation method based on a single reference state, J. Phys. Chem. B. **104**:360 (2000).
16. X. Kong and C. L. Brooks, Lambda-dynamics: a new approach to free energy calculations, J. Chem. Phys. **105**:2414 (1996).
17. S. T. Wlodek, J. Antosiewicz, J. A. McCammon, T. P. Straatsma, M. K. Gilson, J. M. Briggs, C. Humblet, and J. L. Sussman, Binding of tacrine and 6–chlorotacrine by acetylcholinesterase, Biopolymers **38**:109 (1996).
18. H. Resat, T. J. Marrone, and J. A. McCammon, Enzyme–inhibitor association thermodynamics: Explicit and continuum solvent studies, Biophys. J. **72**:522 (1997).
19. S. T. Wlodek, J. Antosiewicz, and J. A. McCammon, Prediction of titration properties of structures of a protein derived from molecular dynamics trajectories, Protein Sci. **6**:373 (1997).
20. I. Massova and P. A. Kollman, Combined molecular mechanical and continuum solvent approach (MM-PBSA/GBSA) to predict ligand binding, Perspect. Drug Discov. **18**:113 (2000).
21. S. Boresch and M. Karplus, The Jacobian factor in free energy simulations, J. Chem. Phys. **105**:5145 (1996).
22. M. Holst, N. Baker, and F. Wang, Adaptive multilevel finite element solution of the Poisson-Boltzmann equation I: algorithms and examples, J. Comp. Chem. **21**:1319 (2000).

23. N. Baker, M. Holst, and F. Wang, Adaptive multilevel finite element solution of the Poisson-Boltzmann equation II: Refinement schemes based on solvent accessible surfaces, J. Comp. Chem. **21**:1343 (2000).
24. N. Baker, D. Sept, M. Holst, and J. A. McCammon, The adaptive multilevel finite element solution of the Poisson-Boltzmann equation on massively parallel computers, IBM J. Res. Dev. in press (2001).
25. K. A. Xavier, S. M. McDonald, J. A. McCammon, and R. C. Willson, Association and dissociation kinetics of bobwhite quail lysozyme with monoclonal antibody HyHEL-5, Prot. Eng. **12**:79 (1999).

Section One

Theory

Chapter 2

Free Energy Calculations: Methods for Estimating Ligand Binding Affinities

David A. Pearlman
Vertex Pharmaceuticals, Incorporated, Cambridge, MA 02139-4242

1. INTRODUCTION

Nearly two decades have now passed since the first macromolecular free energy calculations were published.[1-3] These calculations drew on a statistical framework that was first described by Zwanzig years earlier[4], but which until the advent of fast computers, molecular sampling techniques applicable to complex macromolecular systems, and reliable force fields could not be put into practice. Groundbreaking work in the late 1970s demonstrated that a molecular dynamics (MD) approach could be used to perform configurational sampling for complex systems,[5] and by the early 1980s, computers had become fast and cheap enough that such calculations were within the grasp of most well-funded research groups. Extension of MD (or Monte Carlo [MC])[6] sampling to the calculation of free energy differences as per Zwanzig was a natural one. By the mid 1980s, a series of promising and exciting results reported in early free energy studies had sparked a flurry of research in the area.[7,8]

It is not hard to understand the interest. Free energy is the property that dictates almost every physical process. Understand the free energy behavior for any molecular system, and you can reliably predict how that system will behave. Solvation, diffusion, binding, folding, and many other properties that are of critical interest to scientists can all be understood and (more importantly) predicted if we know the underlying free energy profiles. It is not an exaggeration to say that an ability to reliably and rapidly predict these properties in the general case would revolutionize such endeavors as drug design.

Given the general feelings of euphoria that followed the early, promising papers in this field, one can ask what happened to the revolution? The answer, simply, is that free energy prediction turned out to be significantly more difficult than first thought. While the statistical mechanics foundation is straightforward, as is integration with MD or MC sampling, issues related to sufficient sampling and to the adequacy of the force field quickly emerge when performing these calculations.[9-17] With regard to sampling, we know what we need to do, but, outside of select amenable systems, current computer systems (which are many times faster than those used in the early free energy studies) are still orders of magnitude too slow to allow the kind of full conformational space exploration required to perform enough sampling to reliably predict free energy in the general case. We thus confine ourselves to questions that fall within the class of systems for which we can hope to perform the requisite sampling. While this is sub-optimal, there are still many questions of interest that can be addressed. Much of the development in the free energy field over the past couple of decades has been in areas that attempt to better characterize the convergence characteristics of these calculations, and how to best carry them out to optimize the convergence.[18-25] Major improvements have also been made in various procedural areas that make the models and equations used more correct.[9, 14, 26-31]

The tremendously promising results of early calculations in the field have, with hindsight, turned out to have been largely fortuitous. We now know that those calculations, often performed with 10-40 ps of sampling, cannot possibly have yielded the kind of predictability they appeared to offer.[9-17] The good news is that, after two decades of development and nearly unbelievable increases in available computer resources, we can now--for judiciously chosen questions--obtain predictions with quality that is truly as good as suggested by those first publications.

In this chapter, we shall review the various methods and protocols now available to perform free energy calculations.

2. EXACT FREE ENERGY CALCULATIONS

There now exist several methods for predicting the free energy associated with a compositional or conformational change.[7] These can be crudely classified into two types: "exact" and "approximate" free energy calculations. The former type, which we shall discuss in the following sections, is based directly on rigorous equations from classical statistical mechanics. The latter type, to be discussed later in this chapter, starts with statistical mechanics, but then combines these equations with assumptions and approximations to allow simulations to be carried out more rapidly.

The most commonly reported exact free energy simulations are based on the following equation, which can be derived in a straightforward fashion from elementary classical statistical mechanics:

$$\Delta G = G_B - G_A = -RT \ln < e^{-(V_B - V_A)/RT} >_A \qquad (1)$$

G_B and G_A are the free energies of configurations or molecules B and A, respectively, V_B and V_A are the potential energies of configurations or molecules B and A, respectively, R is the universal gas constant, T is the temperature and $<>_A$ means we evaluate the average of the enclosed quantity from a thermodynamic ensemble generated for state A. Here and throughout this article, we use the potential energy V(**x**) in place of the more general Hamiltonian H(**x**, p), making the typical assertion that the momentum contribution to the free energy difference is zero.

The ensemble is generated using either MD (a numerical integration of Newton's equation) or else a Monte Carlo walk. Each step of MD, or each configuration in MC, requires significant computer resources to evaluate, so the amount of sampling that can reasonably be performed using these techniques is limited. Presently, normal simulations are limited, at best, to total simulation times on the nanosecond timescale. Depending on the nature of systems A and B, the amount of sampling we can carry out may be insufficient to properly evaluate the requisite ensemble average. The severity of this problem usually increases as the difference in states A and B grows larger.

In practice, many physically interesting questions result in states A and B that are so different, with corresponding orthogonality between their respective potential surfaces, that it is practically impossible to carry out enough sampling to properly evaluate the ensemble in Equation 1 (Figure 1). For that reason, implementation of Equation 1 is usually carried out by defining a series of non-physical intermediate states that connect the physical states A and B. As we progress among these intermediates, the system gradually begins to look more like B and less like A. Since free energy is a state function, the total free energy can be rigorously calculated as the sum of the free energies between these similar intermediates (for which, presumably, the required ensemble will converge more quickly). In practice, the potential functions $V_A(\mathbf{x})$ and $V_B(\mathbf{x})$ are replaced by $V(\lambda,\mathbf{x})$. λ is a variable introduced to the potential energy function such that $V(\lambda=0,\mathbf{x}) = V_A(\mathbf{x})$ and $V(\lambda=1,\mathbf{x}) = V_B(\mathbf{x})$.

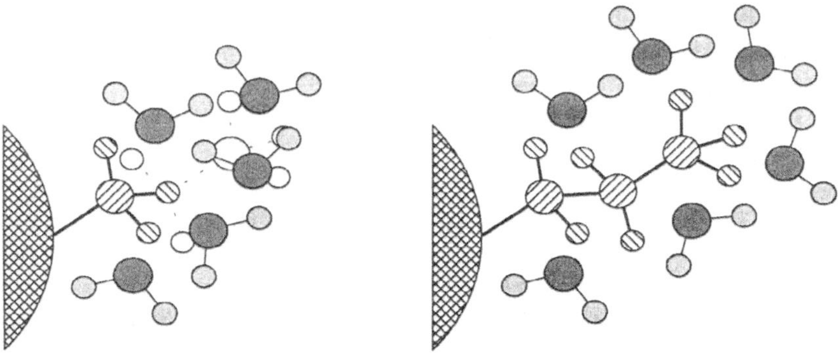

Figure 1. Schematic view of the fundamental difficulty in running a free energy perturbation calculation. The initial endpoint is shown on the left, corresponding to a pendant methyl group attached to the molecular scaffold. On the right is the final endpoint, which corresponds to a propyl group on this scaffold. If we were to carry out this free energy simulation in one window starting with the methyl endpoint, we would need to sample water configurations that are favorable for both groups, but the chances of sampling configurations where the water has moved out to to provide a favorable cavity for the propyl sidechain are vanishingly small. As seen on the left, water molecules will typically overlap the positions that correspond to the second and third carbons (and attached hydrogens) in an ensemble generated using the potential function corresponding to the methyl. Appropriately low energy configurations for the propyl group (right) will be very rare in the methyl ensemble. Atoms with hash lines are carbons and hydrogens. Atoms with no hash lines are "dummy" non-interacting atoms. Bonds to dummy atoms are represented by dashed lines. Atoms with solid shading belong to water molecules.

For example, in the simplest case (though not the case usually used in practice), one could define

$$V(\lambda, \mathbf{x}) = (1-\lambda)\, V_A(\mathbf{x}) + \lambda\, V_B(\mathbf{x}) \qquad (2)$$

With the introduction of the λ factor, Equation 1 is replaced by the equations

$$G_{\lambda(i-1)\rightarrow\lambda(i)} = -RT\ln\left\langle e^{-[V(\lambda(i),\mathbf{x})-V(\lambda(i-1),\mathbf{x})]/RT}\right\rangle_{\lambda(i-1)} \qquad (3)$$

$$\Delta G_{tot} = \sum_{i=1}^{NWINDOW} \Delta G_{\lambda(i-1)\rightarrow\lambda(i)} \qquad (4)$$

where we have used λ(i) to refer to the ith λ point in the series of NWINDOW *points* that starts with λ(0)=0.0 and ends with λ(NWINDOW)=1.0. Each free energy calculation between adjacent λ states is termed a "window." A free energy calculation carried out using Equations 3–4 is usually termed a Free Energy Perturbation (FEP) calculation.

Although historically less common, free energy calculations based on a different equation from classical statistical mechanics have grown in popularity in recent years. These calculations, termed Thermodynamic Integration (TI), are based on the integral

$$\Delta G = \int_0^1 < \frac{dV(\lambda, \mathbf{x})}{d\lambda} >_\lambda d\lambda \qquad (5)$$

where λ has the same meaning as above. In practice, numerical integration is used to evaluate the integral. This requires that the integrand (ensemble) be evaluated at a series of λ intermediates. From these values of the integrand, a method such as the Trapezoidal rule can be used to approximate the integral[32]:

$$\Delta G \approx \sum_0^N \left[< \frac{\partial V(\mathbf{x}, \lambda)}{\partial \lambda} >_{\lambda(i)} + < \frac{\partial V(\mathbf{x}, \lambda)}{\partial \lambda} >_{\lambda(i+1)} \right] \left[\frac{\lambda(i+1) - \lambda(i)}{2} \right] \qquad (6)$$

Both FEP and TI are carried out by systematically varying λ from the initial state 0 to the final state 1. At each λ point, equilibration of the system is performed, followed by data collection to determine the value of the ensemble for the equilibrated system.

Note that the free energy pathway between physical endpoints A and B is divided into a series of λ states for different reasons with FEP and TI. In FEP, we use the λ intermediates to reduce the difference between adjoining states when applying Equation 1. This improves the convergence profile for the required ensemble. In TI, the λ intermediates are required to approximate the continuous integral in Equation 5. The number of intermediates required when using TI will depend on the shape of the accumulated free energy versus λ profile. The greater the variation in the curvature in this profile, the more points that will be required to correctly approximate the required integral.

Another approach to free energy calculations, Slow Growth, has also been employed. Slow Growth is simply the limiting case of either FEP or TI where the number of λ states is extremely large. The assertion is that in such a case, the system will be changing so slowly with each progressive λ state that the ensemble average can be approximated by its instantaneous value (a sample over one step) at each window. This reduces both Equations 1 and 5 to $\Delta G_{tot} = \sum_{i=1}^{NWINDOW} \left[V(\lambda(i),\mathbf{x}) - V(\lambda(i-1),\mathbf{x}) \right]$. This assertion cannot be rigorously proven, and in fact it can be demonstrated that the configuration will systematically lag changes in the potential energy function as the simulation progresses.[33-35] Thus, the validity of this approach has been questioned,[10, 33, 36] although recent work has suggested that the method may have use in bounding the error on free energy simulations.[37]

Other variations on these basic free energy methods have been published, although for various reasons they have not yet been widely adopted. These methods include MD/MC methods,[38] the acceptance ratio method,[39, 40] the weighted histogram method,[41] the particle insertion method,[42, 43] and the energy distribution method.[39] The reader is referred to the original publications for additional discussion of these approaches.

3. FREE ENERGY CALCULATIONS IN PRACTICE

The above equations allow us to calculate the free energy difference between any two configurations or molecules. In general, free energy differences between molecules are *substantially* easier to calculate than those between configurations: free energy differences between configurations require one to postulate an interconversion pathway, which is frequently not a straightforward exercise. Choosing, the wrong interconversion pathway can lead to very poor convergence and unreliable results. For this reason, free energy simulations are most frequently carried in the context of a thermodynamic cycle. For example, to compare the relative binding energies of two inhibitors to a enzyme, we can use the following cycle:

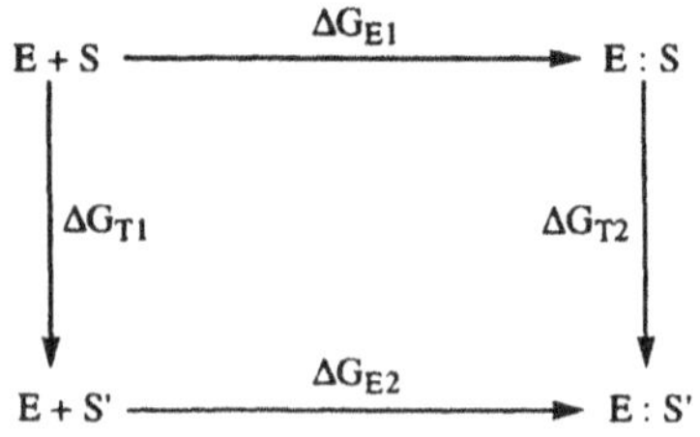

E is the enzyme, S is one inhibitor, S' is the modified inhibitor, E+S represents the unbound state and E:S represents the bound state. In this cycle, we calculate the free energies corresponding to the non-physical "mutation" processes represented by the vertical arrows. From the state function nature of free energy, it follows from this thermodynamic cycle that

$$\Delta\Delta G = \Delta G_{E2} - \Delta G_{E1} = \Delta G_{T2} - \Delta G_{T1} \tag{7}$$

In other words, the relative free energy difference in binding between the two inhibitors is equal to the difference in the free energies calculated for the non-physical mutations.

The equations and methods discussed allow one to determine the free energy difference between two configurations or systems. One might wonder why these calculations are not carried out to determine absolute free energies, which would allow both the differences to be determined and would also allow direct calculation of derivative parameters such as binding constants. The answer is that that direct calculation of the absolute free energy is generally impractical from a convergence standpoint. Refer again to Equation 1. There it is seen that the quantity we need to sample is the potential energy difference between the endpoints. This difference will tend to be relatively modest, even when the absolute potential energies of the endpoint systems are large. In contrast, if we expand the absolute free energy in terms of the potential energy, we get

$$G = \langle V \rangle - \frac{\langle V \rangle^2}{2RT} + \ldots \tag{8}$$

In other words, to calculate G we must be able to determine a converged value of $\langle V \rangle$ (and its higher moments), in contrast to ΔG, which is derived from $\langle \Delta V \rangle$. $\langle V \rangle$ converges much more slowly than $\langle \Delta V \rangle$, which makes calculation of the absolute free energy impractical in most cases.

That said, for select systems it is possible to attempt to calculate an absolute free energy. This can be done by running two FEP or TI simulations and summing the results.[44] For example, to calculate the absolute free energy of binding of a substrate to an enzyme, we perform the following simulations:

$$E{:}S \rightarrow E \qquad \Delta G_a \tag{9}$$

$$S \rightarrow nothing \qquad \Delta G_b \tag{10}$$

Summing the reverse of the first simulation with the second we get the desired net process

$$E + S \rightarrow E{:}S \qquad \Delta G_{EI} \tag{11}$$

where

$$\Delta G_{E1} = -\Delta G_a + \Delta G_b \tag{12}$$

It is not a trivial matter to get a converged value for these simulations, since in both we are forcing the substrate to vanish from the system a substantial mutation. But if one has copious computer time available and is careful, one has the potential for calculating such a value provided the substrate is not too large and there are not appreciable large-scale changes in the protein active site upon binding.

The net free energy is, in the end, the thermodynamic quantity that dictates molecular behavior. However, to understand why the free energy profile for a system looks as it does, it is valuable to also determine the potential and entropic components of the net free energy:

$$\Delta G = \Delta H - T\Delta S \tag{13}$$

If we can obtain an idea of why the free energy behaves as it does, we can often better attempt to make compositional changes to the system that will (hopefully) result in desired changes in binding, solubility, etc.

Unfortunately, it is significantly more difficult to determine the component potential and entropic components of the total free energy, than it is to calculate the free energy itself. Equations that allow the entropy difference (and hence enthalpy difference via Equation 13) to be calculated at the same time we are determining the free energy have been reported. For example, for TI, the following expression can be used:[9]

$$T\Delta S_{\lambda\rightarrow\lambda+\delta\lambda} = \frac{1}{RT}\int_{\lambda}^{\lambda+\delta\lambda} <V(\lambda,\mathbf{x})>_\lambda <\frac{dV(\lambda,\mathbf{x})}{d\lambda}>_\lambda - <V(\lambda,\mathbf{x})\frac{dV(\lambda,\mathbf{x})}{d\lambda}>_\lambda d\lambda \tag{14}$$

Note that in this expression for entropy, one must determine not only averages that depend on the derivative of the potential energy with respect to λ, but also terms that depend on the total potential energy. The derivative terms will converge much more rapidly, since they depend only on interactions of a small number of atoms with the rest of the system. As noted earlier, terms that depend on the net potential energy of the system converge very slowly.

The much slower convergence of the entropy relative to the net free energy can be understood from the following simple model for calculating entropy. From the same equation of state that leads directly to Equation 13, it follows that

$$\Delta S = -d\Delta G/dT \tag{15}$$

If we assume that the heat capacity $\Delta C_v = T(dS/dT)$ is independent of temperature (a reasonable assumption for small temperature ranges), we can expand Equation 15 to its differential approximation

$$\Delta S(T) \approx \frac{-[\Delta G(T+\Delta T) - \Delta G(T-\Delta T)]}{2\Delta T} \tag{16}$$

If the error associated with each value of ΔG (i.e. the error in a standard free energy simulation) is σ, T=300K, and ΔT is 10 degrees (small enough that C_v is temperature independent over the range T-ΔT to T+ΔT), then the error associated with ΔS as calculated by Equation 16 is roughly 42σ. That is, the error in the entropy is 42 times greater than the error in the free energy. Regardless of how the entropy is calculated, if we are attempting to calculate the entropy from the same basic statistical mechanical equations, a similar relative error will apply. Thus, since the error only decreases with the square root of the amount of sampling we perform, one would need to perform between two and three orders of magnitude more sampling to reduce the error in the calculated entropy to the same level as that in the calculated free energy. For this reason, any entropy (or enthalpy) calculations performed in this manner should be considered at this time to be qualitative in nature.

It should be noted that nothing in the derivations of these free energy methods restricts their application to changes in composition (mutations). They can also be applied to conformational changes by associating λ with these changes in the conformation of the system. If we carry out a simulation where λ reflects a conformational constraint (or restraint) on the system, then the free energies we will calculate define a profile of the free energy with respect to the conformational variables defined by λ. Such a profile is termed a Potential of Mean Force (PMF). One can carry out a PMF within the context of a FEP (or TI) simulation in one of two ways. In the first, the conformational variables are rigidly defined by λ through the use of coordinate constraints. For example, to calculate a PMF corresponding to the distance between two molecules, one can, in MD, use holonomic constraints (such as the well-known SHAKE algorithm) to keep the chosen distance fixed at a value that is defined by λ.[45, 46] (In MC, one can simply disallow moves that would change the constrained internal coordinates.)[47] The constraint is imposed without otherwise significantly affecting the conformational ensemble. As λ changes, so does the value. But for MD carried out at any fixed value of λ, the distance does not vary. Methods have been derived that allow the free energy resulting from a λ-dependent constraint to be determined during a free energy simulation.[9, 14, 26, 48, 49] The appeal of these methods is that they are easy to implement, and very simple to carry out. Once a method has been coded into the free energy program, the only difference with a standard free energy calculation is that one defines what internal variables shall be constrained with λ. The remainder of the simulation is performed as usual. The downside to these methods is that, depending on the pathway defined by the λ dependent constraints, convergence can be difficult to attain.

The second general method for performing PMF calculations relies on the use of Umbrella Sampling.[50] In its simplest form, Umbrella Sampling adds a bias restraint (umbrella) term to the standard potential function

$$V_{total} = V_{potential} + V_{bias} \tag{17}$$

where V_{bias} can take a form such as

$$V_{bias} = K(I - I(\lambda))^2 \tag{18}$$

with I the internal variable being restrained. The statistics accumulated during a simulation run with such a biasing term(s) included must be corrected. It is simple to show that for FEP, the corrected master equation is[50]

$$\Delta G = -RT \ln \left[\frac{<e^{-\Delta V/RT} e^{+V_{bias}}>_{A'}}{<e^{+V_{bias}/RT}>_{A'}} \right] \tag{19}$$

where $<>_{A'}$, means we evaluate the averages from ensembles generated using the biased total potential. The corresponding bias-corrected equation for TI is[26]

$$\Delta G = \int_a^b <dV/d\lambda>_a d\lambda + RT \ln <e^{+V_{bias}/RT}>_b \tag{20}$$

Note that for both FEP and TI, the umbrella restraint introduces a term that depends on $e^{+V_{bias}/RT}$, which may (since V_{bias} is always ≥ 0) fluctuate widely, especially if the biasing function is attempting to restrain the system to a conformation far from a local minimum. As a result, use of the umbrella term Equations 19 and 20 is often problematic. This has led to the development of alternate (but related) approaches to Umbrella Sampling.[41, 51, 52] Many of these derive from the following equation, which relates the work function W to the probability of states, corrected for use of the biasing potential:

$$W(I) = -RT \ln \rho^{*}(I) - V(I) - RT \ln \langle e^{+V_{bias}/RT} \rangle_b \tag{21}$$

Here $\rho^*(I)$ is the distribution of conformational states that arises from a simulation using the biased potential. The tricky point with this method comes from the fact that we ultimately need to integrate the work function over a series of windows, and the integration constant for each window is undefined. In practice, this problem is addressed using clever approaches that attempt to match up the probability distributions on consecutive intervals.

Yet another new method for calculating PMFs has recently appeared, which appears promising in initial tests.[53] In this method, an adiabatic separation between the reaction coordinate and the remaining degrees of freedom is imposed. This allows improved sampling while alleviating the need for (often difficult) post processing.

4. CONVERGENCE AND ERRORS

As must be clear by now, the ultimate difficulty in performing free energy simulations--regardless of which approach is chosen--is achieving convergence. The equations are either exact (FEP) or accurate enough (TI) that this is not a major factor in obtaining precise results. But whether we can obtain precise results will depend on evaluating various ensemble averages. (Note that whether we can obtain *accurate,* as opposed to merely precise results will also depend on the force field; a detailed discussion of issues related to force field development is beyond the scope of this chapter-please refer to Chapter 3).

The majority of free energy calculations in the literature have relied on very crude methods to estimate the error in the free energy results. Basically, a simulation is repeated several times, sometimes in the "forward" ($0 \rightarrow 1$) and "reverse" ($1 \rightarrow 0$) λ directions, sometimes only in one direction or the other. Each simulation is performed with a different (but equivalent) starting configuration, e.g. with a different random velocity distribution. The variance in the free energy results over the redundant simulations is taken as a measure of the error in the simulation. Unfortunately, there are several shortcomings to this approach. First, if the simulation is performed very quickly (not much sampling per window), one can encounter a situation where the change in the system is happening much more quickly than the system can relax to reflect it. In this case,[54] one can get very repeatable results over multiple simulations that are completely wrong. Another problem with these crude error estimates is that--even in the best case--they are merely a lower bound on the error.[29, 55] Typically, not enough redundant simulations are performed to have any chance of truly estimating the variance (error) in the simulation. Error estimates derived from double wide sampling (comparing the sums of $\lambda+\delta\lambda$ and $\lambda-\delta\lambda$ windows along a FEP trajectory)[3] are highly correlated and even less reliable.

A tremendous example of the potential folly of estimating error in this fashion can be ascertained by examining certain of the early publications in the field of free energy. These papers presented quite good agreement with experiment for free energy simulations that reflected total MD sampling of only 10-40 ps. They also, by-and-large, reported very good associated errors, as estimated from 1-2 redundant simulations. As has subsequently been shown, 10-40 ps is nowhere near enough MD sampling to precisely calculate the free energy for most changes. Current state-of-the-art free energy simulations are generally run for, at minimum, 150-200 ps, and often for greater than a nanosecond. And, in fact, when some of these early calculations have subsequently been repeated using more sampling, the results have differed considerably from both those obtained previously and from experiment.[56]

What happened? Probably two things. First, the simulations were run too quickly for the environment to respond to the change. And second, because researchers were not using good objective, statistically rigorous measures of convergence, a natural tendency is to accept results that seem to mesh with experiment as good (and to find reasons to dismiss results that do not agree with experiment).

While the general drift toward longer simulation times have ameliorated the problem to some degree, better still would be statistically-based measures of the quality of a free energy simulations. In fact, such measures have been described and implemented within the context of these simulations.[15, 57-59] To determine the error in a calculated free energy, we need to determine the error in the ensemble average upon which it depends. The trick here is to recognize that the data contributing to the average are correlated, and thus to derive a statistical equation which reflects this correlation. The variance in the mean value of an uncorrelated series of data is given as

$$\sigma^2(\bar{X}) = \sigma^2(X_i)/n \qquad (22)$$

where n is the number of data points in the series, $\sigma^2(\overline{X})$ is the variance in the mean, and $\sigma^2(X_i)$ is the variance in the set of data. The error in the mean value of a series of a correlated series can be given by

$$\sigma^2(\bar{X}) = \sigma^2(X_i)/[n/(1+2\tau)] \qquad (23)$$

Here τ is a correlation length, which grows as correlation in the data grows. The net effect of τ is to reduce the effective number of independent data points. τ is calculated from the autocorrelation coefficients for the series of data:

$$\tau = \sum_{k=1}^{n-1}(1 - k/n)\rho_k \qquad (24)$$

with ρ_k the autocorrelation function for two points separated by k-1 data points. Once σ^2 is calculated for the data contributing to the ensemble average, the error in the derived free energy can be calculated by elementary statistical propagation. Accurately estimating the correlation length τ requires that we sample at least 15-20 ps at each window.[15] This puts a lower bound on the length of any simulation for which we would like to use statistically based error estimates. It also eliminates the possibility of using such estimates with the slow growth method (where only one sampling point is obtained at each window). It should also be noted that Equation 23 assumes one is deriving statistics for a stationary series, that is, that the system is in equilibrium. This method will not work properly if the simulation is run so quickly that the system is not close to equilibrium when statistics are being accumulated. This equation will also not reflect any errors that are due to complete failure to sample certain minima. Nonetheless, it is a better measure of the quality of a simulation than a small number of simple repetitions of the experiment.

Statistically-based errors can also be obtained using a block averaging approach. Block averaging essentially places groups of consecutive system configurations into a single block. For example, if we run 100,000 MD steps, these might be placed into 100 blocks of 1000 points each. The average of each block is determined and used as the single observation for that block. Then, the variance for the series of block values is calculated. The idea is that if the blocks are large enough, then there will be no correlation between the average values of the blocks, and we will not have to make any corrections to the simple uncorrelated series statistics (and can use Equation 22). The downside of this approach is that one cannot know, a priori, how large the blocks should actually be. If they are too small, then the assumption that they are independent will not hold. If they are too large, then we effectively waste data and risk not having enough independent blocks to reliably calculate the variance.[59]

5. ISSUES AND TRICKS

Implementation of free energy calculations, in practice, is not quite as simple as the streamlined equations presented above would imply. There are a variety of practical choices that must be made with regard to implementation when a free energy simulation is run. At their root, most of these choices regard how to best ensure that convergence is attained, and that it is attained as efficiently as possible. Here we shall describe some of the most significant options that can be used to hasten convergence (and hence reliability) of a free energy simulation.

The master equations for both FEP and TI (Equations 3-5) are defined in terms of a series of λ intermediates. But nothing in these equations dictates how the series of λ pathways should be chosen. The simplest choice, and the one made in the majority of studies, is to simply define a series of fixed width windows (all $\lambda(i+1)-\lambda(i)$ the same). At each λ point, a pre-chosen fixed amount of equilibration and sampling is carried out. But this is certainly not the optimal choice for all simulations. In the case of FEP, optimal spacing of the windows is dictated by the need to attain reasonable sampling of the quantity $<e^{-(V(\lambda(i+1))-V(\lambda(i)))/RT}>_{\lambda}$. If $\delta\lambda$ is too large, then the potential surfaces of $V(\lambda(i+1))$ and $V(\lambda(i))$ will be too dissimilar, and the required ensemble will converge slowly. For TI, the spacing arises from the need to be able to numerically integrate the ΔG versus λ curve from a finite set of integrand points. More points will be required in regions where the curvature of the graph is changing more quickly. It is clear that fixed $\delta\lambda$ spacing with fixed sampling will not optimize against the requirements of the methods, except in a few fortuitous cases.

Several approaches have been reported which attempt to improve upon the basic fixed $\delta\lambda$, fixed sampling method in an automated fashion. These can be divided into those that modify the window spacing as the simulation progresses, those that modify the functional dependence of the potential on the value of λ, and those that modify the amount of sampling that is performed at each fixed λ point.

An example of the first approach is method of Dynamically Modified Windows (DMW).[18] DMW approximates the slope of the accumulated free energy curve over the past several windows, then adjusts the width of the width of the next window to keep the free energy change per window approximately constant:

$$\Delta\lambda_{\text{next window}} = \Delta G_{\text{target}}/M \tag{25}$$

where ΔG_{target} is the desired free energy change per window and M is the slope of the ΔG versus λ curve over the past several windows. For FEP, this approach will work if the rate of convergence is proportional to the free energy change. Unfortunately, such proportionality only holds for a limited number of systems. This approach is potentially more useful in the context of TI, where the need for more (or fewer) integration points is directly related to the shape (slope) of the free energy curve.

As an alternative to modifying the λ spacing dynamically as the simulation progresses, we can attempt to define a more elaborate λ dependence for the force field that takes into account known sampling issues for the system we are considering. λ dependence can be introduced to the potential function in many different reasonable ways. The most common is to linearly scale the parameters that define the potential function with λ. So, for example, force constants, equilibrium internal coordinates and non-bonded parameters are defined as[60]

$$K_I(\lambda) = (1-\lambda)K_I^A + \lambda K_I^B \tag{26}$$

$$I_{eq}(\lambda) = (1-\lambda)I_{eq}^A + \lambda I_{eq}^B \tag{27}$$

$$R_i^*(\lambda) = (1-\lambda)R_i^{*A} + \lambda R_i^{*B} \tag{28}$$

where K_I is the force constant for an internal coordinate (bond, angle, torsion) term, I_{eq} is the target value of an internal coordinate, R* is the mixed van der Waals radius for a non-bonded interaction, and similar equations are used for charges, non-bonded well depths, and so on. Such a formulation meets, in the simplest fashion possible, the requirements of the λ dependence, namely, that $V(\lambda=0)=V_a$, $V(\lambda=1)=V_b$, and that the function is continuous and differentiable along the entire interval [0,1].

In some cases, we know before we even start the simulation that certain ranges of λ are going to present a greater convergence challenge than others. For example, it is well known that if we are removing a highly charged solvent-accessible group, the simulation will frequently become unstable near the endpoint where the charge is removed. This arises from a combination of the fact that in standard water models there are no van der

Waals parameters on the hydrogens, and that near the endpoint, the van der Waals parameters on the disappearing charged group can become small enough that the hydrogens of the solvent can, occasionally, get close enough to a positively charged group to see the infinitely negative potential singularity at r=0 in the electrostatic term $q_i q_j / \varepsilon r$. A simple procedure, termed electrostatic decoupling, has been used to moderate this problem[61]. In essence, the simulation is run in two parts. In the first part, the charge is removed while keeping the van der Waals parameters fixed. Then, in the second part, the van der Waals parameters are removed. Since the van der Waals parameters on the disappearing group never get small when there is still a charge on the associated atoms, the water molecules can never get close enough to sample the r=0 singularity. Electrostatic decoupling can be implemented as a single simulation, where the electrostatic parameters on the group that is being removed are reduced to 0 (with van der Waals fixed) as λ varies $0 \rightarrow 0.5$ and then the van der Waals parameters are reduced to 0 as λ varies $0.5 \rightarrow 1$. (In typical practice, two separate simulation "legs" are used, but it amounts to the same thing).

A more sophisticated and generalized version of the ideas in electrostatic decoupling has been described.[22-24] Multiple λ values, (λ_1, λ_2, λ_3, ...) are introduced into the potential function, replacing the single λ value that has been described, and subject to the boundary conditions

$$V(\lambda_1 = 0, \lambda_2 = 0, \ldots \mathbf{x}) \equiv V(\lambda = 0, \mathbf{x}) \tag{29}$$

$$V(\lambda_1 = 1, \lambda_2 = 1, \ldots \mathbf{x}) \equiv V(\lambda = 1, \mathbf{x}) \tag{30}$$

Each λ parameter can be used to modify a different aspect of the potential function, and as many λ parameters can be added as one requires. The problem with this approach is that it is often difficult to postulate, a priori, a generalized multi-λ path that will result in greater efficiency. The examples in the literature attempting to utilize this approach have, thus far, been relatively simple.[23] For example, the convergence of free energy simulations on butane-like molecules was improved by reducing the rotational barrier (using one λ) then mutating the non-bonded parameters on the attached atoms (using a second λ) then bringing the rotational barrier back to its normal value (using the first λ parameter again). A more elaborate variant on this approach has been described.[25, 62] In this method, the lowest energy pathway between the two endpoints of the free energy simulation is approximated by determining this pathway for a gas phase simulation. This pathway is imposed on the change between states A and B using appropriately chosen λ dependence of either the internal coordinates or of the atomic coordinates.

An alternative to modifying the λ profile is offered by approaches that dynamically modify the amount of sampling performed at each λ point.[55] A statistical estimate of convergence (Equation 23) is used to determine whether the error at a given point is below a user-specified threshold. While this approach won't work if the λ sampling is too sparse, provided a reasonable number of λ points are used, this method should allow much better convergence for the same total amount of sampling. In fact, this method appears to work quite well.[55] The primary caveat for using this approach is that statistical convergence measures are unreliable unless a reasonable amount of sampling is performed at each λ point. Thus, this method is best suited for simulations using a modest number of λ point with significant sampling at each (subject to a minimum of, say, 10-20 ps sampling to generate reliable statistics regarding convergence).

One generally finds that when running a free energy simulation where groups are being annihilated or created at one/both endpoints, the greatest convergence problems occur at the endpoints. This is because the qualitative change in the system on the first λ step in going from "nothing" to "something" (creation) or vice-versa (deletion) is largest. Consider this issue in the context of FEP for the case of creation and refer again to Figure 1. In the first window, we sample the system using the potential function corresponding to methyl in a particular site, but we also need to sample states where the solvent has moved out of the way to allow the propyl group to be inserted at this site. Subsequent changes in λ only require incremental movements in the solvent (provided δλ is reasonably small), but on the first step to a non-zero λ, the change to the system can be huge. (The same problem is manifest in a non-converging derivative for the first integrand point of TI).

Several approaches have been developed to try to minimize such endpoint problems. Probably the most widely used technique is "bond shrinking".[9, 63] This procedure takes advantage of the fact that, for groups that are disappearing from the system, the lengths of bonds to atoms of the group at the point where it disappears do not need to be physical. (They are basically irrelevant, since the group is non-interacting). Of course, at the other end of the simulation, where the group is fully interacting with the system, physical bond lengths are required. Thus, we can shrink the bonds of the group to small values (typically 0.2-0.4 Å) as the group disappears from the system. The idea is that by making the group much more compact at the endpoint where it first becomes visible to the system, we can reduce endpoint sampling issues. A small group is easier to insert than a larger one, since the chances that the solvent will open up a hole that would accommodate the group is larger. In practice, it has been seen that while this approach is sometimes successful, there are other cases where shrinking the bond actually *slows* convergence.[55] This will have to be evaluated on a case-by-case basis.

Another approach that has been used to attempt to reduce the endpoint problem is to use non-linear λ scaling for the non-bonded interactions. In a general sense this can be represented as[19-21]

$$V_{non-bond}(\lambda) = \lambda^N V_{non-bond}(\lambda = 1) + (1-\lambda)^N V_{non-bond}(\lambda = 0) \tag{31}$$

The scaling can also be performed on the individual parameters that are used to evaluate the non-bonded energy. The effect of non-linear scaling is to reduce the effective physical difference between λ states for small values of λ. If the group is disappearing/appearing at λ = 0, then we will sample more carefully at this endpoint using non-linear scaling. However, even for a small initial value of λ, this does not really solve the issue of the first step, where we go from a non-interacting group to one that has a finite interaction with the system.

A better approach than non-linear scaling is to attempt to reduce and/or eliminate the singularity in the function that occurs on the step when a non-interacting group starts to interact. A clever approach has been described that reduces the problem by modifying the Lennard-Jones van der Waals term in the potential function.[30, 31] For a pair of atoms where one group vanishes at the λ =1 endpoint, the modified Lennard-Jones 6-12 function takes the form:

$$V_{6-12}(\lambda) = (1-\lambda)\left[\frac{A}{(R+S\lambda)^{12}} - \frac{B}{(R+S\lambda)^{6}}\right] + \lambda\left[\frac{A}{R^{12}} - \frac{B}{R^{6}}\right] \tag{32}$$

where A and B are the van der Waals coefficients (which may also be λ dependent), S is a shift coefficient, and R is the interatomic distance. This function is $A/R^{12} - B/R^{6}$ at both λ endpoints (as it must be). But at values of λ near 1 (where the group/atoms are disappearing), the shift coefficient ensures that the effective R value does not get too small, and that, consequently, the van der Waals term does not get too large. This function has been demonstrated to work very effectively for groups that are disappearing from a water bath.

6. CHOOSING SIMULATION CONTROL

All the methods described above can be used to try to improve the convergence profile for a free energy simulation. Statistical methods can be used to estimate error in the calculated values, and simulation times can be increased appropriately, depending on the precision required. In an ideal world, each simulation is run as long as is necessary, and no further discussion is required. In the real world, however, free energy simulations are competing both with other (more approximate) approaches to answering the same questions, and with the speed and cost with which the actual experiments can be carried out. Thus, one would like to be able to answer, in at least a general sense, the question "how much sampling is 'enough'?" With caveats duly noted regarding system-dependent specifics, the following general observations can be made, based on work that has been published and personal experience.[7, 15]

One should assume that at least 200 ps of equilibration+sampling will be required for any reliable simulation in explicit water solvent. Since each simulation should be run at least twice (or forwards and backwards) to ensure a reproducible result, this means a floor of 400 ps simulation time will be required. Note that 200 ps (400 ps) is a lower bound, and that many simulations will need to be run considerably longer. It is not unusual to run protein-based simulations for a nanosecond or more to achieve convergence. For a large (protein based) system, this requires a substantial investment of computer time on today's computers.

In general, a minimum of 10-20 windows are used. It is usually not a good idea to use fewer windows than this (the energy surfaces between adjacent λ intermediates will be too dissimilar for windows, too few integration points will be used for TI). Of course, more windows can be used, but in terms of efficiency, it is best to keep the number of windows to around the minimum number required to address the sampling issues. The reason for this can be seen from the following equation, which gives the total statistical error σ_{total} in the simulation as a function of the number of windows N_w and the amount of sampling S_w [14]:

$$\sigma_{total} = A_{err} / (2 N_w S_w)^{1/2} \tag{33}$$

Here, A_{err} is the proportionality constant from Equation 23, $A_{err} = (\sigma^2 (X_i)(1 + 2\tau))^{1/2}$ which depends on variance and correlation in the data (and which, for simplicity, is assumed in this analysis to be constant over all windows). Thus, we can reduce the total error by either increasing the number of windows or increasing the amount of sampling. However, recall that each time we skip to a new window, we first need to perform some equilibration before statistics can be accumulated. This equilibration phase does not contribute to reducing the net error according to Equation 33. Therefore, with respect to sampling that reduces the error, we "waste" (N_w x equilibration-steps) simulation steps. The greater N_w is, the greater the number of "wasted" steps. On the other hand, there is no hidden overhead with increasing s_w. Therefore, where we have the option, we should increase the sampling per window rather than increase the number of windows.

7. APPROXIMATE FREE ENERGY CALCULATIONS

Despite all the progress in free energy methodology and the rapid increases in available computer resources available, free energy simulations are still too slow and/or system limited to use for many real-world problems. For example, in a commercial setting, a team of bench chemists may synthesize many molecules per day, and combinatorial synthesis can be used to rapidly generate *thousands* of screening candidates. While there is always a need for detailed reliable predictions, even if they are costly and slow to produce, design teams also need access to modeling approaches that can keep up with the rapid synthesis and screening methods. Exact free energy calculations are not acceptably fast for such use.

To address these needs, a variety of methods have been developed allow approximate free energies to be calculated. These methods are based, in one way or another, on the precise free energy methods described above. But they make various assumptions or simplifications that allow them to be carried out much more quickly. All of these methods have shown promise on limited data, but as of yet, all are still in the development stage.

The linear interaction energy method allows the approximation of the free energy of binding using the relationships[64]

$$\Delta G_{bind} = \alpha < \Delta V_{vdw} > + \beta < \Delta V_{coulomb} > \tag{34}$$

where the averages are evaluated from a reference state ensemble. This expression is derived from the master equation of FEP assuming linear response theory holds in the regime of interest. In the original publications using this method,[64] β was set to 0.5 (based on the quadratic response of the free energy to changes in the electric field for polar solutions), and α was set to 0.161, based on empirical best-fit to experimental data for endothiapepsin inhibitors. Subsequent work has demonstrated that the equation, with coefficients as originally developed, is not generally applicable. Better agreement can be obtained by empirically fitting both α and β.[65-67] Further improvement is possible by adding additional terms to this equation, such as those based on solvent exposed surface area.[68, 69]

Another approach that derives from FEP is the free energy grid.[70-72] This method attempts to generate a free energy scoring grid around a molecule of interest. A rectilinear grid is defined about a molecule of interest. This could be a ligand (if one wishes to make changes to the ligand) or the binding site of a protein. A solvated MD simulation is then run, during which a one step FEP calculation is performed at each grid point:

$$\Delta G_{grid} = -RT \ln \langle e^{-Vprobe/RT} \rangle_{nothing} \tag{35}$$

$\langle e^{-Vprobe/RT} \rangle_{nothing}$ indicates that the reference system at each grid point is "nothing" at that point. The other endpoint is a probe atom with user-specified non-bonded parameters (usually similar to a united atom methyl group). In essence, at each point, we are calculating the free energy for going from nothing at the grid point to having a probe atom at that grid point. Since the reference state for each grid point is the same--no probe group at any grid point--we can simultaneously calculate the one-window FEP energy for all grid points from a single simulation. As has been discussed, the orthogonality of the potential energy spaces corresponding to "nothing" and to the probe atom means we will rarely achieve good convergence of Equation 35 at any grid point. But, if we calculate enough grid points, we may be able to infer from the resulting *approximate* free energy grid how favorable it is to introduce a probe atom at different regions of space. Initially, this approach was applied to the question of where we might best modify ligands to improve their binding or solubility.[70, 71] More recently, this method has been extended to the creation of a scoring grid within the binding site of a protein.[72] In this latter guise, it shows promise as a rapid database-friendly free energy-based screening function that improves upon traditional potential energy based methods.

Another clever approach to determining multiple free energies at reduced cost involves the use of "soft sites" in the reference ensemble.[31] This method is applicable when one wishes to determine the free energies associated with a number of modest perturbations to fixed core molecule, e.g. if one wishes to determine the free energy changes for substitutions about a known ligand. In this approach, one adds terms of the following form to the potential energy function:

$$Vij = 4\varepsilon_{ij}\left[\frac{\sigma_{ij}^{12}}{(\alpha\sigma_{ij}^{6} + r_{ij}^{6})^{2}} - \frac{\sigma_{ij}^{6}}{(\alpha\sigma_{ij}^{6} + r_{ij}^{6})}\right] \tag{36}$$

where σ_{ij} is a specified van der Waals interaction distance for atom i to the "soft core" nucleation site, r_{ij} is the distance from atom i to the "soft core" nucleation site, α is an empirical parameter that specifies the softness of the interaction and ε_{ij} is the effective well depth. This function is added for each of the nucleation (possible substitution) sites about the root molecule. This function provides a small repulsion force to the potential centered on the nucleation sites. The result is that occupation of the soft sites by solvent is less likely (but not so large a force that water is entirely prohibited). By so doing, we can hope to calculate--at least qualitatively--the free energy for placing a group in the soft repulsion site in a one window FEP simulation. Since the reference state is the same for all these simulations, we can carry out several FEP simulations around the molecule simultaneously. We can also calculate the free energies of changing to a number of different substituents at the same nucleation site using only a single FEP simulation. The net effect is that in one single window FEP calculation, we determine the free energies for a number of potential changes. All free energy differences calculated in this fashion will be relative to a non-physical reference state, but this is not a problem because the non-physical reference state drops out if we construct thermodynamic cycles to compare the relative free energies of the substituents to each other. Though this method is not applicable to high-throughput screening (unlike, say, the free energy grid), this approach has yielded promising results for amenable systems.[73]

Finally, surprisingly good initial results have been obtained with a simple approach that makes use of the Poisson-Boltzmann approach to estimating solvation free energies. In this method,[74-76]

$$G = V_{mm} + G_{PB} - TS_{mm} \tag{37}$$

A solvated MD simulation is performed to determine an ensemble of conformations for the molecule of interest. This ensemble is then used to calculate the terms in this equation. V_{mm} is the standard molecular mechanics energy for each member of the ensemble (calculated after removing the solvent water). G_{PB} is the solvation free energy calculated by numerical integration of the Poisson-Boltzmann equation plus a simple surface energy term to estimate the nonpolar free energy contribution. T is the absolute temperature. S_{mm} is the entropy, which is estimated using either quasi-harmonic analysis of the trajectory or else normal-modes analysis.[74] Given the substantial approximations and assumptions inherent in this approach, the results that have been obtained using it for select nucleic acid and protein systems are surprisingly good.

8. CONCLUSIONS

Though the basic equations have stayed the same, the field of free energy calculations have made marked gains in terms of reliability and precision over the nearly two decades since these calculations first came to prominence. In fact, it is not without irony that the promise of many of those early short calculations was--with hindsight--due in part to luck, but that with progress in methodology and computer speed, we are now at a point that is, in reality, where we thought we were 20-years ago! Which is not really a bad thing, given that initial expectations for free energy simulations were very high. It is expected that free energy calculations will remain a vital part of the computational toolkit for a long time to come.

Of course, there are many problems that are not amenable to precise calculations, either because these calculations are (still) too slow, or because they simply require throughput that is unlikely to *ever* be practical for exact simulations. For example, modern drug design in the industrial setting starts with high-throughput methods that require scoring methods faster than precise free energy calculations can hope to be. (Precise free energy calculations are more useful during the later, during bench-based optimization phase of drug design). To address these cases, approximate free energy methods are required. A number of such methods have now appeared in the literature, and it is expected that still more will appear, as these methods attain ever increasing importance.

Overall, free energy calculations continue to evolve--they have gotten more reliable, faster, and (with the approximate methods) more universally applicable. As such, they remain, and will continue to remain, a vital part tool in the modeler's arsenal.

9. REFERENCES

1. J. P. M. Postma, H. J. C. Berendsen, and J. R. Haak, Thermodynamics of cavity formation in water, Faraday Symp. Chem. Soc. **17**:55 (1982).
2. B. L. Tembe and J. A. McCammon, Ligand-receptor interactions, Comput. Chem. **8**:281 (1984).
3. W. L. Jorgensen and C. Ravimohan, Monte Carlo simulation of differences in free energies of hydration, J. Chem. Phys. **83**:3050 (1985).
4. R. W. Zwanzig, High-temperature equation of state by a perturbation method. I. Nonpolar gasses, J. Chem. Phys. **22**:1420 (1954).
5. M. Karplus and J. A. McCammon, Dynamics of proteins: Elements and function, Annu. Rev. Biochem. **52**:263 (1983).
6. W. L. Jorgensen, Theoretical studies of medium effects on conformational equilibria, Phys. Chem. **87**:5304 (1983).
7. D. A. Pearlman and B. G. Rao, Free energy calculations: Methods and applications, in: Encyclopedia of Computational Chemistry, P. v. R. Schleyer, N. L. Allinger, T. Clark,

J. Gasteiger, P. A. Kollman, H. F. Schaefer, III, and R. P. Schreiner, eds., John Wiley & Sons, New York (1998), pp. 1036-1061.
8. M. R. Reddy, M. D. Erion, and A. Agarwal, Free energy calculations: Use and limitations in predicting ligand binding affinities, in: Reviews in Computational Chemistry, K. B. Lipkowitz and D. B. Boyd, eds., VCH Publishers, New York (2000), vol. 16, pp. 217-304.
9. D. A. Pearlman and P. A Kollman, The overlooked bond-stretching contribution in free energy perturbation calculations, J. Chem. Phys. **94**:4532 (1991).
10. M. J. Mitchell and J. A. McCammon, Free energy difference calculations by thermodynamic integration--difficulties in obtaining a precise value, J. Comp. Chem. **12**:271 (1991).
11. M. Mazor and B. M. Pettitt, Convergence of the chemical potential in aqueous simulations, Mol. Simul. **6**:1 (1991).
12. S. H. Fleischman and D. A Zichi, Free energy simulations of methane solvation--a study of integrand convergence properties using thermodynamic integration, J. Chim. Phys. **88**:2617 (1991).
13. A. Hodel, T. Simonson, R. O. Fox, and A. T. Brunger, Confrontational substates and uncertainty in macromolecular free energy calculations, J. Phys. Chem. **97**:3409 (1993).
14. D. A. Pearlman, Determining the contributions of constraints in free energy calculations: Development, characterization, and recommendations, J. Chem. Phys. **98**:8946 (1993).
15. D. A. Pearlman, A. free energy derivatives: A new method for probing the convergence problem in free energy calculations, J. Comp. Chem. **15**:105 (1994).
16. V. Helms and R. C. Wade, Free energies of hydration from thermodynamic integration: Comparison of molecular mechanics force fields and evaluation of calculation accuracy, J. Comp. Chem. **18**:449 (1997).
17. P. M. King, Free energy via molecular simulation: A primer, in: Computer Simulation of Biomolecular Systems: Theoretical and Experimental Applications, vol. 2., W. F. van Gunsteren, P. K. Weiner, and A. J. Wilkinson, eds., Escom Science Publishers, Netherlands (1993), pp. 101-119.
18. D. A. Pearlman and P. A. Kollman, A new method for carrying out free energy perturbation calculations: Dynamically modified windows, J. Chem. Phys. **90**:2460 (1989).
19. A. J. Cross, Evaluation of molecular mechanics and calculation accuracy, Chem. Phys. Lett. **128**:198 (1986).
20. U. C. Singh, F. K. Brown, P. A. Bash, and P. A. Kollman, An Approach to the application of free energy perturbation methods using molecular dynamics: Applications to the transformations of $CH_3OH \rightarrow CH_3CH_3$, $H_3O^+ \rightarrow NH_4^+$, Glycine → Alanine, and Alanine → Phenylalanine in aqueous solution and $H_3O^+(H_2O)_3 \rightarrow NH_4(H_2O)_3$ in the gas phase, J. Am. Chem. Soc. **109**:1607 (1987).
21. H. Resat and M. Mezei, Studies on free energy calculations. 2. A theoretical approach to molecular solvation, J. Chem. Phys. **101**:6126 (1994).
22. S. Yoneda and H. Umeyama, Free energy perturbation calculations of multiple mutation bases, J. Chem. Phys. **97**:6730 (1992).
23. A. E. Mark, W. F. van Gunsteren, and H. J. C. Berendsen, Calculation of relative free energy via indirect pathways, J. Chem. Phys. **94**:3808 (1991).
24. T. P. Straatsma and J. A. McCammon, Treatment of rotational isomeric states. 3. The use of biasing potentials, J. Chem. Phys. **101**:5032 (1994).
25. R. Elber, Calculation of the potential of mean force using molecular dynamics with linear constraints: An application to a conformational transition in a solvated dipeptide, J. Chem. Phys. **93**:4312 (1990).

26. T. P. Straatsma, M. Zacharias, and J. A. McCammon, Holonomic constraint contributions to free energy differences from thermodynamic integration molecular dynamics simulations, Chem. Phys. Lett. **196**:297 (1992).
27. D. L. Severance, J. W. Essex, and W. L. Jorgensen, Generalized alteration of structure and parameters: A new method for free energy perturbations in systems containing flexible degrees of freedom, J. Comp. Chem. **16**:311 (1995).
28. M. Prevost, S. J. Wodak, B. Tidor, and M. Karplus, Contribution of the hydrophobic effect to protein stability: Analysis based on simulations of the Ile-96 → Ala mutation in barnase, Proc. Natl. Acad. Sci. USA **88**:10880 (1991).
29. D. A. Pearlman, A Comparison of alternative approaches to free energy calculations, J. Phys. Chem. **98**:1487 (1994).
30. W. Zacharias, T. P. Straatsma, and J. A. McCammon, Separation-shifted scaling, a new scaling method for lennard-jones interactions in thermodynamic integration, J. Chem. Phys. **100**:9025 (1994).
31. T. C. Beutler, A. E. Mark, R. C. van Schaik, P. R. Gerber, and W. F. van Gunsteren, Avoiding singularities and numerical instabilities in free energy calculations based on molecular simulations, Chem. Phys. Lett. **222**:529 (1994).
32. W. H. Press, B. P. Flannery, S. A. Teukolsky, and W. T. Vetterling, Numerical Recipes, Cambridge University Press, New York (1989).
33. D. A. Pearlman and P. A. Kollman, The lag between the Hamiltonian and the sytem configuration in free energy perturbation calculations, J. Chem. Phys. **91**:7831 (1989).
34. R. H. Wood, Estimation of errors in free energy calculations due to the lag between the Hamiltonian and the system configuration, J. Phys. Chem. **95**:4838 (1991).
35. R. H. Wood, W. C. F. Muhlbauer, and P. T. Thompson, Systematic errors in free energy perturbation calculations due to a finite sample of configuration space: Sample size hysteresis, J. Phys. Chem. **95**:6670 (1991).
36. J. Hermans, Simple analysis of noise and hysteresis in (slow-growth) free energy simulations, J. Phys. Chem. **95**:9029 (1991).
37. H. Hu, R. H. Yun and R. H. J. Hermans, Reversibility of free energy simulations: Slow growth may have a unique advantage, Mol. Sim. in press (2001).
38. F. Guarnieri and W. C. Still, A rapidly convergent simulation method: Mixed Monte Carlo/stochastic dynamics, J. Comp. Chem. **15**:1302 (1994).
39. C. H. Bennett, Efficient estimation of free energy differences from Monte Carlo data, Comp Phys. **22**:245 (1976).
40. D. M. Ferguson, On the use of acceptance ratio methods in free energy calculations, J. Chem. Phys. **99**:10086 (1993).
41. S. Kumar, D. Bouzida, R. H. Swendsen, and P. A. Kollman, The weighted histogram analysis method for free-energy calculations on biomolecues, J. Comp. Chem. **13**:1011 (1992).
42. J. Widom, Some topics in the theory of fluids, J. Chem. Phys. **39**:2808 (1963).
43. J. G. Powles, W. A. B. Evans, and N. Quirke, Non-destructive molecular dynamics simulation of the chemical potential of a fluid, Mol. Phys. **46**:1347 (1982).
44. W. L. Jorgensen, J. K. Buckner, S. Boudon, and J. Tirado-Rives, Efficient computation of absolute free energies of binding by computer simulations. Application to the methane dimer in water, J. Chem. Phys. **89**:3742 (1988).
45. J. P. Ryckaert, G. Ciccotti and H. J. C. Berendsen, Numerical integration of the cartesian equations of motion of a system with constraints: Molecular dynamics of n-alkanes, J. Comput. Phys. **23**:327 (1977).
46. D. J. Tobias and C. L. Brooks, Thermodynamics of amide hydrogen bond formation in polar and apolar solvents, J. Chem. Phys. **89**:5115 (1988).
47. W. L. Jorgensen and J. Tirado-Rives, Free energies of hydration for organic molecules from Monte Carlo simulations, Perspec. Drug Disc. and Design **3**:123 (1995).

48. S. Boresch and M. Karplus, The Jacobian factor in free energy simulations, J. Comp. Chem. **105**:5145 (1996).
49. W. K. den Otter and W. J. Briels, The calculation of free-energy differences by constrained molecular dynamics, J. Chem. Phys. **109**:4139 (1998).
50. J. P. Valleau and G. M. Torrie, in: Modem Theoretical Chemistry, Vol. 5, B. Beme, ed., Plenum, New York, pp.137.
51. E. M. Boczko and C. L Brooks, Constant-temperature free energy surfaces for physical and chemical processes, J. Phys. Chem. **97**:4509 (1993).
52. S. Kumar, J. M. Rosenberg, D. Bouzida, R. H. Swendsen, and P. A. Kollman, Multidimensional free-energy calculations using the weighted histogram analysis method, J. Comput. Chem. **16**:1339 (1995).
53. L. Rosso and M. E. Tuckerman, An adiabatic molecular dynamics method for the calculation of free energy profiles, Mol. Sim. in press (2001).
54. A. E. Mark, S. P. van Helden, P. E. Smith, L .H. M. Janssen, and W. F. van Gunsteren, Convergence properties of free energy calculations: Alpha-cyclodextrin complexes as a case study. J. Am. Chem. Soc. **116**:6293 (1994).
55. D. A. Pearlman and P. R. Connelly, Determination of the differential effects of hydrogen bonding and water release on the binding of FK506 to native and Tyr82 → Phe82 FKBP-12 proteins using free energy simulations, J. Mol. Biol. **248**:696 (1995).
56. D. A. Pearlman and P. A. Kollman, A Free energy perturbation calculations: Problems and pitfalls along the gilded road, in: Computer Simulation of Biomolecular Systems: Theoretical and Experimental Applications, W. F. van Gunsteren and P. K. Weiner, P. K., eds., Escom Science Publishers, Netherlands (1989), pp101-119.
57. W. B. Davenport, Probability and Random Processes, McGraw-Hill, New York (1970).
58. S. K. Schiferl and D. C. Wallace, Statistical Errors in Molecular Dynamics Averages, J. Chem. Phys. **83**:5203 (1985).
59. T. P. Straatsma, H. J. C. Berendsen, and A. J. Stam, Estimation of statistical errors in molecular simulation calculations, Mol. Phys. **57**:89 (1986).
60. D. A. Pearlman, D. A. Case, J. C. Caldwell, W. S. Ross, T. E. Cheatham III, S. DeBolt, D. M. Ferguson, G. L. Seibel, and P. A. Kollman, AMBER, a package of computer programs for applying molecular mechanics, normal mode analysis, molecular dynamics and free energy calculations to simulate the structural and energetic properties of molecules, Comp. Phys. Comm. **91**:1 (1995).
61. P. A. Bash, U. C. Singh, R. Langridge, and P. A. Kollman, Free energy calculations by computer simulation, Science **236**:564 (1987).
62. Y. X. Sun and P. A. Kollman, Determination of salvation free energy using molecular dynamics with solute cartesian mapping: An application to the salvation of 18-crown-6, J. Chem. Phys. **97**:5108 (1992).
63. L. Wang and J. Hermans, Change of bond lengths in free-energy simulations: Algorithmic improvements, but when is it necessary?, J. Chem. Phys. **100**:9129 (1994).
64. J. Aqvist, C. Medina, and J. E. Samuelsson, New method for predicting binding affinity in computer-aided drug design, Protein Eng. **7**:385 (1994).
65. T. Hansson, J. Marelius, and J. Aqvist, Ligand binding affinity prediction by linear interaction energy methods, J. Comput. Aided Mol. Des. **12**:27 (1998).
66. J. Wang, R. Dixon, and P. A. Kollman, Ranking ligand binding affinities with avidin: A molecular dynamics-based interaction energy study, Proteins **34**:69 (1999).
67. I. D. Wall, A. R. Leach, D. W. Salt, M. G. Ford, and J. W. Essex, Binding constants of neuraminidase inhibitors: An investigation of the linear interaction energy method, J. Med. Chem. **42**:5142 (1999).
68. D. K. Jones-Hertzog and W. K. Jorgensen, Binding affinities for sulfonamide inhibitors with human thrombin using Monte Carlo simulations with a linear response method, J. Med. Chem. **40**:1539 (1997).

69. H. A. Carlson and W. L Jorgensen, An extended linear response method for determining free energies of hydration, J. Phys. Chem. **99**:10667 (1995).
70. R. J. Radmer and P. A. Kollman, The application of three approximate free energy calculations methods to structure based ligand design: Trypsin and its complex with inhibitors, J. Comp.-Aided Mol. Des. **12**:215 (1998).
71. D. A. Pearlman, Free energy crids: A practical qualitative application of free energy perturbation to ligand design using the OWFEG method. J. Med. Chem. **42**:4313 (1999).
72. D. A. Pearlman and P. S. Charifson, Improved scoring of ligand-protein interactions using OWFEG free energy grids, J. Med. Chem. **44**:502 (2001).
73. H. Liu, A. E. Mark, and W. F. van Gunsteren, Estimating the relative free energy of different molecular states with respect to a single reference state, J. Phys. Chem. **100**:9485 (1996).
74. J. Srinivasan, T. E. Cheatham, P. Cieplak, P. A. Kollman, and D. A. Case, Continuum solvent studies of the stability of DNA, RNA and phosphoamidate-DNA helices, J. Am. Chem. Soc. **120**:9401 (1998).
75. Y. N. Vorobjev, J. C. Almagro, and J. Herrnans, Discrimination between native and intentionally misfolded conformations of proteins: ES/IS, a new method for calculating conformational free energy that uses both dynamics simulations with an explicit solvent and an implicit solvent continuum model, Proteins: Struct. Funct. Genet. **32**:399 (1998).
76. B. Jayaram, D. Sprous, M. A. Yong, and D. L. Beveridge, Free energy analysis of the conformational preferences of A and B forms of DNA in solution, J. Am. Chem. Soc. **120**:10629 (1998).

Chapter 3

Molecular Mechanics Force Field Development and Applications

Pat Metthe Todebush and J. Phillip Bowen*
Computational Center for Molecular Structure and Design, Department of Chemistry, University of Georgia, Athens, Georgia 30602

1. INTRODUCTION

The use of computational methods for the calculation of molecular properties has been a perennial goal of chemists. In recent years, the field of computational chemistry has become a firmly established discipline. Computational chemists have made impressive contributions to almost every aspect of chemistry, ranging from structural organic and inorganic chemistry to the prediction of polymer properties and the design of medicinally important therapeutic agents. While many computer-based methods are robust and widely utilized, the continued development and refinement of software and the underlying theory remains an active area of research.[1, 2]

Energy-based calculations on structures of biological relevance are a challenge. Typically, structures of pharmaceutical interest are a complex array of diverse functional groups and heterocycles with specific stereochemical relationships. From the quest to predict protein folding to the design and study of drug-receptor interactions, the ability to perform meaningful calculations on large complex systems has been a somewhat frustrating problem, even with the ever-increasing power and performance of a new generation of computers. Computer-assisted drug design (CADD) is more than traditional computational chemistry (structure- and energy-based calculations). It includes a host of diverse computer-based pharmacophore perception methods (database searching, virtual screening, quantitative structure activity relationships, combinatorial library design, as well as chemo- and bioinformatics).[3] All of these approaches have been successfully applied in drug discovery. Computational chemistry and the other methods are applicable to many different areas of chemistry in addition to pharmaceutically related applications.

Computational chemistry may be defined as any method derived from a set of existing equations and assumptions which calculates molecular structure, its corresponding energy, and other molecular properties. The discipline may be divided into three broad areas: ab intio quantum mechanics, semi-empirical quantum mechanics, and force field or molecular mechanics. In the past, considerable debate arose over which method was superior. These arguments led to needless friction, antagonistic claims, and misinformation. Today, quantum mechanics (ab initio, semi-empirical, and density functional theory) and molecular mechanics are accepted widely as having unique roles to play in structure-energy calculations. This is particularly true in industrial research settings. Typically, the nature of a specific problem needing a solution dictates the appropriateness of the theoretical approach selected.

Ab initio quantum mechanics is based on a rigorous treatment of the Schrödinger equation (or equivalent matrix methods)[4-7] which is intellectually satisfying. While there are a number of approximations made, it relies on a set of equations and a few physical constants.[8] The use of ab initio methods on large systems is limited if not impossible, even with the fastest computers available. Since the size of an ab initio calculation is defined by the number of basis functions in the system, ab initio calculations are extremely costly for anything past the second row in the periodic table, and for all systems with more than 20 or 30 total atoms.

The term "semi-empirical" has been reserved commonly for electronic-based calculations which also starts with the Schrödinger equation.[9-31] Due to the mathematical complexity, which involve the calculation of many integrals, certain families of integrals have been eliminated or approximated. Unlike ab initio methods, the semi-empirical approach adds terms and parameters to fit experimental data (e.g., heats of formation). The level of approximations define the different semi-empirical methods. The original semi-empirical methods can be traced back to the CNDO,[12, 13] NDDO,[14] and INDO.[15] The success of the MINDO,[16] MINDO/3,[17-21] and MNDO[22-27] level of theory ultimately led to the development of AM1[28] and a reparameterized variant known as PM3.[29, 30] In 1993, Dewar et al. introduced SAM1.[31] Semi-empirical calculations have provided a wealth of information for practical applications.

Density functional theory (DFT),[32] also a semi-empirical method, is capable of handling medium-sized systems of biological interest, and it is not limited to the second row of the periodic table. DFT has been used in the study of some small protein and peptide surfaces. Nevertheless, it is still limited by computer speed and memory. DFT offers a quantum mechanically based approach from a fundamentally different perspective, using electron density with an accuracy equivalent to post Hartree-Fock theory. The ideas have been around for many years,[33, 34] but only in the last ten years have numerous studies been published. DFT, compared to ab initio quantum mechanics, is still being evaluated in terms of its overall reliability. This will be less so in the future as more DFT calculations accumulate.

Molecular mechanics[35] is a mathematical approach used for the prediction of molecular structure, energy, vibrational spectra, dipole moments, and other physical properties. The ideas have existed prior to computers.[36-38] It is a routinely used calculational technique for the study of macromolecules and structures not amenable to quantum mechanics. Like all computer-based approaches, molecular mechanics is only a model, not necessarily a complete description of physical reality. Any deviation from *accurate* experiment data implies that something is wrong with the model. (It should be pointed out that if there is a disagreement between calculations and experiments one should not automatically assume the former is wrong. There are numerous examples where theory gave the correct answer which was later verified by repeating the experiments. The reverse has also been true.) Molecular mechanics has proven itself to be a reliable predictive method when (a) the force field equations are accurate and (b) the constants or parameters in the equations describing the force field have been determined properly. Force field calculations are able to handle many diverse chemical systems (e.g., proteins, large crystal structures, solvated systems. This method is by far the fastest available for the calculation of molecular structures and energies. Importantly, however, unlike quantum mechanics molecular mechanics is limited by the determination of parameters, most typically the vast number of unique torsion angles present in structurally diverse molecules.

2. FOUNDATIONS OF MOLECULAR MECHANICS

Molecular mechanics force fields rest on four fundamental principles. The first principle is derived from the Born-Oppenheimer approximation. Electrons have much lower mass than nuclei and move at much greater velocity. The velocity is sufficiently different that the nuclei can be considered stationary on a relative scale. In effect, the electronic and nuclear motions are uncoupled, and they can be treated separately. Unlike quantum mechanics, which is involved in determining the probability of electron distribution, molecular mechanics focuses instead on the location of the nuclei. Based on both theory and experiment, a set of equations are used to account for the electronic-nuclear attraction, nuclear-nuclear repulsion, and covalent bonding. Electrons are not directly taken into account, but they are considered indirectly or implicitly through the use of potential energy equations. This approach creates a mathematical model of molecular structures which is intuitively clear and readily available for fast computations. The set of equations and constants is defined as the force field. The mathematical formalism, compared to ab initio quantum mechanical calculations, is much less complex which lends itself to macromolecules (including solvation) simply beyond the scope of electronic based calculations.

The second principle is founded on the premise that each type of bond (e.g., Csp^3-Csp^3 or Csp^2=O) have "natural" bond lengths and angles. Any deviation from these natural bond lengths and angles is penalized by an increase in the steric energy of the system. The mathematical model of a molecule will adjust its nuclear positions to yield a new geometry with lower energy according to the force field. In strained systems, the molecule will deform in predictable ways. This steric energy of the molecule is the summation of the individual energy terms (bond length energy, bond angle energy etc.), and is the subject of the discussion in Section 4.

The third principle relates to the set of equations which describe the potential energy surface of the molecule. These potential energy equations, derived primarily from classical physics, contain parameters optimized to obtain the best match between experimental data and/or theoretical results for a training set of compounds. Once the parameters are evaluated for a set of structures (as diverse as possible), they are fixed and then used unmodified for other similar (and usually larger) compounds. As a first approximation, these parameters must be transferable from one structure to another for this method to work and be generally applicable.

The fourth principle requires every atom in the molecular model to be *classified* or *typed* according to the element, electronic characteristics, and hybridization. This is unlike quantum mechanics where hybridization and bonding can be inferred at the conclusion of a calculation. In molecular mechanics, the atom must be given a specific designation related to its hybridization at the outset of a calculation, and the element will always be in this state throughout the calculation. If parameters do not exist for all of the unique interaction terms, the calculation will fail, default to generalized parameters, or select parameters through some extrapolation algorithm.

3. MOLECULAR MECHANICS FORCE FIELDS

There are many different molecular mechanics force fields available. Many of them were originally developed in academic laboratories to solve specific problems. For example, some were designed to handle small molecules while others were developed to deal with protein structures. Today, the original demarcation between macromolcules and small molecules has become blurred, and they now are commercially available. Initially, many molecular mechanics programs were distributed at nominal costs, but due to the lack of federal funding for most molecular mechanics methods and software development, researchers were forced to find alternative ways to fund these efforts. Otherwise, many of the programs in currently in use would not exist.

It is beyond the scope of this short review to list every available molecular mechanics program. Only a selected few programs are mentioned here, without descriptive details of the potential functions, minimization algorithms, or comparative evaluations. Both the CHARMM[39-41] and AMBER[42-47] force fields use harmonic potential functions to calculate protein structures. They were developed in the laboratories of Karplus and Kollman, respectively, and work remarkably well. The CFF[48-50] and MMFF[51-56] force fields use more complex potential functions. Both force fields were developed in commercial settings and based extensively or exclusively on results obtained from quantum mechanics. Unlike the other molecular mechanics methods, the OPLS[56, 57] force field was parameterized by Jorgensen to simulate solution phase phenomena.

A series of small molecule force fields[58-66] were developed by Allinger and coworkers. The popular MM2 program,[58] widely used and distributed along with its source code, was imitated by a number of other groups. The latest commercial release of its successor is MM3(2000),[59-61] which is the focus of this review. Efforts directed toward another version, MM4,[62-67] have been reported. Presently MM4 is not commercially available, and there are no firm release dates as of this writing. These programs are excellent hydrocarbon force fields. The major problem, however, with the MM series of programs, in particular the unreleased MM4, is the lack of diverse functional groups. Each version of the MM series was developed to improve the results of previous versions by reducing the known errors and updating the potential functions in light of new experimental data or high level ab initio calculations. The complexity of the MM3 force field[68] allows for further predictions, such as vibrational frequencies and the vibrational energy levels within a given energy minimum for a molecule as well as thermodynamic information.

The MM3(2000) force field is the basis of this chapter. The program includes an induced dipole calculation that allows for the treatment of induction.[69] This improvement in the electrostatics yields better predicted dipole moments than in previous versions of MM3. It should be pointed out that most other force fields use point charges whereas the MM series of programs is based on point dipoles.

4. MM3 FORCE FIELD

4.1 Overview

The MM3 total steric energy is the summation of the individual energies for each type of interaction in the molecular model (Equation 1).[68] $E_{stretch}$ is the stretching energy, E_{bend} is the bending energy, $E_{torsion}$ is the torsional energy, $E_{nonbonded}$ is the van der Waals and London dispersion energies, and $E_{electrostatics}$ is the dipole-dipole, charge-dipole, and charge-charge interaction energies. Most force fields have theses terms included in their description of the total energy. $E_{cross\text{-}terms}$ describes the energies attributed to more than one variable: bend-bend, stretch-bend, and torsion-stretch interactions (Equation 2).

$$E_{total}^{MM3} = E_{stretch} + E_{bend} + E_{nonbonded} + E_{torsion} + E_{electrostatics} + E_{cross-terms} \quad (1)$$

$$E_{cross-terms}^{MM3} = E_{bend-bend} + E_{stretch-bend} + E_{torsion-stretch} \quad (2)$$

4.2 Bond Stretching Function

Molecular mechanics treats atoms as point masses connected together by harmonic forces, analogous to a classical picture of balls connect by springs.[35, 67, 68] The simplest mathematical description used to describe this type of system (the ball and spring model) uses harmonic potentials or Hook's law (see Equation 3). The change in bond length squared, Δl^2, from its natural bond length, l_i^0, to its distorted value, l_i, gives rise to a steric energy. As a first approximation quadratic terms are reasonable descriptions for bond stretching and compression, but they are not accurate for many complex molecules. Hooke's law is not valid for extreme distortions from equilibrium value. The true one-dimensional potential energy for molecular stretching and compression is anharmonic, and this results in a breakdown of the quadratic term approximation. A Morse potential describes the energy associated with the stretching and compression of bond lengths, accounting for anharmonicity.

$$E_{stretch} = \frac{1}{2}\sum_{i=1}^{n} k_{s,i}\left(l_i - l_i^0\right)^2 \quad (3)$$

Although quadratic expressions are reasonable approximations, as discussed above, they only works for small deviations from the natural state of the molecule. Some successful force fields continue to use these terms for macromolecules in order to decrease computational time. Most errors are not readily discernable because high quality experimental data are not available for large systems compared with the data available for small molecules. The logical step to rectify the problem is introduction of a Morse function. Such a solution results in excessive computational time, and introduces other mathematical problems. A more efficient way to describe bond stretching is through the mathematical stratagem of a power series expansion. The first nonvanishing term in a Taylor series expansion is the harmonic or quadratic potential. In MM3, the stretching term is described by a three-term truncated power series expansion (Equation 4). The third and fourth power terms introduce anharmonicity into the stretching potential. If the series was truncated with a cubic term, however, large bond length distortions would cause the energy to plummet toward negative infinity with a correspondingly dramatic increase in the bond length. Basically, the bond dissociates. To avoid this, the series was truncated with an even power term.

$$E_{stretch}^{i-j} = 143.88\frac{k_s}{2}\left[(\Delta l)^2 - c_s(\Delta l)^3 + \frac{7}{12}(c_s)^2(\Delta l)^4\right] \tag{4}$$

Some additional comments regarding Equation 4 are in order. The factor 143.88 converts the units to kcal/mol. There are two additional constants. The first is k_s, which is the stretching force constant parameter in units of md $Å^{-1}$. The second constant is c_s, which is the cubic term with a unitless value of 2.55. When a Morse potential is expanded in a power series, the factor 7/12 is obtained.

4.3 Angle Bending Function

As with bond stretching, a simple harmonic term can be used to describe angle bending deformations (Equation 5). Usually, a quadratic term gives an accuracy of approximately 10 degrees in angle variations. For similar reasons outlined above for bond stretching, a power series expansion is a more accurate way to describe angle compression and angle opening.

$$E_{bend} = \frac{1}{2}\sum_{i=1}^{n} k_{\theta,ij}\left(\theta_i - \theta_i^0\right)^2 \tag{5}$$

In the MM3 program, the bending motions are divided into two separate descriptions: in-plane and out-of-plane bending. The in-plane bending is shown in Equation 6. Depending on the geometry of the central atom in the bond angle, the angle can have as many as three different equilibrium bending parameters or values. If the central atom of the bond is either an sp^2 or sp^3 type carbon atom, there are three bond angle types available depending on the substitution pattern. The bending constant parameter, k_b, is in units of md Å $radian^{-2}$ $molecule^{-1}$. The value 0.043828 converts the units md Å $radian^{-2}$ $molecule^{-1}$ to kcal deg^{-2} mol^{-1}. The c_c, c_q, c_p, and c_s are the cubic, quartic, pentic, and sextic angle bending constants, with values of -1400.00×10^{-5}, 5.60×10^{-5}, -0.07×10^{-5}, 0.0022×10^{-5}, respectively.

$$E_{bend}^{i-j-m} = 0.043828\frac{k_b}{2}(\Delta\theta)^2\left[1 + c_c(\Delta\theta) + c_q(\Delta\theta)^2 + c_p(\Delta\theta)^3 + c_s(\Delta\theta)^4\right] \quad (6)$$

The out-of-plane bending expression is identical to Equation 6. A value of 0.00 is assigned for the natural bond angles, θ_0. It was necessary to introduce out-of-plane bending to reproduce trigonal planar systems. For example, to keep the Csp^2 centers (found in aldehydes, ketones, and alkenes) from distorting to nonplanar geometries, a force constant was added to the Csp^2 center and an imaginary point in the plane defined by the three attached atoms. This approach works well. Other alternative approaches have used improper torsion angles to achieve the same results. These out-of-plane or improper torsion angles are necessary mathematical formulations for resonance effects to be incorporated into force field models of trigonal planar hybridized atoms.

4.4 Non-Bonded Interaction Functions

4.4.1 Overview

The discussion thus far has focused on the forces between an array of atoms connected together through covalent bonds and their angles. Important interactions occur between atoms not directly bonded together. The theoretical explanation for attractive and repulsive forces for nonbonded atoms i and j is based on electron distributions. The motion of electrons about a nucleus creates instantaneous dipoles. The instantaneous dipoles on atom i induce dipoles of opposite polarity on atom j. The interactions between the instantaneous dipole on atom i with the induced instantaneous dipole on atom j of the two electron clouds of nonbonded atoms are responsible for attractive interactions. The attractive interactions are know as London Dispersion forces,[70] which are related to r^{-6}, where r is the distance between nonbonded atoms i and j. As the two electron clouds of nonbonded atoms i and j approach one another, they start to overlap. There is a point where electron-electron and nuclear-nuclear repulsion of like charges overwhelms the London Dispersion forces.[33] The repulsive interactions are known as van der Waals interactions. The London Dispersion and van der Waals interactions may be taken into account by simple potential functions, again introducing quantum mechanical effects.

4.4.2 London Dispersion and van der Waals Interactions

In MM3, the two types of nonbonded interactions which are included in the program are handled separately. The first includes attractive and repulsive forces between nonbonded atoms, their origins are described above, and the second is hydrogen bonding.

Many force fields use a Lennard-Jones 6-12 potential[71] to reproduce nonbonded interactions, see Equation 7. As two atoms approach one another, the steepness or hardness of the energy curve is proportional to r^{-12}. The use of an exponential term instead of the r^{-12} term in force field equations better reproduces experimental data for organic structures, and it is more consistent with quantum chemical calculations.

$$E_{vdw} = \sum_{i<j}\sum\left[\frac{A}{r^{12}} - \frac{B}{r^{6}}\right] \tag{7}$$

Potential exponential functions have been known for some time. In fact, the MM series of programs use a modified Buckingham potential or Hill equation,[72] represented by Equation 8. The $\varepsilon^{\#}$ represents the depth of the potential energy well, and is defined by $(e_i \bullet e_j)^{1/2}$, where ε_i and ε_j are the hardness of each atom. The variable P is defined as the ratio of the sum of the van der Waals radii of atoms i and j (Å) divided by the sum of the effective interatomic distances between the two atoms. The potential becomes increasingly attractive and drops to negative infinity when P goes to zero. A second expression, Equation 9, is needed to introduce a repulsive potential when this happens. Otherwise the two atoms would fuse together if they came close enough (in silica cold fusion). Equation 9 is most important when dealing with a poorly built starting geometry.

$$E_{vdw}^{A...B} = \varepsilon^{\#}\left(1.84\times10^{5} e^{-12.0/P} - 2.25P^{6}\right) \tag{8}$$

$$E_{vdw}^{A...B} = \varepsilon^{\#}192.270P^{2} \tag{9}$$

4.4.3 Hydrogen Bonding Interactions

In MM3, the hydrogen bonding energy primarily arises from electrostatics, described below in Section 4.6, but an extra contribution for the hydrogen bond energy is calculated from a modified version of the modified Buckingham potential used in nonbonded calculations.[72, 73] This hydrogen bonding term takes into account charge transfer in the hydrogen bond and allows for the directionality of the hydrogen bond to be calculated, see Equation 10. As described above regarding nuclear superposition of nonbonded atoms, the second equation (Equation 11) used in hydrogen bonding prevents the two atoms from fusing. The directionality comes from the cosθ term, which tends to weaken as the bond deviates from the ideal bond angle.

$$E_{NB} = \varepsilon^{\#}\left[1.84 \times 10^{5} e^{-12.0/P} - 2.25\frac{P^{6}}{D}\left(\frac{l}{l_{0}}\right)\cos\theta\right] \quad (10)$$

$$E_{HB} = \varepsilon^{\#}\left[192.270P^{2} + 1706.96 - 1706.96\left(\frac{l}{l_{0}}\right)\frac{\cos\theta}{D}\right] \quad (11)$$

4.5 Torsion Function

It is not possible to reproduce rotational barrier phenomena with only stretching, bending, and nonbonded terms.[35, 68] The explicit inclusion of a torsional potential energy function was required. This is another mathematical stratagem for introducing quantum mechanical effects into molecular mechanics. Equation 12 is the MM3 torsion potential, which is a three-term truncated Fourier series. The summation is over every dihedral angle. Through the appropriate selection of dihedral or torsion parameters, V_1, V_2, and V_3, a rotational energy profile about single and double bonds can be generated. The dihedral or torsion parameters are in kcal/mol, and ω is the actual dihedral angle.

$$E_{tor} = \sum_{i=1}^{n}\left[\frac{1}{2}V_{i,1}(1+\cos\omega) + \frac{1}{2}V_{i,2}(1-\cos 2\omega) + \frac{1}{2}V_{i,3}(1+\cos 3\omega)\right] \quad (12)$$

4.6 Electrostatic Interactions

4.6.1 Overview

Traditionally, the Allinger MM series of molecular mechanics programs have used dipole-dipole interactions to represent the electrostatic interactions. Each polar covalent bond is assigned a moment rather than point charges being assigned to every atom. This was the method of choice for historic reasons, but limited the calculations to non-ionic species. Most other force fields use point charges. MM3 has three electrostatic terms: dipole-dipole, charge-dipole, and charge-charge terms described in Equations 13, 14, and 15, respectively.

4.6.2 Dipole-dipole term

The dipole-dipole interaction energy is computed between two bond moments using Jean's equation.[35] The use of dipole-dipole interactions in the MM series of programs is historical. Originally, the goal of earlier versions of MM3 was to study small organic molecules. The results were reasonable, but it became necessary to introduce additional terms for charged species. In Equation 13, the value 14.39418 is a conversion factor to convert the units from ergs molecule^{-1} to kcal/mol. The two dipoles, μ and μ', are the two bond moments in units of Debyes, and the angle between them is χ. In the denominator, the distance between the midpoints of the two dipoles is R and the dielectric constant is D. In MM3, the dielectric constant has a default value of 1.5 which may be varied to mimic other solvents.

$$E_{\mu\mu'} = \frac{14.39418\left[\mu\mu'(\cos\chi - 3\cos\alpha\cos\beta)\right]}{R^3 D} \qquad (13)$$

4.6.3 Charge-dipole term

The charge-dipole interaction energy is computed between and atom with a formal charge and a polar bond. It is only computed when a molecule has a charged segment (ion) along with an inherent dipole moment. In Equation 14, the variable q_A is the formal charge on atom A and μ, defined above, is the bond moment. The angle α, defined by the two vectors μ, and **r** attenuate the interaction energy as a cosine function. The vector **r** goes from atom A to the midpoint of the bond moment μ. The dielectric constant is D.

$$E_{Q\mu} = 69.12\left[\frac{q_A\mu\cos\alpha}{r^2 D}\right] \tag{14}$$

4.6.4 Charge-charge term

Coulomb's law describes the charge-charge interaction energy (Equation 15). It is used in MM3 for the calculation of two charges interacting with one another. This term is used to calculate ionic interactions. The variables q_A and q_B are the formal charges on atoms A and B, respectively. The distance between the two atoms is r, and the dielectric constant is D.

$$E_Q = 14.39418(4.80298)^2\left[\frac{q_A q_B}{rD}\right] \tag{15}$$

4.6.5 Bond Polarizabilities

MM3(2000) has also included a new approach to obtain bond polarizabilities and induced dipole moments.[69] A general formula based on the original MM3 force constants and bond polarizabilities was derived and is used to compute bond polarizabilities, and then molecular polarizabilities by an additive model.

As a demonstration of the power and versatility of the MM3(2000) force field, a comparative study of dipole moments was computed on forty-four small organic molecules. A segment of those results are discussed here with an emphasis on the improvement of the MM3(2000) force field due to the inclusion of the induced dipole moments.

4.7 Cross-Term Function

4.7.1 Overview

The stretching, bending, and torsional terms need to be coupled to each other in order to reproduce subtle yet important structural features. For example, it is a well known molecular effect that when bond angles are compressed, the bonds stretch to relieve the strain. Also, bond stretching is known to be a function of the dihereal angle. The maximum bond length is achieved during an all-eclipsed conformation. Stretching, bending, and torsional terms along with the nonbonded and electrostatic expressions are not able in general to take into account coupling interactions. Thus, explicit incorporation of cross terms are necessary to reproduce experimental data and high level ab initio calculations.

4.7.2 Stretch-Bend Interaction

As the bond angle θ_{ijk} is distorted, the MM3 stretch-bend interaction describes what happens to the two bond lengths, l_{ij} and l_{jk}. This term accounts for the lengthening of the bonds as the angle is decreased. As defined above, the change in the bond angle and the bond lengths are symbolized by $\Delta\theta$ and Δl and represent the differences between the altered and natural values in units of degrees and Å, respectively. The force constant k_{sb} is the stretch bend constant parameter for the angle ijk in md rad^{-1} molecule^{-1}. The value 2.51118 is the conversion factor to convert md rad^{-1} molecule^{-1} to kcal deg^{-1} Å^{-1}.

$$E_{stretch-bend} = 2.51118 k_{sb}^{i-j-k}(\Delta\theta)\left[\left(\Delta l_{ij} - \Delta l_{jk}\right)\right] \qquad (16)$$

4.7.3 Bend-Bend Interaction

The MM3 bend bend interaction connects two adjacent angles with the same common central atom. The bend-bend constant is a geometric mean of the two individual angles, and it is a pre-determined value which depends only on the atom type arrangement of the angle. The bend-bend constant parameter k_{bb} is in units of md Å rad^{-2} molecule^{-1}, which is converted to kcal degree^{-2} mol^{-1} by the conversion factor -0.02191418. The two terms $\Delta\theta_{ijk}$ and $\Delta\theta_{ijm}$ represent the differences in altered and natural bond angles in units of degrees.

$$E_{bend-bend} = -0.02191418 k_{bb}\left[\left(\Delta\theta_{ijl}\right)\left(\Delta\theta_{ijm}\right)\right] \qquad (17)$$

4.7.4 Torsion-Stretch Interaction

The bond length is a function of its dihedral angle, which can, in part, be described by Equation 4. The stretching term alone cannot account for bond stretching as a function of torsion angle without the introduction of a torsion stretch term. In the early releases of MM3, the first two constants were set to zero (Equation 19), but in subsequent releases the expression was expanded to include the three-term Fourier series used to describe torsion energy. Two corrections have been included. The first correction is for the Bohlmann effect, where the C-H bond length has to be adjusted after the minimization of the torsion angle for proper fitting of the molecular geometry. The second correction is made for systems with hyperconjugation, where one torsion stretch term is not sufficient to model the system properly.

$$E_{tor} = -\frac{\Delta l}{2}\left[k_{ts1}(1+\cos\omega) + k_{ts2}(1-\cos 2\omega) + k_{ts3}(1+\cos 3\omega)\right] \quad (18)$$

$$E_{tor} = -\frac{\Delta l}{2}\left[k_{ts3}(1+\cos 3\omega)\right] \quad (19)$$

4.8 Chemical Effects

The mechanical, nonbonded, and electrostatic expressions described above are not sufficient to describe some structural and spectroscopic effects. Three specific structural and spectroscopic phenomena have been incorporated into MM3. They are the electronegativity, anomeric, and Bohlmann effects, which essentially can be traced to molecular orbital origins. These chemical effects are a part of the MM3 program.[74, 75]

5. PARAMETERIZATION

MM3 is a better force field for predicting molecular structure and vibrational frequencies than its predecessor. The parameterization of the force field was carried out in MM3 in a similar fashion to MM2.[68] Experimental data for vibrational frequencies, structures, dipole moments, and heats of formation are collected from experimental and/or ab initio studies. The bond length and angle parameters are determined first; they are usually taken as a first approximation from previous parameterization. The next stage is spectral fitting, where the force constants are determined and assigned to the appropriate vibrational frequencies. The moments of inertia, the rotational energy barriers, the dipole moments, and the heats of formation are fit last. Each modification to specific parameters requires review and revision of all parameters until no further adjustments can be made. The parameterization process is tedious and demands extensive study until a valid parameter set is determined. Automated parameter development programs have been devised to aid in the parameterization process.[76]

6. DIPOLES

For many years, chemists have been using theory to determine molecular properties. One property of particular interest is the dipole moment. The dipole moment is a measure of the overall electronic charge separation within a molecular geometry. Any neutral molecule is made up of a system of electric charges distributed in space around the nuclei. In many molecules, however, the "center of gravity" of the total positive charge and the total negative charge do not coincide, and the distance (defined as a vector quantity) between the charges is defined as the dipole moment (Equation 20).

$$\text{Dipole } (\vec{\mu}) = \text{charge} * \text{distance} \tag{20}$$

The next section in this chapter provides a brief comparison of the dipole moment (magnitude and direction) for a set of simple alcohols. Experimental gas phase dipole moments[45] are compared to ab initio and as molecular mechanics computed values. It is important to note that the direction of the vector dipole used by chemists is defined differently in classical physics. In the former definition, the vector points from the positive to the negative direction, while the latter has the orientation reversed.

Since the parameters used in molecular mechanics contain all of the electronic interaction information to cause a molecule to behave in the way that it does, proper parameters are important for accurate results. MM3(2000), with the included calculation for induced dipole interactions, should model more accurately the polarization of bonds in molecules. Since the polarization of a molecular bond does not abruptly stop at the end of the bond, induced polarization models the pull of electrons throughout the molecule. This changes the calculation of the molecular dipole moment, by including more polarization within the molecule and allowing the effects of polarization to take place in multiple bonds. This should increase the accuracy with which MM3(2000) can reproduce the structures and energies of large molecules where polarization plays a role in structural conformation.

7. METHODS

7.1 Overview

In this next section the dipole magnitude and directionality from MM3(2000) is compared to results obtained by MM3(96), Hartree-Fock and Møller-Plesset minimized structures (calculated using GAUSSIAN94),[77] as well as experimental dipole moment measurements.[78] For the molecular mechanics geometry optimizations, full matrix energy minimizations were carried out, and ground state structures were verified by the vibrational frequencies. The HF/6-31G** basis set was used throughout for all Hartree-Fock (HF) calculations; the frequency calculations were examined for the optimized structure to assure that the ground state geometry had indeed been located. The Møller-Plesset (MP) optimizations used the MP2/6-31G** basis set with the HF output as a starting geometry (MP2/6-31G**/MP2/6-31G**//HF/6-31G**/HF/6-31G**).[8] Frequency calculations were not performed with this level of theory due to computer limitations.

Also included in this work is a comparison of the direction of the dipole moments for each compound using two different measurement techniques. The first technique involves comparing the vector components against a standard Cartesian coordinate system, where the center of mass of the molecular model rests at the origin and the inertial axes of the molecule are oriented along X, Y, and Z axes, respectively. This represents the dipole in coordinate space, as shown in Figure 1.

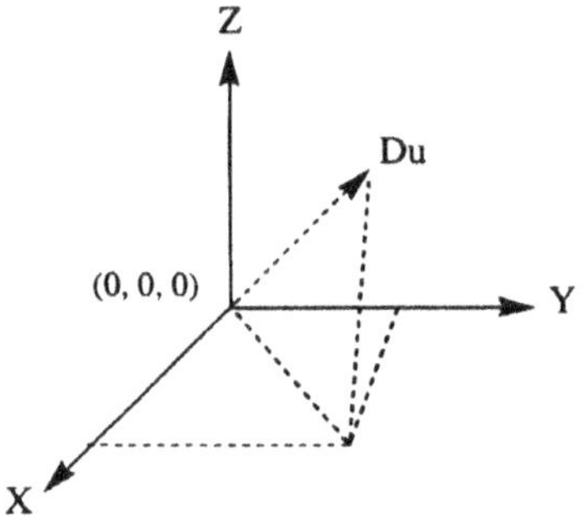

Figure 1. A representation of the dipole moment of a molecule (specified as a dummy atom, Du) in regards to the coordinate axis, (X,Y,Z).

The second technique measures the angle of separation between the dipole of a molecule calculated with two different computational methods. In this case, the molecules are superimposed upon each other and then the angle is measured, see Figure 2.

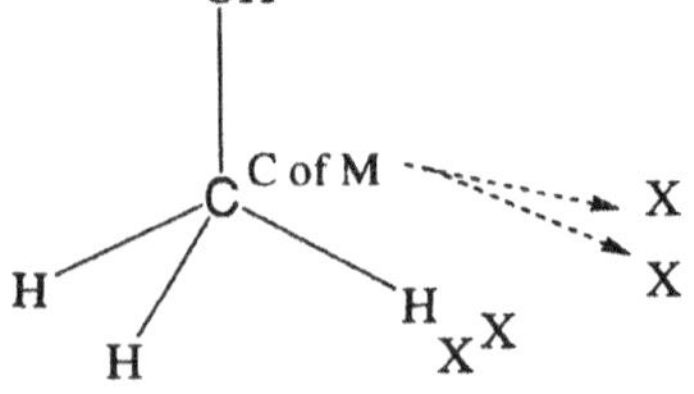

Figure 2. A representation of the dipole moment of a molecule in regards to the direction of the dipole computed by the other computational methods.

These two measurements give a clear picture of the orientation of the dipole in 3D space and how the dipole moment of a molecule moves when different computational methods are used.

7.2 Results and Discussion

Tables 1, 2, and 3 present a set of five alcohols. In Table 1, it should be noted that while MM3(96) calculates the magnitude of the dipole moment to be essentially the same for the entire set of molecules, MM3(2000) is superior in reproducing the experimental dipole moments. This is demonstrated by comparing the root mean squared deviation of 0.0878 Debye in MM3 to the 0.0524 Debye deviation in MM3(2000). (All of the experimental values except where notes are stark effect measurements determined from microwave spectra.)

It is known that ab initio methods are not accurate in reproducing or predicting molecular dipole moments. For example, a typical basis set minimization with no additional keywords was carried out, and the results show that the computed magnitude of the dipole moment is not particularly accurate when compared with experimental values. For alcohols, MP2 has a root mean squared deviation of 0.146 Debye, while HF had a deviation of 0.0734 Debye when measured against the experimental values.

The ability to predict the direction of the dipole moment is important since this value cannot be experimentally measured. Chemists typically use intuition to determine where the positive and negative parts of the molecule are located, but a good tool for predicting this is very helpful in large complex molecules where partial charges may determine important but subtle chemical effects, such as in a protein binding site. All of the methods used in this study did a reasonably good job in predicting the direction of the dipole moment. This is an important finding since consistency within the computational methods is the only measure of accuracy. Figure 1 is a pictorial display of one set of results in Table 2. It shows the coordinate axis system onto which the dipole moment of a molecule is superimposed. A dummy atom represents the termination of the dipole moment vector. This is a complete description of where the dipole moment is located in space, but

Table 1. Magnitude of the dipole moment for the alcohol family.

	HF	MP2	MM3	MM3(2000)	EXP[6]
Methanol	1.836	1.921	1.713	1.798	1.700
Ethanol	1.702	1.783	1.712	1.611	1.680
Propanol	1.616	1.698	1.712	1.569	1.550
2-Propanol	1.678	1.777	1.709	1.618	1.580
t-Butanol	1.625	1.734	1.701	1.708	1.670*

* liquid phase measurement

Table 2. Direction of dipole moment compared with the coordinate axis for the alcohols.

	HF			MP2			MM3			MM3(2000)		
	X	Y	Z	X	Y	Z	X	Y	Z	X	Y	Z
Methanol	119	29	90	121	31	90	109	19	90	106	16	90
Ethanol	98	8	90	87	3	90	73	17	90	72	18	90
Propanol	86	4	90	39	51	90	83	7	90	83	7	90
2-Propanol	47	89	43	46	63	56	51	77	42	47	74	47
t-Butanol	38	90	52	40	90	50	33	90	57	38	90	52

it is not a particularly informative presentation of where the dipole moment is in relation to its molecular environment. It is important to note that due to the symmetry of some of the molecules, a value of 90 or 180 degrees will be observed no matter which computational method is used.

Table 3 represents the direction of the dipole moment. Instead of comparing it to the coordinate axis, the direction is compared with the dipole moment as it was computed by another computational method. In Figure 2, the molecules are superimposed upon each other and the dipole moment represented by a dummy atom at each vector termini is displayed for each molecule, *i*. The angle, θ_{ij}, between the two vectors μ_i and μ_j is then measured using the center of mass as the vertex of the angle. From Table 3, using ethanol as an example, the angle between MM3 and MM3(2000) is one degree, therefore the difference between the angles of HF/MM3 and HF/MM3(2000) is also one degree, whereas the angle between HF and MM3 is 9 degrees and HF and MM3(2000) is 10 degrees.

Both of these techniques display features of the dipole moment's direction which are very important. Since there are no experimental data available to compare with these techniques, rendering a value judgement on the accuracy of one technique over another is really impossible. Yet, the comparison itself is of interest.

Table 3. Direction of dipole moment compared with other computed moments for alcohols.

	HF/MM3	MP2/MM3	MM3 (2000)/ MM3	HF/MM3 (2000)	MP2/ MM3 (2000)	HF/ MP2
Methanol	11	12	3	15	16	1
Ethanol	9	11	1	10	12	2
Propanol	9	12	0	9	12	3
2-Propanol	16	21	6	12	16	4
t-Butanol	12	15	4	7	10	3

8. CONCLUSIONS

Molecular mechanics is a useful and reliable computational method for structure, energy, and other molecular properties. The mathematical basis for molecular models in MM3 has been described, along with the limitations of the method. One of the major difficulties associated with molecular mechanics, in general, and MM3 in particular is the lack of accurately parameterized diverse functional groups. This lack of diverse functional groups has severely limited the use of MM3 in pharmaceutical applications.

The differences between ab initio and molecular mechanics generated dipole moments were discussed. The MM3(2000) force field is better able to reproduce experimental dipole moments for a set of forty-four molecules with a root mean squared deviation (rmsd) of 0.145 Debye compared with Hartree-Fock (rmsd 0.236 Debye), Møller-Plesset 2 (rmsd 0.263 Debye) or MM3(96) force field (rmsd 0.164 Debye). The orientation of the dipole moment shows that all methods give comparable angle measurements with only small differences for the most part. This consistency within methods is important information and encouraging since the direction of the dipole moment cannot be measured experimentally.

9. REFERENCES

1. K. Lipkowitz and D. Boyd, eds. Review of Computational Chemistry, VCH, New York (1991).
2. P. v. R. Schleyer, N. L. Allinger, T. Clark, J. Gasteiger, P. A. Kollman, H. F. Schaefer, III, and P.R. Schreiner, eds., Encyclopedia of Computational Chemistry, John Wiley & Sons, Chichester, UK, (1998).
3. J. P. Bowen and M. Cory, Computer-assisted drug design, in: Encyclopedia of Pharmaceutical Technology, Ja. Swarbrick and J. C. Boylan, eds., Marcel Dekker, Inc., New York (2001).
4. E. Schrödinger, Quantisierung als Eigenwertproblem, Ann. Phys. **79**:361 (1926).
5. E. Schrödinger, Quantisierung als Eigenwertproblem, Ann. Phys. **80**:437 (1926).
6. E. Schrödinger, Quantisierung als Eigenwertproblem, Ann. Phys. **81**:109 (1926).
7. W. Z. Heisenberg, The translation of kinematical and mechanical relations into terms of the quantum theory, Phys. **33**:879 (1927).
8. W. J. Hehre, L. Radom, P. Schleyer, and J. Pople, Ab Initio Molecular Orbital Theory, John Wiley & Sons, New York (1986).
9. J. A. Pople and D. Beveridge, Approximate Molecular Orbital Theory, McGraw-Hill, New York (1970).
10. M. J. S. Dewar, The Molecular Orbital Theory of Organic Chemistry, McGraw-Hill, New York (1969).
11. J. N. Murrell and A. J. Harget, Semiempirical Self-Consistent-Field Molecular Orbital Theory of Molecules, Wiley-Interscience, London (1972).
12. J. A. Pople and G. A. Segal, Approximate self-consistent molecular orbital theory. II. Calculations with complete neglect of differential overlap, J. Chem. Phys. **43**:136 (1965).

13. J. A. Pople and G.A. Segal, Approximate self-consistent molecular orbital theory. III. CNDO results for AB_2 and AB_3 systems, J. Chem. Phys. **44**:3289 (1966).
14. J. A. Pople, D. P. Santry, and G. A. Segal, Approximate self-consistent molecular orbital theory. I. Invariant procedures, J. Chem. Phys. **43**:S129 (1965).
15. J. A. Pople, D. L. Beveridge, and P. A. Dobosh, Approximate self-consistent molecular orbital theory. V. Intermediate neglect of differential overlap, J. Chem. Phys. **47**:2026 (1967).
16. M. J. S. Dewar and W. Thiel, Ground states of molecules. The MNDO method. Approximations and parameters, J. Am. Chem. Soc. **99**:4899 (1977).
17. R. C. Bingham, M. J .S. Dewar, and D. H. Lo, Ground states of molecules. XXV. MINDO/3. An improved version of the MINDO semiempirical SCF-MO method, J. Am. Chem. Soc. **97**:1285 (1975).
18. R. C. Bingham, M. J. S. Dewar, and D. H. Lo, Ground states of molecules. XXVI. MINDO/3 calculations for hydrocarbons, J. Am. Chem. Soc **97**:1294 (1975).
19. R. C. Bingham, M. J. S. Dewar, and D. H. Lo, Ground states of molecules. XXVII. MINDO/3 calculations for CHON species, J. Am. Chem. Soc. **97**:1302 (1975).
20. R. C. Bingham, M. J. S. Dewar, and D. H. Lo, Ground states of molecules. XXVIII. MINDO/3 calculations for compounds containing carbon, hydrogen, fluorine, and chlorine, J. Am. Chem. Soc. **97**:1307 (1975).
21. M. J. S. Dewar, D. H. Lo, and C. A. Ramsden, Ground states of molecules. XXIX. MINDO/3 calculations of compounds containing third row elements, J. Am. Chem. Soc. **97**:1311 (1975).
22. M. J .S. Dewar and W. Thiel, Ground states of molecules. 39. MNDO results for molecules containing hydrogen, carbon, nitrogen, and oxygen, J. Am. Chem. Soc. **99**:4907 (1977).
23. M. J. S. Dewar and M. L. McKee, Ground states of molecules. 41. MNDO results for molecules containing boron, J. Am. Chem. Soc. **99**:5231 (1977).
24. M. J. S. Dewar and H. S. Rzepa, Ground states of molecules. 45. MNDO results for molecules containing beryllium, J. Am. Chem. Soc. **100**:777 (1978).
25. L. P. Davis, R. M. Guidry, J. R. Williams, M. J. S. Dewar, and H. S. Rzepa, MNDO calculations for compounds containing aluminum and boron, J. Comput. Chem. **2**:433 (1988).
26. M. J. S. Dewar, M. L. McKee, and H. S. Rzepa, MNDO parameters for third period elements, J. Am. Chem. Soc. **100**:3607 (1978).
27. M. J. S. Dewar and E. Healy, Ground states of molecules. 64. MNDO calculations for compounds containing bromine, J. Comput. Chem. **4**:542 (1983).
28. M. J. S. Dewar, E. G. Zoebisch, E. F. Healy, and J. J. P. Stewart, AM1: A new general purpose quantum mechanical molecular model, J. Am. Chem. Soc. **107**:3902 (1985).
29. J. J. P. Stewart, Optimization of parameters for semi-empirical methods 1. Method, J. Comput. Chem. **10**:209 (1989).
30. J. J. P. Stewart, Optimization of parameters for semi-empirical methods 1. Applications, J. Comput. Chem. **10**:221 (1989).
31. M. J. S. Dewar, C. Jie, and J. Yu, SAM1: The first of a new series of general purpose quantum mechanical molecular models, Tetrahedron **49**:5003 (1993).
32. R. G. Parr and W. Yang, Density Functional Theory of Atoms and Molecules. Oxford University Press, Oxford (1989).
33. W. Kohn and L. J. Sham, Self-consistent equations including exchange and correlation effects, Phys. Rev. **A140**:1133 (1965).
34. P. Hohenberg and W. Kohn, Inhomogeneous electron gas, Phys. Rev. **B136**:864 (1964).
35. U. Burkert and N. L. Allinger, Molecular Mechanics, American Chemical Society, Washington, DC, ACS Monograph 177, (1982).
36. F. H. Westheimer and J. E. Mayer, The theory of the racemization of optically active derivatives of biphenyl, J. Chem. Phys. **14**:733 (1946).

37. T. L. Hill, Steric effects, J. Chem. Phys. **14**:465 (1946).
38. I. Dostrovsky, E. D. Hughes, and C. K. Ingold, XXXII. The role of steric hindrance. magnitued of steric effect, range of occurrence of steric and polar effects, and place of the Wagner rearrangement in nucleophilic substitution and elimination, Chem. Soc. 173 (1946).
39. B. R. Brooks, R. E. Bruccoleri, B. D. Olafson, D. J. States, S. Swaminathan, and M. Karplus, CHARMM: Program for macromolecular energy, minimization, and dynamics calculations, J. Comput. Chem. **4**:187 (1983).
40. A. D. MacKerell, J. Wiórkiewicz-Kuczera, and M. Karplus, An all-atom empirical energy function for the simulation of nucleic acids, J. Am. Chem. Soc. **117**:11946 (1995).
41. J. C. Smith and M. Karplus, Empirical force field study of geometries and confromational transitions of some organic molecules, J. Am. Chem. Soc. **114**:801 (1982).
42. S. J. Weiner, P. A. Kollman, D. A. Case, U. C. Singh, C. Ghio, G. Alagona, S. Profeta, Jr., and P. J. Weiner, A new force field for molecular mechanical simulation of nucleic acids and proteins, J. Am. Chem. Soc. **106**:765 (1984).
43. S. J. Weiner, P. A. Kollman, D. T. Nguyen, and D. A. Case, An all atom force field for simulation of protein and nucleic acids, J. Comput. Chem. **7**:230 (1986).
44. D. A. Pearlman, D. A. Case, J. W. Caldwell, W. S. Ross, T. E. Cheatham, S. Debolt, D. Ferguson, G. Seibel, and P. Kollman, AMBER, a package of computer programs for applying molecular mechanics, normal-mode analysis, molecular dynamics and free energy calculations to simulate the structural and energetic properties of molecules. Comput. Phys. Commun. **91**:1 (1995).
45. W. D. Cornell, P. Cieplak, C. I. Bayly, I. R. Gould, K. M. Merz, D. M. Ferguson, D. C. Spellmeyer, T. Fox, J. W. Caldwell, and P. A. Kollman, A 2nd generation force field for the simulation of proteins, nucleic acids, and organic molecules, J. Am. Chem. Soc. **117**:5179 (1995).
46. W. D. Cornell, P. Cieplak, C. I. Bayly, and P. A. Kollman, Application of RESP charges to calculate confromational energies, hydrogen bond energies, and free energies of solvation, J. Am.Chem. Soc. **115**:9620 (1993).
47. P. A. Kollman, Advances and continuing challenges in achieving realistic and predictive simulations of the properties of organic and biological molecules, Acc. Chem. Res. **29**:461 (1996).
48. J. R. Maple, M. J. Hwang, T. P. Stockfisch, U. Dinur, M. Waldman, C. S. Ewig, and A. T. Hagler, Derivation of class II force fields. 1. Methodology and quantum force field for the alkyl functional group and alkane molecules, J. Comput. Chem. **15**:162 (1994).
49. M. J. Hwang, T. P. Stockfisch, and A. T. Hagler, Derivation of class II force field. 2. Derivation and characterization of a class II force field, CFF93, for the alkyl functional group and alkane molecules, J. Am. Chem. Soc. **116**:2515 (1994).
50. J. R. Maple, M. J. Hwang, T. P. Stockfisch, and A. T. Hagler, Derivation of class II force fields. 3. Characterization of a quantum force field for alkanes, Isr. J. Chem. **34**:195 (1994).
51. T. A. Halgren, Merck molecular force field. I. Basis, form, scope, parameterization, and performance of MMFF94, J. Comput. Chem. **17**:490 (1996).
52. T. A. Halgren, Merck molecular force field. II. MMFF94 van der Waals and electrostatic parameters for intermolecular interactions, J. Comput. Chem. **17**:520 (1996).
53. T. A. Halgren, Merck molecular force field. III. Molecular geometries and vibrational frequencies for MMFF94, J. Comput. Chem. 17:553 (1996).
54. T. A. Halgren and R. B. Nachbar, Merck molecular force field. IV. Conformational energies and geometries for MMFF94, J. Comput. Chem. **17**:587 (1996).
55. T. A. Halgren, Merck molecular force field. V. Extension of MMFF94 using experimental data, additional computational data, and empirical rules, J. Comput. Chem. **17**:616 (1996).

56. W. L. Jorgensen, Optimized intermolecular potential functions for liquid hydrocarbons, J. Am. Chem. Soc. **106**:6638 (1984).
57. W. L. Jorgensen and J. Tirado-Rives, The OPLA potential functions for proteins: Energy minimizations for crystals of cyclic peptides and crambin, J. Am. Chem. Soc. **110**:1666 (1989).
58. N. L. Allinger, Conformational analysis 130. MM2. A hydrocarbon force field utilizing V_1 and V_2 torsional terms, J. Am. Chem. Soc. **99**:8127 (1977).
59. N. L. Allinger, Y. H. Yuh, and J. H. Lii, Molecular Mechanics. The MM3 force field for hydrocarbons. 1, J. Am. Chem. Soc. **111**:8551 (1989).
60. J. H. Lii and N. L. Allinger, Molecular Mechanics. The MM3 force field for hydrocarbons. 2. Vibrational frequencies and thermodynamics, J. Am. Chem. Soc. **111**:8566 (1989).
61. J. H. Lii and N. L. Allinger, Molecular Mechanics. The MM3 force field for hydrocarbons. 3. The van der Waals' potentials and crystal data for aliphatic and aromatic hydrocarbons, J. Am. Chem. Soc. **111**:8576 (1989).
62. N. L. Allinger, K. S. Chen, and J. H. Lii, An improved force field (MM4) for saturated hydrocarbons, J. Comp. Chem. **17**:642 (1996).
63. N. Nevins, K. S. Chen, and N. L. Allinger, Molecular mechanics (MM4) calculations on alkenes, J. Comp. Chem. **17**:669 (1996).
64. N. Nevins, J. H. Lii, and N. L. Allinger, Molecular mechanics (MM4) calculations on conjugated hydrocarbons, J. Comp. Chem. **17**:695 (1996).
65. N. Nevins and N. L. Allinger, Molecular mechanics (MM4) vibrational frequency calculations for alkenes and conjugated hydrocarbons, J. Comp. Chem. **17**:730 (1996).
66. N. L. Allinger, K. S. Chen, J. A. Katzenellenbogen, S. R. Wilson, and G. M. Anstead, Hyperconjugative effects on carbon-carbon bond lengths in molecular mechanics (MM4), J. Comp. Chem. **17**:747 (1996).
67. J. P. Bowen, New vistas in molecular mechanics calculations, in: Drug Design, Marcel Dekker, Inc., New York (1997), pp. 495-538.
68. J. P. Bowen and N. L. Allinger, Molecular mechanics: The art and science of parameterization in: Reviews in Computational Chemistry, K. B. Lipkowitz and D. B. Boyd, eds.; VCH, New York, NY, (1991) pp. 81-97.
69. M. Buyong, J. H. Lii, and N. L. Allinger, Molecular polarizabilities and induced dipole moments in molecular mechanics, J. Comp. Chem. **21**:814 (2000).
70. F. London, Zur theori und systematik der molekularkräfte, Zeitschrift für Physik. **63**:245 (1930).
71. J. E. Lennard-Jones, Cohesion, Proc. R. Soc. (London, A.) **106**:463 (1924).
72. T. L. Hill, Steric effects. I. van der Waals potential energy curves, J. Chem. Phys. **16**:399 (1948).
73. J. H. Lii and N. L. Allinger, Directional hydrogen bonding in the MM3 force field. II, J. Comput. Chem. **19**:1001 (1998).
74. N. L. Allinger, M. R. Imam, M. R. Frierson, Y. H. Yuh, and L. Schafer, The effect of electrone gativity on bond lengths in molecular mechanics calculations, in: Mathematics and Computational Concepts in Chemistry, N. Trinajstic, London (1986), p. 8.
75. L. Norskov-Lauritsen and N. L. Allinger, A molecular mechanics treatment of the anomeric effect, J. Comput. Chem. **5**:326 (1984).
76. G. Liang, P. C. Fox, and J. P. Bowen, The parameter analysis and refinement toolkit system (PARTS) and its application in MM3 parameterization for phosphine and its derivatives, J. Comput. Chem. **17**:940-953 (1996).
77. M. J. Frisch, G. W. Trucks, H. B. Schlegel, P. M. W. Gill, B. G. Johnson, M. A. Robb, J. R. Cheeseman, T. Keith, G. A. Peterson, J. A. Montgomery, K. Raghavachari, M. A. Al-Laham, V. G. Zakrzewski, J. V. Ortiz, J. B. Foresman, J. Cioslowski, B. B. Stefanov, A. Nanayakkara, M. Challacombe, C. Y. Peng, P. Y. Ayala, W. Chen, M. W. Wong, J. L. Andres, E. S. Replogle, R. Gomperts, R. L. Martin, D. J. Fox, J. S. Binkley, D. J.

Defrees, J. Baker, J. P. Stewart, M. Head-Gordon, C. Gonzalez, and J. A. Pople, Gaussian 94, Revision E.2, Gaussian, Inc., Pittsburgh (1995).

78. (a) A. L. McClellan, Tables of Experimental Dipole Moments, Vol. 1., W. H. Freeman and Co., San Francisco (1963). (b) A. L. McClellan, Tables of Experimental Dipole Moments, Vol. 2. Rahara Enterprises, El Cerrito (1973), (c) A. L. McClellan, Tables of Experimental Dipole Moments, Vol. 3., Rahara Enterprises, El Cerrito (1989).

Section Two

Molecular Properties

Chapter 4

Solvation Thermodynamics and the Treatment of Equilibrium and Nonequilibrium Solvation Effects by Models Based on Collective Solvent Coordinates

Christopher J. Cramer and Donald G. Truhlar
Department of Chemistry and Supercomputer Institute, University of Minnesota, Minneapolis, MN 55455-0431

1. INTRODUCTION

Computing the free energy of solvation is important in rational drug design for both pharmacokinetics (drug transport to the site of action) and pharmacodynamics (drug interactions at the site of activity). In the former case, it is well recognized that bioavailability depends on the availability of a particular drug to have a favorable interaction with both water (for transport in the blood) and lipophilic media (for transport through cell membranes). The solubility and partitioning of a molecule in and between various media are thus critical physicochemical parameters that correlate strongly with biological activity.[1] Even though such effects are nonspecific with regard to receptor structure, they must be taken into account in rational drug design. Then, at the receptor, desolvation is a major contributor to ligand-protein binding, and the estimation of this effect plays an important role in structure-based drug design.[2] A third area where solvation effects must be considered is drug metabolism, since the kinetics and thermodynamics of the enzymatic biotransformations of drugs may be very dependent on solvation effects. Both the ability of a drug to survive deactivation and elimination and the nature of the potentially toxic metabolites of a drug may depend on solvation effects on kinetics.

For both nonspecific and structure-based approaches, physicochemical solvation parameters may be used directly, or they may be embedded in quantitative structure-activity relationships.[3] This chapter starts with a review of the thermodynamic equations that may be used for a quantitative description of the free energy of solutes in fluid media. Then it provides an overview of statistical mechanical models for solvation effects that are based on collective coordinates, both macroscopic coordinates such as the electric polarization of the solvent and microscopic collective coordinates such as the surface area of the solute-solvent interface. Such models are sometimes called implicit models (because the individual atomic coordinates of the solvent molecules are not treated explicitly) or continuum models (because the fine grained detail of the solvent is replaced by an averaged description in which the solvent is treated mathematically as a continuous medium).

2. MOLAR FREE ENERGY

An ideal gas obeys the equation of state

$$PV = nRT \tag{1}$$

where P is the pressure, V is the volume, n is the number of moles, R is the universal gas constant, and T is the absolute temperature. In convenient units, R takes a value of 0.08206 L atm mol^{-1} K^{-1}.

The chemical potential μ of an ideal gas is its molar Gibbs free energy. In mechanics and thermodynamics, absolute values of energies depend on how the zero of energy is defined, but physical observables only depend on energy differences. In order to standardize the tabulation of quantities useful for the calculation of energy differences, a system of standard states has been defined. The most common standard state for gases is an ideal gas at one atmosphere of pressure and the temperature of interest. At 298° K this corresponds to an ideal gas at a concentration of 0.04089 mol L^{-1}. This is not always the most convenient standard state for discussing gases in equilibrium with liquid solutions. Another useful choice of standard state is the ideal gas at a concentration of 1 mol L^{-1}; at 298° K this corresponds to a pressure of 24.45 atm = 18583 torr. It is always important to understand what standard state is being used when one uses tabulated thermodynamic data.

The chemical potential of an ideal gas at temperature T depends on pressure according to the following relation:

$$\mu = \mu^{\mathrm{O}} + RT \ln\left(\frac{P}{P^{\mathrm{O}}}\right) \tag{2}$$

where μ^{O} and P^{O} are the chemical potential and pressure in the standard state. The significance of this equation is that if μ^{O} is tabulated for a given temperature for a known value of P^o, we can calculate the molar free energy of an ideal gas at any other pressure at this temperature by using the tabulated values. If we change the standard state, i.e., if we change P^o, then a different μ^o must be tabulated, but the resulting μ calculated from Equation 2 must be independent of the choice of standard state. Consider two possible standard states 1 and 2. We must have

$$\mu^o(1) + RT \ln \frac{P}{P^o(1)} = \mu^o(2) + RT \ln \frac{P}{P^o(2)} \tag{3}$$

from which it follows that

$$\mu^o(1) = \mu^o(2) + RT \ln \frac{P^o(1)}{P^o(2)} \tag{4}$$

Let standard state 1 be the 1 atm ideal gas, and standard state 2 be the 1 mol L^{-1} ideal gas. Then, using the values of P^o given above we have, at 298° K,

$$\mu^o(1 \text{ atm}) = \mu^o(1\ M) - RT \ln 24.45 = \mu^o(1\ M) - 1.89 \text{ kcal/mol} \tag{5}$$

That is, the magnitude of the chemical potential changes by 1.89 kcal/mol on going from one standard state to the other.

Equations 1 and 5 apply only to ideal gases. For nonideal gases one writes[4a]

$$\mu = \mu^o + RT \ln\left(\frac{f}{f^o}\right) \tag{6}$$

where f is the fugacity of the gas. The fugacity is defined by Equation 6. In other words, we keep the simple form of Equation 2 by hiding the complicated behavior of the real gas (as opposed to an ideal gas) in the fugacity function.

It is conventional to define fugacity so that in the limit of a dilute gas, it becomes the pressure. Thus fugacity has units of pressure. When one deals with condensed phases and with gases in equilibrium with condensed phases, it becomes convenient to introduce a unitless generalization of fugacity, which is called activity. The activity is defined by[4b]

$$a = f/f^o \tag{7}$$

and therefore

$$\mu = \mu^{o} + RT \ln a \tag{8}$$

At equilibrium, all components of a mixture have the same molar free energy, i.e., the same chemical potential, in any phase in which they are present, and they have the same chemical potential as all other components. However it is not always convenient to use the same standard state for all components or even for the same component in all phases. Just as Equation 6 defines fugacity, Equation 7 or 8 defines activity. Furthermore, Equations 6–8 define f and a for all substances, not just gases. However we should keep in mind that we do not use the same standard state for a substance in all the phases, mixtures, or pure states in which it may occur or for all components of a mixture.

To proceed we will first summarize the treatment of ideal mixtures. Then we will consider nonideal mixtures.

3. IDEAL MIXTURES

An ideal mixture is one for which[4c]

$$f_A = f_A^{\bullet} X_A \tag{9}$$

for every component A, where f_A is the fugacity of A in the mixture, X_A is the mole fraction of A in the mixture, i.e.,

$$X_A = \frac{n_A}{n_A + n_B} \tag{10}$$

and $f_A^{\bullet}$ is the fugacity of pure A at the same temperature and pressure of the mixture. Note that one can postulate ideal mixtures of ideal gases, but one can also postulate ideal mixtures of nonideal components, such as real gases and liquids.

Consider an ideal mixture of ideal gases, A and B. Equation 9 yields

$$P_A = P_A^{\bullet} X_A \tag{11}$$

where $P_A^{\bullet}$ is the pressure of pure A at the same pressure and temperature as the mixture. The pressure of the mixture is

$$P = (n_A + n_B)RT/V \tag{12}$$

Substituting $P_A^\bullet = P$ along with Equations 10 and 12 into Equation 11 gives

$$P_A = n_A RT/V \tag{13}$$

which is the expected result.

Now apply Equation 9 to a liquid solution in equilibrium with its vapor. We assume that both the solute and the solvent are liquids in their pure state at the temperature and pressure of interest. Substituting Equation 9 into Equation 6 yields

$$\mu_A = \mu_A^o + RT \ln\left(\frac{f_A^\bullet X_A}{f_A^o}\right) \tag{14}$$

When X_A is unity, the left-hand side becomes the chemical potential of the pure substance:

$$\mu_A^\bullet = \mu_A^o + RT \ln\left(\frac{f_A^\bullet}{f_A^o}\right) \tag{15}$$

Note that the standard state fugacity now carries a subscript, because for A it is defined by the nonideal behavior of A. Substituting Equation 15 into Equation 14 yields[4d, 5]

$$\mu_A = \mu_A^\bullet + RT \ln X_A \tag{16}$$

This is independent of standard state, but we can also view it as a version of Equation 14 in which we conveniently choose the standard state as the pure liquid (we shall abbreviate this as liq.s.s. to denote the (pure) liquid standard state). Therefore

$$f_A^o(\text{liq.s.s}) = f_A^\bullet \tag{17}$$

This illustrates the statement made earlier that the most convenient choice of standard state may depend on the problem. For gas-phase problems involving A, it is convenient to choose the standard state for A as an ideal gas at 1 atm pressure. But, where the vapor of A is in equilibrium with a solution, it is sometimes convenient to choose the standard state as the pure liquid. Since f_A is the same for the pure liquid and the vapor in equilibrium with the pure liquid, we may equivalently equate $f_A^\bullet$(liq.s.s.) to either the fugacity of the pure liquid or the vapor in equilibrium with it. In general this vapor does not have a pressure of 1 atm nor does it have a concentration of 1 mol L^{-1}. Thus this standard state is not equivalent to either of the two gaseous standard states mentioned above.

Suppose that the vapor of A in equilibrium with the liquid is an ideal gas. Then Equation 2 yields

$$\mu_A = \mu_A^o + RT \ln \frac{P_A}{P^o} \tag{18}$$

where P_A is the vapor pressure of A in equilibrium with the liquid, and

$$\mu_A^\bullet = \mu_A^o + RT \ln \frac{P_A^\bullet}{P^o} \tag{19}$$

where $P_A^\bullet$ is the vapor pressure of pure A. Substituting Equations 18 and 19 into Equation 16 yields

$$RT \ln \frac{P_A}{P^o} = RT \ln \frac{P_A^\bullet}{P^o} + RT \ln X_A \tag{20}$$

which simplifies to

$$P_A = P_A^\bullet X_A \tag{21}$$

which is Raoult's law.[4e] Thus Equation 9 may be considered to be a generalization of Raoult's law.[7] Note that Equations 18–21 are equally applicable to a solution of one or more solutes B, C, D, etc. in liquid A. Raoult's law says that the partial pressure of each will be proportional to its concentration.

4. NONIDEAL SOLUTIONS

In actuality, Raoult's law is only an approximation for real systems. Although it is a good approximation for many solvents, for which $X_A \approx 1$ if the solution is dilute, it is often a very poor approximation for solutes, for which $X_A < 0.5$. In the limit as $X_A \to 0$, though, there is still a linear relationship

$$P_A \xrightarrow[X_A \to 0]{} k_A^X X_A \tag{22}$$

but k_A^X does not equal $P_A^\bullet$. Equation 22 is called Henry's law,[4f] and k_A^X is called the Henry's law constant. For small X_A, the molarity M_A and molality m_A are linear functions of X_A; thus there is also a linear relation between P_A and molarity or molality:

$$P_A \xrightarrow[M_A \to 0]{} k_A^M M_A \tag{23}$$

$$P_A \xrightarrow[m_A \to 0]{} k_A^m m_A \tag{24}$$

Equations 23 and 24 are alternative forms of Henry's law, and k_A^M and k_A^m are alternative forms of the Henry's law constant.

We can equate the chemical potential of the solute to the chemical potential of the vapor in equilibrium with it. Assume the vapor is an ideal gas:

$$\mu_A(\text{solute}) = \mu_A(\text{vapor}) = \mu_A^o(\text{g.s.s.}) + RT \ln P_A / P^o(\text{g.s.s.}) \tag{25}$$

$$= \mu_A^o(\text{g.s.s.}) + RT \ln \frac{k_A^X}{P^o(\text{g.s.s.})} + RT \ln X_A \tag{26}$$

where quantities referring to the gaseous standard state are labeled g.s.s. In this case, it may be convenient to use Equation 8 and let

$$a_A = X_A \tag{27}$$

Comparing Equations 26 and 27 to Equation 8 implies a new standard state (to be denoted as the Henry's Law standard state or HL s.s.) whose chemical potential is related to that for the gaseous standard state by

$$\mu_A^o(\text{HL s.s.}) = \mu_A^o(\text{g.s.s.}) + RT \ln \frac{k_A^X}{P^o(\text{g.s.s.})} \tag{28}$$

With the liquid standard state we have

$$\mu_A = \mu_A^o + RT \ln X_A \tag{29}$$

or

$$\mu_A = \mu_A^o + RT \ln a_A \tag{30}$$

Both Equations 29 and 30 are valid in the region where the limit of Equation 22 holds, but at higher concentrations of A, Equation 29 fails to hold. However, Equation 30 is a special case of Equation 8 and as such it defines the activity μ for any value of X_A. Thus,

$$a_A \neq X_A \tag{31}$$

but

$$a_A \xrightarrow[X_A \to 0]{} X_A \tag{32}$$

It is then convenient to define an activity coefficient γ_A such that[9]

$$a_A = X_A X_A \tag{33}$$

at all X_A. Equation 31 then implies

$$\gamma_A \xrightarrow[X_A \to 0]{} 1 \tag{34}$$

The activity coefficient measures the deviation from ideality. Substituting Equation 33 into Equation 30 yields

$$\mu_A = \mu_A^o + RT \ln \gamma_A X_A \tag{35}$$

Equation 29 implies that μ_A^o is the chemical potential of a hypothetical solution in which $X_A = 1$, but the vapor pressure over the solution still obeys Henry's law as extrapolated from infinite dilution. Thus the standard state is a hypothetical Henry's law solution of unit mole fraction.

The numerical value of the activity clearly depends upon the standard state, and one often encounters other choices for the standard state for solutes. For example, just as we obtained Equations 29 and 30 from Equation 22, we could have obtained similar looking equations from Equations 23 or Equation 24. However, the derivation requires a mention of one more thermodynamic convention if we wish to avoid nonsensical logarithms of quantities with units. Notice that Equation 22 implies that k_A^X has units of pressure; thus the logarithms in Equations 26 and 28 are well defined. At first, it might appear that the constants in Equations 23 and 24 have different units. However, the convention that is followed in thermodynamics is that M_A and m_A are the unitless numerical values of the molarity and molality; thus one can take their logarithms.

If we take the standard state as the hypothetical 1 molar Henry's law solution (sometimes shortened to "hypothetical ideal 1 molar solution," where the ideality referred to is Henry's law ideality in molarity units, that is, the proportionality of partial pressure and molarity, not Raoult's law ideality) we get

$$\mu_A^o(1\,M\ \text{sol.s.s.}) = \mu_A^o(\text{g.s.s.}) + RT\ln\frac{k_A^M}{P_A^o(\text{g.s.s.})} \tag{36}$$

$$\mu_A = \mu_A^o + RT\ln a_A \tag{37}$$

$$a_A = \gamma_A M_A \tag{38}$$

$$\gamma_A \xrightarrow[M_A \to 0]{} 1 \tag{39}$$

where 1 *M* sol.s.s. denotes the 1 molar hypothetical solute standard state. Furthermore if we take a hypothetical 1 molal Henry's law solute as the standard state (1 *m* sol.s.s.) we get

$$\mu_A^o(1\,m\ \text{sol.s.s.}) = \mu_A^o(\text{g.s.s.}) + RT\ln\frac{k_A^m}{P_A^o(\text{g.s.s.})} \tag{40}$$

$$\mu_A = \mu_A^o + RT\ln a_A \tag{41}$$

$$a_A = \gamma_A m_A \tag{42}$$

$$\gamma_A \xrightarrow[m_A \to 0]{} 1. \tag{43}$$

One cannot emphasize too often that the numerical values of μ_A^o, a_A, and γ_A depend on the choice of standard state. The usual thermodynamic convention is to say what standard state is used in words, not in the equation itself, but in many cases we have indicated it in the equation for clarity.

Before closing this section we note that even in nonideal solutions we can use the standard state of Equation 16 for the solute. Since Equation 16 only holds for ideal solutions, one generalizes to obtain[4g]

$$\mu_A = \mu_A^\bullet + RT \ln a_A \tag{44}$$

with

$$a_A = \gamma_A m_A / m_A^\bullet \tag{45}$$

where $m_A^\bullet$ is the molality of the standard state. Now, however, γ_A does not tend to unity as X_A tends to 0. However, comparing Equations 44 and 45 to Equations 34 and 35 yields

$$\lim_{X_A \to 0} \gamma_A(\text{liq.s.s.}) = e^{[\mu_A^o(\text{HL s.s.}) - \mu_A^\bullet]/RT} \tag{46}$$

Thus, γ_A of Equation 45 tends to a constant value in dilute solutions. This constant value is sometimes called the limiting activity coefficient γ_A^∞. Consideration of Equations 21 and 22 allows us to evaluate this limit:

$$\gamma_A^\infty \equiv \lim_{X_A \to 0} \gamma_A(\text{liq.s.s.}) = k_A^X / P_A^\bullet \tag{47}$$

Alternatively, we may write:

$$k_A^X = P_A^\bullet \gamma_A^\infty \tag{48}$$

This shows that knowledge of the vapor pressure of A and its limiting activity coefficient allows us to calculate the Henry's law constant.[7]

5. ELECTROLYTES

Electrolytes are solutes that carry an electrical charge. As charged species typically have negligible vapor pressures, it is convenient to introduce yet another standard state for their description.[8, 9] In general, the same conditions of concentration, temperature, and pressure are assumed as for a non-electrolyte (e.g., the HLss standard state), but the chemical potential of an anion is defined as

$$\mu_{A^-} = \mu_A + \Delta G^o_{(A/A^-)} \text{(e.s.s.)} \tag{49}$$

where $\Delta G^o_{(A/A^-)}$ (e.s.s.) denotes the standard-state free energy change for the reaction

$$A_{(sol)} + 1/2\ H_{2(g)} \rightarrow A^-_{(sol)} + H^+_{(sol)} \tag{50}$$

In Equation 50 the chemical potential of non-electrolyte A depends on the usual choice of standard-state conventions described above, and the chemical potentials of both $H_{2(g)}$ and $H^+_{(sol)}$ are taken to be zero (this defines e.s.s., the electrolyte standard state). By setting the standard-state free energy of the solvated proton equal to zero, this standard-state convention incorporates the absolute potential, ΔG^o_{NHE}, of the hydrogen electrode process producing one proton in solution,

$$1/2\ H_{2(g)} \rightarrow e^-_{(g)} + H^+_{(sol)} \tag{51}$$

into the magnitude of the chemical potential of the electrolyte. In water as solvent, the absolute potential of this electrode is 4.44 eV.[10]

For a cation, the analogous equations defining the standard-state chemical potential for A^+ are

$$\mu_{A^+} = \mu_A + \Delta G^o_{(A^+/A)} \text{(e.s.s.)} \tag{52}$$

$$A^+_{(sol)} + 1/2\ H_{2(g)} \rightarrow A_{(sol)} + H^+_{(sol)} \tag{53}$$

where $\Delta G^o_{(A^+/A)}$ (e.s.s.) in Equation 52 refers to the standard-state free energy change for Equation 53.

Note, in using Equations 50 and 53 above, that tabulations of thermodynamic data for electrolytes tend to employ a 1 molar ess concentration for all species in solution. For situations defined to have a standard-state pH value different from 0 (which corresponds to a 1 molar concentration of solvated protons), the standard-state chemical potentials for anions and cations are determined as

$$\mu_{\mathrm{A}^-}(\text{e.s.s.; pH} = k) = \mu_{\mathrm{A}^-}(\text{e.s.s.}) - kRT \quad (54)$$

$$\mu_{\mathrm{A}^+}(\text{e.s.s.; pH} = k) = \mu_{\mathrm{A}^+}(\text{e.s.s.}) + kRT \quad (55)$$

Note the analogy between Equations 54 and 55 and Equation 4.

An example of the use of these standard states for working with solvation effects on one-electron oxidation potentials is provided elsewhere.[11]

6. SOLVATION

We are now in a position to consider the difference in chemical potential between a solute in its (hypothetical) standard state in a liquid solution and that same chemical species in its gas-phase standard state. This difference is the standard-state molar free energy of solvation $\Delta G_{\mathrm{S}}^{\mathrm{o}}$ (also referred to as the standard state molar free energy of transfer from the vapor phase to a liquid solution). We define

$$\Delta G_{\mathrm{S}}^{\mathrm{o}} = \mu_{\mathrm{A}}^{\mathrm{o}}(\text{solute}) - \mu_{\mathrm{A}}^{\mathrm{o}}(\text{vapor}) \quad (56)$$

where we have used different standard states for each phase. Thus the superscript on $\Delta G_{\mathrm{S}}^{\mathrm{o}}$ unlike all previous standard-state superscript in these notes refers not to a particular standard state of one substance, but rather to the fact that the dissolution process being considered is a hypothetical one involving a transfer from one standard state to another. (This may be compared to the situation for heat of formation, which is the enthalpy of forming one mole of a substance in its standard state from the elements, each in their own standard state. In that case, there are different standard states for different substances in the initial and final states of the transfer process.)

Now substitute Equations 2 and 35 into Equation 56:

$$\Delta G_{\mathrm{S}}^{\mathrm{o}}(\text{HL s.s.}) = \mu_{\mathrm{A}}(\text{solute}) - \mu_{\mathrm{A}}(\text{vapor}) - RT\ln\gamma_{\mathrm{A}}X_{\mathrm{A}} + RT\ln\frac{P_{\mathrm{A}}}{P^{\mathrm{o}}} \quad (57)$$

where the parenthetical notation on the left-hand side indicates that we are using a Henry's law standard state for the solute. If the vapor is in equilibrium with the solution, the first two terms on the right-hand side of Equation 57 cancel, and we have

$$\Delta G_{\mathrm{S}}^{\mathrm{o}}(\mathrm{HL\ s.s.}) = RT \ln \frac{P_{\mathrm{A}}}{P^{\mathrm{o}} X_{\mathrm{A}} \gamma_{\mathrm{A}}} \tag{58}$$

Using Equation 34 allows us to write this as

$$\Delta G_{\mathrm{S}}^{\mathrm{o}}(\mathrm{HL\ s.s.}) = RT \lim_{X_{\mathrm{A}} \to 0} \ln \frac{P_{\mathrm{A}}}{P^{\mathrm{o}} X_{\mathrm{A}}} \tag{59}$$

To evaluate the logarithm, we must measure the vapor pressure P_{A} of A in equilibrium with a solution where its mole fraction is X_{A} in the limit where the solution becomes infinitely dilute. That is, in the limit of infinite dilution where γ is 1, the free energy of solvation can be obtained from measurement of the solute vapor pressure (in the appropriate standard state units) over a solution of known concentration.

Substituting Equation 22 into Equation 59 yields

$$\Delta G_{\mathrm{S}}^{\mathrm{o}}(\mathrm{HL\ s.s.}) = RT \ln k_{\mathrm{A}}^{X} / P^{\mathrm{o}} \tag{60}$$

Now substituting Equation 48 yields

$$\Delta G_{\mathrm{S}}^{\mathrm{o}}(\mathrm{HL\ s.s.}) = RT\left(\ln \frac{P_{\mathrm{A}}^{\bullet}}{P^{\mathrm{o}}} + \ln \gamma_{\mathrm{A}}^{\infty} \right) \tag{61}$$

Thus the free energy of solvation may be calculated from the Henry's law constant or from the vapor pressure of the pure substance and the limiting activity coefficient. Thus, if the deviation of the solution from Raoult's law behavior is known, calculation of the standard state free energy of solvation requires only the vapor pressure of the pure substance (in the standard state units). For an ideal solution that behaves according to Raoult's law, γ^{∞} would be 1, leading to the observation that $\Delta G_{\mathrm{S}}^{\mathrm{o}}$ would depend simply on the vapor pressure of the pure solute.

In our quantum mechanical solvation modeling,[12-27] we take the standard state of the vapor to be a 1 molar ideal gas at 298° K and the standard state of the solute to be a hypothetical 1 molar Henry's law solute at the same temperature and pressure. Free energies of solvation, $\Delta G_{\mathrm{S}}^{\mathrm{o}}$, for this choice of standard states, may be derived by employing the theory given above. First, we combine Equations 28 and 36 to get

$$\mu_{\mathrm{A}}^{\mathrm{o}}(1\,M\ \text{sol.s.s}) = \mu_{\mathrm{A}}^{\mathrm{o}}(\text{HL s.s}) + RT\ \ln\frac{k_{\mathrm{A}}^{M}}{k_{\mathrm{A}}^{X}} \tag{62}$$

$$= \mu_{\mathrm{A}}^{\mathrm{o}}(\text{HL s.s.}) - RT\ \ln M_{\mathrm{A}}^{\bullet} \tag{63}$$

Equation 63 follows from Equation 62 by combining Equations 22 and 23. Clearly, the ratio of Henry's law constants is the molarity of the solution in which the mole fraction of solute A is unity, i.e., the molarity of pure liquid A. We next note that Equation 56 implies

$$\Delta G_{\mathrm{S}}^{\mathrm{o}}(1\,M\ \text{sol.s.s.}) = \Delta G_{\mathrm{S}}^{\mathrm{o}}(\text{HL s.s.}) + \mu_{\mathrm{A}}^{\mathrm{o}}(1\,M\ \text{sol.s.s.}) - \mu_{\mathrm{A}}^{\mathrm{o}}(\text{HL s.s.}) \tag{64}$$

and using Equation 63 then yields

$$\Delta G_{\mathrm{S}}^{\mathrm{o}}(1\,M\ \text{sol.s.s.}) = \Delta G_{\mathrm{S}}^{\mathrm{o}}(\text{HL s.s.}) - RT\ \ln M_{\mathrm{A}}^{\bullet} \tag{65}$$

Combining this with Equations 60 and 61 yields

$$\Delta G_{\mathrm{S}}^{\mathrm{o}}(1\,M\ \text{sol.s.s.}) = RT\ \ln\frac{k_{\mathrm{A}}^{X}}{P^{\mathrm{o}} M_{\mathrm{A}}^{\bullet}} \tag{66}$$

and

$$\Delta G_{\mathrm{S}}^{\mathrm{o}}(1M\,\text{sol.s.s.}) = RT\ \ln\frac{P_{\mathrm{A}}^{\bullet}\gamma_{\mathrm{A}}^{\infty}}{P^{\mathrm{o}} M_{\mathrm{A}}^{\bullet}} \tag{67}$$

where, for a 1 M gaseous standard state, P^{o} should be set to 24.45 atm, as discussed between Equations 1 and 2. Other transformations of standard states may be accomplished equivalently.

Let's do a numerical example: 1,2-ethanediol. We will use Equation 67 since $\gamma_{\mathrm{A}}^{\infty}$ is known to be 0.8.[28] The density is[29] 1.113 g cm^{-1}; therefore

$$M^{\bullet} = \frac{1.113\text{g}}{\text{m}^3}\,\frac{10^3\,\text{cm}^3}{\text{L}}\,\frac{\text{mol}}{62.07\,\text{g}} = 17.93 \tag{68}$$

The vapor pressure is[30] 0.010 kPa; therefore

$$P^{\bullet} = 0.010 \text{ kPa} \frac{\text{atm}}{101.325 \text{ kPa}} = 9.87 \times 10^{-5} \tag{69}$$

Finally, Equation 67 yields ($T = 298°$ K):

$$\Delta G_S^o = 0.592 \text{ kcal mol}^{-1} \ln \frac{(9.87 \times 10^{-5})(0.8)}{(24.45)(17.93)} = -9.19 \text{ kcal mol}^{-1} \tag{70}$$

Notice that if we had used the Henry's law standard state for the solute, we would have obtained –7.48 kcal/mol (which can be derived by combining Equations 63, 68, and 70). Note that if $\gamma_A^{\infty} = 0.9$, the free energy of solvation would be less negative by only 0.07 kcal/mol. Thus, when the solution is nearly ideal ($\gamma_A^{\infty} \approx 1$), the free energy of solvation is primarily determined by the vapor pressure.

Note that we previously[16] quoted ΔG_S^o as –9.6 kcal/mol and as[17] –9.3 kcal/mol. These values differ from the present value by 0.4 kcal/mol (apparently a math error) and 0.1 kcal/mol (apparently a round-off error), respectively.

A special case of Equation 67 concerns the "free energy of solvation of A in A." Since a solution of A in A obviously satisfies Raoult's Law, we have $\gamma_A^{\infty} = 1$. Then Equation 67 can be used to find the solvation energy of A in A from the vapor pressure of A and its density.[31]

7. SOLUBILITY

Next consider the relationship between the free energy of solvation and the solubility of a solute. First, in keeping with all the developments above, we consider a solute that is a liquid in its pure state. By combining of Equations 55 and 61, we may write

$$\Delta G_S^o = RT \ln\left(\frac{P_A^{\bullet}}{P^o}\right) + RT\left(\frac{k_A^X}{P_A^{\bullet}}\right) \tag{71}$$

Then Equation 22 yields

$$\Delta G_S^o = RT \ln\left(\frac{P_A^{\bullet}}{P^o}\right) + \lim_{X_A \to 0} RT \ln\left(\frac{P_A}{P_A^{\bullet} X_A}\right) \tag{72}$$

$$= RT \ln\left(\frac{P_A^\bullet}{P^o}\right) + RT \ln\left(\frac{P_A}{P_A^\bullet X_A^\infty}\right) \tag{73}$$

where X_A^∞ refers to the mole fraction of solute in the infinitely dilute regime in which Henry's law is obeyed. In the special case of a liquid solute which is saturated in the solvent at such very dilute concentrations, we must have a case where the equilibrium chemical potentials of the pure liquid solute, the solute in solution, and the solute vapor over both of the phases are all equal. For the vapor over the pure liquid solute, we have

$$\mu_A^\bullet(\text{gas}) = \mu_A^o(\text{g.s.s.}) + RT \ln\left(\frac{P_A^\bullet}{P^o}\right) \tag{74}$$

while for the vapor over the solution we have

$$\mu_A^\bullet(\text{gas}) = \mu_A^o(\text{g.s.s.}) + RT \ln\frac{P_A}{P^o} \tag{75}$$

Since these two chemical potentials must be equal (given the equilibrium between the pure liquid and the saturated solution), it must be the case that $P_A^\bullet = P_A$, in which case Equation 72 becomes

$$\Delta G_S^o = RT \ln\left(\frac{P_A^\bullet}{P^o}\right) - RT \ln X_A^\infty \tag{76}$$

This equation has the expected behavior that ΔG_S^o becomes more positive with decreasing solubility of the solute. However, free energies of solvation for different solutes cannot be related to their relative solubilities unless the vapor pressures of the different solutes are similar or one takes account of this via Equation 76. Furthermore, if the solubility is high enough that Henry's law does not hold, then one must consider finite-concentration activity coefficients, not just the infinite-dilution limit.

The situation is more complicated for saturated solutions of solid solutes, since there is a free energy of fusion term associated with leaving the pure solid in order to dissolve into solution (or, in the other direction, we must take account of the free energy of crystallization). Since this term, like vapor pressure, will be different for different solutes, it is in general not appropriate to assign relative solubilities based on relative free energies of solvation. Furthermore, molecular modeling techniques for estimating crystallization energies are not as well developed as those for estimating solvation energies. For a discussion of solubility of solids in terms of infinite dilution activity coefficients, see Grant and Higuchi.[32]

A similar multiphase complication that should be kept in mind when discussing solutions at finite concentrations is possible micelle formation. It is well known that for many organic solutes in water, when the concentration exceeds a certain solute-dependent value, called the critical micelle concentration (cmc), the solute molecules are not distributed in a random uncorrelated way but rather aggregate into units (micelles) in which their distances of separation and orientations with respect to each other and to solvent molecules have strong correlations. Micelle formation, if it occurs, will clearly have a major effect on the apparent activity coefficient but the observation of the phenomenon requires more sophisticated analytical techniques than observation of, say, liquid-liquid phase separation.

8. MODELING: EQUILIBRIUM PROPERTIES

The reason we prefer to use 1 *M* for the standard state in both the gas phase and in liquid solution is that using the same concentration in the gas phase and solution eliminates an entropic term in the statistical mechanical free energy and allows us to focus on the interaction terms coupling the solute to the solvent.[33-35] In particular, using the standard state of Equation 67, we can write[33-35] the free energy of solvation of a rigid, non-rotating solute as

$$\Delta G_S^{o}(\mathbf{x}) = -RT \ln \left\langle \exp\left(-B/\tilde{k}T\right)\right\rangle_{\text{solvent}} \tag{77}$$

where **x** denotes the set of vibrational coordinates of the solute, $\tilde{k}$ is Boltzmann's constant, *B* is the potential energy of interaction between the solute and the solvent, and $\langle\cdots\rangle_{\text{solvent}}$ denotes an average over all possible solvent configurations. In practice we are interested in the free energy of solvation of a nonrigid, rotating solute, and this is given by

$$\Delta G_S^{o} = \left\langle \Delta G_S^{o}(\mathbf{x})\right\rangle_{\mathbf{x}} + \Delta G_{\text{int}} \tag{78}$$

where $\langle\cdots\rangle_{\mathbf{x}}$ denotes an average over vibrational coordinates, G_{int} is the internal (i.e., conrovibronic, i.e., conformational-rotational-vibrational-electronic) free energy of the solute, and

$$\Delta G_{\text{int}} = G_{\text{int}}(\text{sol.}) - G_{\text{int}}(\text{g}). \quad (79)$$

In practice we often neglect the distinction between ΔG_S^o and $\Delta G_S^o(\mathbf{x})$, although sometimes it is important to optimize the geometry in solution[21] or to at least include the conformational part.[14] (If one did try to include the rotational part, one would run into the problem that the 3 gas-phase rotations are converted in liquid solution into low-frequency librations that are strongly coupled to low-energy solvent motions). In the rest of this section we focus on $\Delta G^o(\mathbf{x})$.

There are two main approaches to calculating Equation 77. In the brute-force or atomistic approach the solvent molecules are treated explicitly, and the average is calculated by using Monte Carlo[36] or molecular dynamics[38, 39] methods to sample the solvent configurations. In the so-called continuum approach, the solvent is modeled by a field, i.e., a collective solvent coordinate, that does not depend on the instantaneous positions of the nuclei.[12, 15, 21, 22] The advantages of the atomistic approach are that it provides a detailed picture of the solvent and it can readily be applied to arbitrary systems provided a potential energy function is available. The advantages of the collective-solvent-coordinate model are lower cost, and earlier route to including quantum mechanical effects, and elimination of the need for potential energy functions for individual solute-solvent interactions. Since the atomistic approach is treated in a separate chapter of this text, the present chapter will only consider collective-solvent-coordinate approaches.

In general, collective-coordinate approaches separate $\Delta G_S^o(\mathbf{x})$ into two parts: bulk electrostatics (henceforth called just electrostatic or ΔG_{elec}) and the rest. This is an extrathermodynamic distinction, and there is no unique way to separate the two kinds of effects in either thermodynamics or statistical mechanics. In the most accurate collective-coordinate approaches, is modeled by self-consistent reaction field (SCRF) ΔG_{elec} theory.[12, 15, 21, 22, 39-49] The reaction field is the field acting on the solute due to the electric polarization of the solvent induced by the solute. The electric polarization of the solvent partially cancels the electric field lines emanating from the charges and partial charges of the solute, thereby reducing the self-energy and charge-charge interactions within the solute. This favors higher partial atomic charges in molecular solutes as compared to their gas-phase charge distributions.

If the solute were simply a collection of point charges surrounded by a continuous dielectric medium with the bulk dielectric constant ε of the solvent, the self-energy and the strength of charge-charge interactions in the solute would be reduced by a factor of ε. This is called dielectric screening. However, the solute itself occupies a finite volume, and solvent is excluded from this volume. This reduces the dielectric screening and is called dielectric descreening. In the early days of solvation modeling, the solute was represented as a sphere or ellipsoid;[39-45] such a model provides only a crude accounting for dielectric descreening. Modern theories represent the solute as a superposition of atomic spheres centered at the nuclei; this is much more realistic.

The electric polarization of the solvent has three components: electronic, "atomic" (i.e., translational and vibrational), and orientational. The polarization of a nonpolar solvent is almost entirely electronic; this leads to $\varepsilon \approx 2$. Polar solvents can have much larger dielectric constants, e.g. ε is 13.9 for 1-pentanol, 37.7 for methanol, and 78.3 for water.[50]

The electrostatic contribution to the free energy of solvation is one half the interaction energy of the solute with the reaction field. The factor of one half comes from the fact that the free energy cost of polarizing the solvent is one half of the favorable interaction energy that one gains; the simplicity of this result is a consequence of assuming linear response of the solvent to the solute.[21, 43, 45]

The chief uncertainties in calculating G_{elec} are (i) the charge distribution of the solute and (ii) the location of the boundary at which one switches between solute screening and descreening. We will consider these in order. In molecular mechanics modeling, one associates standard partial charges to the various atoms of common functional groups. This is only satisfactory for zero-order estimates. Better charges may be obtained from quantum mechanical electronic structure calculations employing a self-consistent field (SCF), i.e., in which each orbital is optimized self-consistently in the field of the others. If the orbitals are optimized not only in the field of the other occupied orbitals but also self-consistently with the reaction field, one obtains the SCRF method. In our own most recent work, the charges used in the SCRF calculations are obtained from the orbitals by what is called a class IV mapping.[51] This mapping contains semiempirical parameters that makes up for the lack of complete electron correlation and other deficiencies of the electronic structure method. Our most recent set of mapping parameters is called Charge Model[52, 53] (CM2).

In most of our own solvation models,[12-17, 19-26] the interaction of the solute partial atomic charges with electric polarization of the solvent dielectric medium outside the overlapping spheres representing the solute is calculated by the generalized Born[42, 54-57] (GB) approximation. We have also developed a semiempirical model[27] based on the conductor-like screening model[58] (COSMO), which uses the $\varepsilon = \infty$ limit of the Poisson equation and then scales the results to finite ε. (We have also used a pairwise descreening model[18] discussed below.) The COSMO model uses the continuous solute electron density function $\rho(\mathbf{r})$ rather than replacing it by a set of partial atomic charges.[18, 58, 59] In principle this allows a more accurate treatment of lone pairs and atomic dipoles, but it cannot take advantage of the improvement in accuracy afforded by the class IV mapping.

A third class of solvation models is based directly on the Poisson equation for finite ε, based either on $\rho(\mathbf{r})$[48, 60] or its multipole expansion.[47, 61] The approach based on $\rho(\mathbf{r})$ is usually called polarized continuum model (PCM). It is unclear if the extra accuracy of using the Poisson equation is sufficient to offset its chief disadvantage, which is sensitivity to the portion of the solute charge that lies outside the set of overlapping solute spheres. This results in a spurious contribution to the solvation energy is variously called outlying charge error or escaped charge error.[60] The outlying charge error is usually compensated by a renormalization procedure,[61] but it can still be severe, especially for anions. The use of a truncated multipole expansion eliminates the outlying charge error in principle, but suffers from the fact that the multipole expansion is slowly convergent for large molecules.

An alternative to the GB, COSMO, and Poisson electrostatic calculations is to model the solution to the Poisson equation in terms of pair potentials between solute atoms; this procedure is based on the physical picture that the solvent screens the intra-solute Coulombic interactions of the solute, except for the critical descreening of one part of the solute from the solvent by another part of this solute. This descreening can be modeled in an average way to a certain level of accuracy by pairwise functions of atomic positions.[18, 64, 65] One can obtain quite accurate solvation energies in this way, and it has recently been shown that this algorithm provides a satisfactory alternative to more expensive explicit-solvent simulations even for the demanding cases of 10-base-pair duplexes of DNA and RNA in water.[66]

The electrostatic methods just discussed suitable for nonelectrolytic solvent. However, both the GB and Poisson approaches may be extended to salt solutions, the former by introducing a Debye-Hückel parameter[67] and the latter by generalizing the Poisson equation to the Poisson-Boltzmann equation.[68] The Debye-Hückel modification of the GB model is valid to much higher salt concentrations than the original Debye-Hückel theory because the model includes the finite size of the solute molecules.

Perhaps the most widely discussed source of uncertainty in electrostatic calculations is the location of the solute/solvent boundary. The most common treatment is to place the boundary at the surface of a set of overlapping spheres centered at the nuclei. But what radius should one use for those spheres? One common answer is van der Waals radii times 1.2.[46] In our own quantum mechanical solvation models,[12-27] and those of several others[59, 69], these radii are empirical parameters. Recently Barone et al.[70] have modified the PCM to use charge-dependent united-atom spheres instead of all-atom spheres, and they optimized the electrostatic radii for a particular normalization scheme, known as the ICOMP = 4 scheme. The resulting model is called the united-atom Hartree-Fock (UAHF) model.

The "correct" radii for electrostatic solvation calculations have been debated for over 50 years. But there can be no correct answer. The model that the dielectric constant changes from unity to ε at a definite point is wrong. The solute boundary region fluctuates as a consequence of solute vibrations and solvent motions, and the dielectric constant, to the extent that it is even defined in a microscopic space, also fluctuates and changes gradually over this region. The latter aspect of solvent polarization is sometimes treated as a form of dielectric saturation, i.e., one notes that at high enough fields (e.g., close to a polar or changed solute), the permittivity (dielectric constant) of the solvent is not constant, it is reduced and depends on the field.[71, 72] Thus we believe that it is futile to hope to find the "best" radii by considering only electrostatics. Instead we recommend finding reasonable radii that clearly define a bulk electrostatic contribution, however arbitrary, and then concentrating on making the non-electrostatic term (mentioned above) be *consistent* with this choice. In our work, we do this by parameterizing the non-electrostatic term against experimental data.

Some of the physical effects that must be included in the non-electrostatic term are:

1) cavitation, i.e., the free energy cost of making a cavity in the solvent to allow room for the solute;

2) dispersion, i.e., the change in solvent-solvent dispersion forces and the introduction of solute-solvent dispersion forces when the solute is placed in the cavity (the change in solvent-solvent dispersion forces due to having made the cavity are in principle in the cavitation term, but one can see that the effects are not neatly separable);

3) exchange repulsion of solute and solvent and the change in intrasolvent exchange repulsion (Exchange repulsion is ultimately due to the Pauli exclusion principle which prevents the charge clouds of atoms from overlapping significantly without a large energy penalty. Those who appreciate advanced quantum mechanics realize that, due to the Hellmann-Feynman theorem,[73-75] all such quantum effects can still be calculated by electrostatics, i.e., once you know the wave function, everything is electrostatics. That is one reason why, when we speak carefully, we say that the so called non-electrostatic term is actually a measure of the deviation of the solvation energy from a bulk electrostatic model):

4) the deviation of other solute-solvent interactions from bulk electrostatics, e.g., the fact that a hydrogen bond is not explainable in terms of bulk electrostatics;

5) other changes in solvent structure due to the introduction of the solute, e.g., those changes that are responsible for the hydrophobic[76-82] effect.

When the non-electrostatic terms are semiempirical, they also make up in an average way for systematic deficiencies in the treatment of electrostatics, e.g., for the truncation of the distributed multipole representation of the solute charge density at the monopole term on each center.

There are three popular ways to treat the nonelectrostatic effects: (i) ignore them, (ii) combine specialized models for cavitation, dispersion, exchange repulsion, and so forth,[46-48, 70] (iii) employ atomic surface tensions.[12-27, 83-86] In the third approach, which is the most accurate in an empirical sense, one writes[22-27]

$$\Delta G_S^O(\mathbf{x}) = G_{\text{elec}}(\mathbf{x}) + \sum_k \sum_i \sum_j \sum_\delta \sigma_{kij\delta} A_{ki}(\mathbf{x}) f_{kij}(\mathbf{x}) S_\delta \tag{80}$$

where $\sigma_{kij\delta}$ is an empirical atomic surface tension, $A_{ki}(\mathbf{x})$ is the solvent-accessible surface area of atom k corresponding to effective solvent radius r_{kij}, $f_j(\mathbf{x})$ is geometrical factor, and S_δ is a solvent descriptor. The use of more than one effective solvent radius allows a more physical treatment of the various separate effects (e.g., dispersion forces extend only a short distance into the solvent whereas solvent structural perturbations may penetrate into the solvent over a longer length scale), the geometrical factor takes account of chemical functionality (e.g., an H bonded to C is hydrophobic whereas an H bonded to O is not), and the use of several solvent descriptors is key to obtaining a universal model that works in any organic solvent. We believe that the solvent descriptors should include measures of the solvent's acidity, basicity, macroscopic surface tension, electronic polarizability (as indicated, e.g., by its index of refraction n), halogenic character, and aromaticity.[22-27] Descriptors representing these solvent characteristics may be thought of as collective solvent coordinates that interact with the solute through Equation 73.

Because surface curvature depends on radius and different atoms have different sizes, and because the atomic surface tension depends on atomic number, the atomic surface tensions also include surface curvature effects, which has recently been studied as a separate effect.[7] Local surface curvature may also correlate with nearest-neighbor proximity and thus may be implicitly included to some extent when semiempirical atomic surface tensions depend on interatomic distances in the solute.

It is actually possible to create a model based entirely on atomic surface tensions, and, at least for species with no net charge, it does quite well.[23-25, 87, 88] Such a model can be quite useful for drug design because of its speed and simplicity, but it is somewhat unsatisfactory theoretically because the correct physics is not *manifest*.

In drug design one often uses an organic solvent as a surrogate for a cell membrane or for the blood-brain barrier in designing the partitioning properties that are essential to bioavailability. For example, partitioning into 1-octanol, *n*-hexadecane, and chloroform have all been used to correlate or predict bioavailability. Table 1 indicates that these three solvents span a range of hydrogen bonding acidity, basicity, and polarity (as measured by Abraham's hydrogen bonding descriptors[89] α and β and the dielectric constant ε, respectively), but have very similar indices of refraction n and macroscopic surface tensions γ. Also included in Table 1 are estimated values for solvent descriptors to characterize phosphatidyl choline (PC) if it were to be a pure liquid (these previously unpublished estimates derive from analysis of various related molecules, long chain esters, phosphates, etc.) Note that, if we assume that partitioning into a PC bilayer (which would represent a reasonable model for a biomembrane) can be well modeled by the solvent descriptors in Table 1 in the same way that solvent/solvent partitioning can be, PC does not really look much like octanol, chloroform, or hexadecane. That is, it is not clear that any one of these solvent/solvent partitioning models should be expected to be terribly predictive of bioavailability when that property is tied to membrane crossing of drug molecules.

A better idea may be to develop specific effective solvent descriptors by using data on actual membranes.[23] Experimental partitioning data are available for the case of water/phosphatidyl choline bilayer for a variety of organic solutes.[23, 90] If we assume that the dielectric constant of the phosphatidyl choline bilayer is 5.0 (an estimate based on the dielectric constant of 1-octanol) and that the α value is zero (there are no hydrogen bond donors for this molecule), and regress the experimental partition coefficients on the remaining three solvent parameters n, γ, and β, the regression provides values of 1.40, 25, and 1.15 for these parameters when using the AM1/SM5.4 model[19, 20] for computing solvation free energies. The similarity between the estimates given above for these descriptors and the values obtained from regression speaks to the physicality of the model . The regression itself has an R value of 0.9 over 19 reasonably diverse solutes, which is high enough that one might anticipate useful performance for screening. The virtue of this approach is that it is quite general. Given any particular membrane model and some initial data for partitioning, one can

Table 1. Solvent descriptors.[a]

	n	α	β	γ	ε[b]
1-octanol	1.43	0.37	0.48	39	10
chloroform	1.45	0.15	0.02	38	5
n-hexadecane	1.43	0	0	39	2
phosphatidyl choline[c]	1.37	0	0.9	27	5

[a] α and β are Abraham's $\sum\alpha_2^H$ and $\sum\beta_2^H$, respectively. [b] In Tables 1–3, dielectric constants are rounded to nearest integer to highlight major trends, but unrounded values were used for all calculations. [c] Estimated.

Table 2. Free energies of solvation (kcal/mol) of three solutes in eight solvents.

	Solvent (dielectric constant)							
Solute	$C_{16}H_{34}$ (2)	CCl_4 (2)	ethyl ether (4)	$HCCl_3$ (5)	butyl acetate (5)	tributyl phosphate (8)	1-octanol (10)	water (78)
1-butanol	−3.5[a]	−4.2	−5.7	−5.3	−5.2	−6.3	−5.7	−4.7
	−3.6[b]	−4.6	−5.4	−5.1	−5.3	−6.4	−6.3	−4.8
butylamine	−3.6	−4.3	−4.4	−5.3	NA[c]	−4.3	−5.4	−4.3
	−3.7	−4.6	−4.6	−4.9	−4.3	−4.3	−5.3	−4.1
aniline	−5.4	−6.1	−6.5	−6.9	−7.3	−7.6	−6.7	−5.5
	−5.4	−6.4	−6.7	−6.5	−7.6	−7.3	−6.9	−5.1

[a]upper value: experiment; [b]lower value: SM5.2R/MNDO; [c]not available

design model-specific descriptors to help evaluate the bioavailability aspect of further drug design efforts.

How well can continuum solvation models distinguish changes in one or another of these solvent properties? This is illustrated in Table 2, which compares solvation energies for three representative solutes in eight test solvents. Three of the test solvents are those shown in Table 1, one is water, and the other four were selected to provide useful comparisons on the basis of their solvent descriptors, which are shown in Table 3. Notice that all four solvents in Table 3 have no acidity, which makes them more suitable, in this respect, than 1-octanol or chloroform for modeling biomembranes. Table 2 shows that the SM5.2R model, with gas-phase geometries and semiempirical molecular orbital theory for the wave function, does very well indeed in reproducing all the trends in the data.

Table 3. Descriptors for more solvents.

	n	α	β	γ	ε
ethyl ether	1.35	0	0.41	24	4
butyl acetate	1.39	0	0.45	36	5
tributyl phosphate	1.42	0	1.21	28	8
carbon tetrachloride	1.45	0	0	38	2

9. NONEQUILIBRIUM PROPERTIES

Self-consistent reaction fields (SCRF) methods[17, 91] and other methods that predict the thermodynamic free energy of solvation are based on the assumption of thermodynamic equilibrium. This is certainly the correct model for calculating free energies of reaction, molecular partition coefficients of stable molecules, molecular solubilities, and ionic solubility products. There are, however, two types of problems where nonequilibrium effects may manifest themselves, namely spectroscopy and chemical dynamics. Nonequilibrium effects in these areas are discussed in two recent reviews.[21, 49] and the status of current understanding is summarized in the rest of this section. Understanding the differences between equilibrium solvation and solvation effects in spectroscopy is important for designing structure-property relationships, and understanding solvation effects on dynamics is important for modeling reactivity in general. For example, the fate of a drug lead in an aqueous biophase or the factors controlling transition state stabilization in a catalyzed reaction may depend significantly on nonequilibrium solvation effects.

In spectroscopy we may distinguish two types of process, adiabatic and vertical. Adiabatic excitation energies are by definition thermodynamic ones, and they are usually further defined to refer to at 0° K. In practice, at least for electronic spectroscopy, one is more likely to observe vertical processes, because of the Franck-Condon principle. The simplest principle for understandings solvation effects on vertical electronic transitions is the two-response-time model in which the solvent is assumed to have a fast response time associated with electronic polarization and a slow response time associated with translational, librational, and vibrational motions of the nuclei.[92] One assumes that electronic excitation is slow compared with electronic response but fast compared with nuclear response. The latter assumption is quite reasonable, but the former is questionable since the time scale of electronic excitation is quite comparable to solvent electronic polarization (consider, e.g., the excitation of a 4.5 eV $n \rightarrow \pi^*$ carbonyl transition in a solvent whose frequency response is centered at 10 eV; the corresponding time scales are 10^{-15} s and 2×10^{-15} s respectively). A theory that takes account of the similarity of these time scales would be very difficult, involving explicit electron correlation between the solute and the macroscopic solvent. One can, however, treat the limit where the solvent electronic response is fast compared to solute electronic transitions; this is called the direct reaction field (DRF).[49,93] The accurate answer must lie somewhere between the SCRF and DRF limits;[94] nevertheless one can obtain very useful results with a two-time-scale version of the more manageable SCRF limit, as illustrated by a very successful recent treatment of the vertical excitation of acetone in nine solvents.[95] In this treatment, solvent electronic polarization is treated by SCRF theory, but solvent nuclear polarization is frozen on the time scale of the electronic transition.

In chemical dynamics, one can distinguish two qualitatively different types of processes: electron transfer and reactions involving bond rearrangement; the latter involve heavy-particle (proton or heavier) motion in the formal reaction coordinate. The zero-order model for the electron transfer case is pre-organization of the nuclear coordinates (often predominantly the solvent nuclear coordinates) followed by pure electronic motion corresponding to a transition between diabatic electronic states.[96] The zero-order model for the second type of process is transition state theory[97] (or, preferably, variational transition state theory[80]) in the lowest adiabatic electronic state (i.e., on the lowest-energy Born-Oppenheimer potential energy surface).

Nonequilibrium considerations for electron transfer are similar to those for vertical photoexcitation discussed above, except that the pre-organization of the solvent prior to the electron transition makes the effective gap at the time of the electron transfer smaller, and thus the assumption of rapid electronic response of the solvent is even better.

It is generally believed that equilibrium solvation is a reasonable assumption or zero-order point for most reactions involving rearrangement of bonds. The most difficult case is probably the case of adiabatic reactions involving rearrangement or transfer of charged species; this includes proton transfer, S_N2 and S_N1 reactions, electrocyclic reactions, etc. In this case the equilibrium SCRF treatment is very reasonable for electronic response, and it should often be approximately valid for the nuclear solvent response as well. The reason for the latter statement is that the solute reaction coordinate motion is slowed down at the critical dynamical bottleneck region (a particle crossing a barrier is moving most slowly when it is at the top of the barrier), whereas the critical solvent motions appear to be very fast, with a time scale on the order of 10–400 fs.[98] In recent years there has been considerable effort devoted to trying to understand these nonequilibrium effects.[49] A general consensus is that they seldom (if ever) exceed a factor of 10, and more typically they contribute a factor of at most 2 or 3, and sometimes considerably less. We should keep in mind that factors of 10, 3, and 2 in the rate correspond to changes of 1.4, 0.6, and 0.4 kcal/mol, respectively, in the phenomenological free energy of activation at room temperature so these effects may become significant when this level of accuracy is required.

There are two major approaches to including nonequilibrium effects in reaction rate calculations. The first approach treats the inability of the solvent to maintain its equilibrium solvation as the system moves along the reaction coordinate as a frictional drag on the reacting solute system.[97, 99, 100] The second approach adds one or more collective solvent coordinate to the nuclear coordinates of the solute.[101-107] When these solvent coordinates are at their classical equilibrium position, the solvent is at equilibrium. But these collective coordinates can couple into the reaction coordinate and take on nonequilibrium values. It can be shown that the two approaches are equivalent.[108-111] The generalized-solvent-coordinate approach has been used successfully to explain kinetic isotope effects for H and Mu addition to benzene in water,[104] to predict the aqueous acceleration effect on the reaction of H with CH_3OH in water,[106] and to explain the frictional effect on the rate of contact ion pair dissociation in water.[107] The chief uncertainty in estimating such nonequilibrium effects is predicting the effective solvent time constant and the relevant "force constants" coupling the solute coordinates to the collective solvent coordinates. Two general approaches may be distinguished. One approach centers on the electrostatics of solvent polarization and therefore makes a strong connection between this process and the frozen-nuclear-polarization approximation in the theory of vertical photoexcitation.[101-103, 105] The other approach attempts to obtain the relevant parameters from macroscopic solvent descriptors such as macroscopic viscosity or macroscopic diffusion coefficients.[104-106] This latter approach may be useful for rough and ready estimates of the approximate size of the nonequilibrium effect but ultimately suffers from uncertainty as to whether the macroscopic frictional forces involved in diffusion or viscosity are the same as those that operate on atomic motions over very short time and distance scales.

Nonequilibrium solvent effects can indeed by significant at the kcal level–maybe even at a greater level, but so far there is no evidence for that when the reaction coordinate involves protonic or heavier motions. Our goal in this section has been to emphasize just how powerful and general the equilibrium model is. In addition, in both the previous section and the present section, we have emphasized the use of models based on collective solvent coordinates for calculating both equilibrium and nonequilibrium solvation properties.

Acknowledgments

This work was supported in part by the National Science Foundation.

10. REFERENCES

1. R. B. Silverman, The Organic Chemistry of Drug Design, Academic Press, San Diego (1992).
2. X. Xou, Y. Sun, and I. Kuntz, Inclusion of solvation in ligand binding free energy calculations using the generalized-Born model, J. Am. Chem. Soc. **121**:8033 (1999).

3. C. Hansch and A. Leo, Exploring QSAR, American Chemical Society, Washington (1995).
4. I. M. Klotz and R. M. Rosenberg, Chemical Thermodynamics, 5th ed., John Wiley & Sons, New York (1994), (a) p. 272, (b) p. 344, (c) p. 307, (d) p. 326, (e) p. 324, (f) p. 342.
5. K. Denbigh, The Principles of Chemical Equilibrium, 4th ed., Cambridge University Press, London (1981) p. 249.
6. A. W. Adamson, A Textbook of Physical Chemistry, Academic Press, New York (1973), p. 342. Or see 3rd ed., (1986).
7. S. R. Sherman, D. B. Trampe, D. M. Bush, M. Schiller, C. A. Eckert, A. J. Dallas, J. Li, and P. W. Carr, Compilation and correlation of limiting activity coefficients of nonelectrolytes in water, Ind. Eng. Chem. Res. **35**:1044 (1996).
8. R. M. Noyes, Thermodynamics of ion hydration as a measure of effective dielectric properties of water, J. Am. Chem. Soc. **84**:513 (1962).
9. D. D. Wagman, W. H. Evans, V. B. Parker, R. H. Schumm, I. Halow, S. M. Bailey, K. L. Churney, and R. L. Nuttall, The NBS tables of chemical thermodynamic properties. Selected values for inorganic and C1 and C2 organic substances in SI units, J. Phys. Chem. Ref. Data **11**/Suppl. No. 2 (1982).
10. S. Trasatti, The absolute electrode potential: An explanatory note, Pure Appl. Chem. **58**:955 (1986).
11. P. Winget, E. J. Weber, C. J. Cramer, and D. G. Truhlar, Computational electrochemistry: Aqueous one-electron oxidation potentials for substituted anilines, Phys. Chem. Chem. Phys. **2**:1231 (2000).
12. C. J. Cramer and D. G. Truhlar, General parameterized SCF model for free energies of solvation in aqueous solution, J. Am. Chem. Soc. **113**:8305 (1991).
13. C. J. Cramer and D. G. Truhlar, AM1-SM2 and PM3-SM3 parameterized SCF solvation models for free energies in aqueous solution, J. Comput.-Aided Mol. Design **6:**629 (1992).
14. C. J. Cramer and D. G. Truhlar, Quantum chemical conformational analysis of glucose in aqueous solution, J. Am. Chem. Soc. **115**:5745 (1993).
15. C. J. Cramer and D. G. Truhlar, Continuum solvation models: classical and quantum mechanical implementations, in: Reviews in Computational Chemistry, Vol. 6, K. B. Lipkowitz and D. B. Boyd, eds., VCH Publishers, New York (1995) pp. 1–72.
16. C. J. Cramer and D. G. Truhlar, Quantum chemical conformational analysis of 1,2-ethanediol: Correlation and solvation effects on the tendency to form internal hydrogen bonds in the gas phase and aqueous solution, J. Am. Chem. Soc. **116:**3892 (1994).
17. C. C. Chambers, G. D. Hawkins, C. J. Cramer, and D. G. Truhlar, Model for aqueous solvation based on class IV atomic charges and first solvation shell effects, J. Phys. Chem. **100**:16385 (1996).
18. G. D. Hawkins, C. J. Cramer, and D. G. Truhlar, Parameterized models of aqueous free energies of solvation based on pairwise descreening of solute atomic charges from a dielectric medium, J. Phys. Chem. **100:**19824 (1996); erratum: to be published.
19. D. J. Giesen, M. Z. Gu, C. J. Cramer, and D. G. Truhlar, A universal organic solvation model, J. Org. Chem. **61**:8720 (1996).
20. D. J. Giesen, G. D. Hawkins, D. A. Liotard, C. J. Cramer, and D. G. Truhlar, A universal solvation model for the quantum mechanical calculation of free energies of solvation in non-aqueous solvents, Theoret. Chem. Acc. **98**:85–109 (1997) erratum: **101:**309 (1999).
21. C. J. Cramer and D. G. Truhlar, Continuum solvation models, in: Solvent Effects and Chemical Reactivity, O. Tapia and J. Bertrán, eds., Kluwer, Dordrecht (1996) pp. 1–80.
22. D. J. Giesen, C. C. Chambers, G. D. Hawkins, C. J. Cramer, and D. G. Truhlar, Modeling free energies of solvation and transfer, in: Computational Thermochemistry, K. Irikura and D. J. Frurip,. eds., American Chemical Society Symposium Series, Vol. 677, Washington, DC (1998) pp. 285–300.

23. C. C. Chambers, D. J. Giesen, G. D. Hawkins, W. H. J. Vaes, C. J. Cramer, and D. G. Truhlar, Modeling the effect of solvation on structure, reactivity, and partitioning of organic solutes: Utility in drug design, in: Rational Drug Design, D. G. Truhlar, W. J. Howe, A. J. Hopfinger, J. M. Blaney, and R. A. Dammkoehler, eds., Springer, New York (1999) pp. 51–72.
24. G. D. Hawkins, J. Li, T. Zhu, C. C. Chambers, D. J. Giesen, D. A. Liotard, C. J. Cramer, and D. G. Truhlar, New tools for rational drug design, in: Rational Drug Design: Novel Methodology and Practical Applications, A. L. Parrill and M. R. Reddy, eds., American Chemical Society Symposium Series, Vol. 719, Washington, DC (1999) pp. 121–140.
25. G. D. Hawkins, T. Zhu, J. Li, C. C. Chambers, D. J. Giesen, D. A. Liotard, C. J. Cramer, and D. G. Truhlar, Universal solvation models, in: Combined Quantum Mechanical and Molecular Mechanical Methods, J. Gao and M. A. Thompson, eds., American Chemical Society Symposium Series, vol. 712, Washington, DC (1998) pp. 201–219.
26. J. Li, T. Zhu, G. D. Hawkins, P. Winget, D. A. Liotard, C. J. Cramer, and D. G. Truhlar, Extension of the platform of applicability of the SM5.42R universal solvation model, Theoret. Chem. Acc. **103**:9 (1999).
27. D. M. Dolney, G. D. Hawkins, P. Winget, D. A. Liotard, C. J. Cramer, and D. G. Truhlar, A universal solvation model based on the conductor-like screening model, J. Comput. Chem. **21**:340 (2000).
28. D. Suleiman and C. A. Eckert, J. Chem. Eng. Data **39**:692 (1994).
29. Aldrich Catalog Handbook of Fine Chemicals, 1998–1999, Aldrich Chemical Co.: Milwaukee, WI (1998), p. 766.
30. Handbook of Chemistry and Physics, 78th ed.; D. R. Lide, ed., CRC Press, Boca Raton, FL (1997) p. 6–71.
31. P. Winget, G. D. Hawkins, C. J. Cramer, and D. G. Truhlar, J. Phys. Chem. B **104**:4726 (2000).
32. D. J. W. Grant and T. Higuchi, Solubility Behavior of Organic Compounds, Chapter 2 Wiley, New York (1990).
33. A. Ben-Naim, Solution Thermodynamics, Plenum, New York (1987).
34. A. Ben-Naim, Statistical Thermodynamics for Chemists and Biochemists, Plenum, New York (1992).
35. D. J. Giesen, C. J. Cramer, and D. G. Truhlar, Entropic contributions to free energies of solvation, J. Phys. Chem. **98**:4141 (1994).
36. J. I. Siepmann, An introduction to the Monte Carlo method for particle simulations, Adv. Chem. Phys. **105**:1 (1999).
37. A. Rahman, Correlations in the motion of atoms in liquid argon, Phys. Rev. **136**:A405 (1964).
38. J. Gao, Methods and applications of combined quantum mechanical and molecular mechanical potentials, in: Reviews in Computational Chemistry, Vol. 7, K. B. Lipkowitz and D. B. Boyd, eds., VCH, New York (1996) p. 119–186.
39. L. Onsager, Electric moments of molecules in liquids, J. Am. Chem. Soc. **58**:1486 (1936).
40. D. Rinaldi and J. L. Rivail, Polarisabilités moléculaires et effet diélectrique de mileau à l'etude théorique de la molécule d'eau et de ses dimères, Theor. Chim. Acta. **32**:57 (1973).
41. O. Tapia and O. Goscinski, Self-consistent reaction field theory of solvent effects, Mol. Phys. **29**:1653 (1975).
42. O. Tapia, Local field representation of surrounding medium effects. From liquid solvent to protein core effects, in: Quantum Theory of Chemical Reaction, Vol. 2, R. Daudel, A. Pullman, L. Salem, and A. Veillard, eds., Reidel, Dordrecht (1980) pp. 25–72.
43. O. Tapia and B. Silvi, Solvent effects on the structure and the electronic properties of simple molecules—a MINDO/3-SCRF-MO study, J. Phys. Chem. **84**:2646 (1980).

44. M. F. Ruiz-Lopez and D. Rinaldi, Electrostatic solvent effect on the circular dichroism of the carbonyl n t π * transition, J. Mol. Struct. Theochem. **10**:277 (1983).
45. R. Constanciel and R. Contreras, Self-consistent field theory of solvent effects representation by continuum models-introduction of desolvation contribution, Theor. Chim. Acta **65**:1 (1984).
46. J. Tomasi and M. Persico, Molecular interactions in solution: An overview of methods based on continuous distributions of the solvent, Chem. Rev. **94**:2027 (1994).
47. J. L. Rivail and D. Rinaldi, Liquid state quantum chemistry: computational applications of the polarizable continuum models, in: Computational Chemistry, Review of Current Trends, J. Leszczynski, ed., World Scientific, New York (1996) pp. 139–174.
48. J. Tomasi, B. Menucci, R. Cammi, and M. Cossi, Quantum mechanical models for reactions in solutions, in: Computational Approaches to Biochemical Reactivity, G. Náray-Szabó and A. Warshel, eds., Kluwer, Dordrecht (1997) pp. 1–102.
49. C. J. Cramer and D. G. Truhlar, Implicit solvation models: equilibria, structure, spectra, and dynamics, Chem. Rev. **99**:2161 (1999).
50. C. Reichardt, Solvents and Solvent Effects in Organic Chemistry, 2nd ed., VCH, Weinheim (1988).
51. J. W. Storer, D. J. Giesen, C. J. Cramer, and D. G. Truhlar, Class IV charge models: a new semiempirical approach in quantum chemistry, J. Comput.-Aided Mol. Design **9**:87 (1995).
52. J. Li, T. Zhu, C. J. Cramer, and D. G. Truhlar, New class IV charge model for extracting accurate partial charges from wave functions, J. Phys. Chem. A **102**:1820 (1998).
53. J. Li, J. Xing, C. J. Cramer, and D. G. Truhlar, Accurate dipole moments from Hartree-Fock calculations by means of class IV charges, J. Chem. Phys. **111**:885 (1999).
54. G. J. Hoijtink, E. de Boer, P. H. Van der Meij, and W. P. Weijland, Potentials of various aromatic hydrocarbons and their univalent anions, Recl. Trav. Chim. Pays-Bas. **75**:4873 (1956).
55. F. Peradejordi, On the Pariser and Parr semiempirical method for computing molecular wave functions. The basic strength of N-heteroatomic compounds and their monomers, Cahiers Phys. **17**:393 (1963).
56. S. C. Tucker and D. G. Truhlar, Generalized Born fragment charge model for solvation effects as a function of reaction coordinate, Chem. Phys. Lett. **157**:164 (1989).
57. W. C. Still, A. Tempczyk, R. C. Hawley, and T. Hendrickson, Semianalytical treatment of solvation for molecular mechanics and dynamics, J. Am. Chem. Soc. **112**:6127 (1990).
58. A. Klamt and G. Schüürmann, COSMO: A new approach to dielectric screening in solvents with explicit expressions for the screening energy and its gradient, J. Chem. Soc. Perkin Trans. **2**:799 (1993).
59. T. N. Truong, U. N. Nguyen, and E. V. Stefanovich, Generalized conductor-like screening model (GCOSMO) for solvation: An assessment of its accuracy and applicability, Int. J. Quantum Chem. **60**:403 (1996).
60. S. Miertus, E. Scrocco, and J. Tomasi, Electrostatic interaction of a solute with a continuum. A direct utilization of ab initio molecular potentials for the prevision of solvent effects, Chem. Phys. **55**:117 (1981).
61. V. Dillet, D. Rinoldi, and J. L. Rivail, Liquid-state quantum chemistry: An improved cavity model, J. Phys. Chem. **98**:5034 (1994).
62. K. Baldridge and A. Klamt, First principles implementation of solvent effects without outlying charge error, J. Chem. Phys. **106**:6622 (1997).
63. C. Amovilli, V. Barone, R. Cammi, E. Cancès, M. Cossi, B. Menucci, C. S. Pomelli, and J. Tomasi, Recent advances in the description of solvent effects with the polarizable continuum model, Adv. Quantum Chem. **32**:227 (1998).

64. M. Schaefer and C. Froemmel, A precise analytical method for calculating the electrostatic energy of macromolecules in aqueous solution, J. Mol. Biol. **216**:1045 (1990).
65. G. D. Hawkins, C. J. Cramer, and D. G. Truhlar, Pairwise solute descreening of solute charges from a dielectric medium, Chem. Phys. Lett. **246**:122 (1995).
66. V. Tsui and D. A. Case, Molecular dynamics simulations of nucleic acids with a generalized born solvation model, J. Am. Chem. Soc. **122**:2489 (2000).
67. J. Srinivasan, M. W. Trevathan, P. Beroza, and D. A. Case, Application of a pairwise generalized Born model to proteins and nucleic acids, Theor. Chem. Acc. **101**:426 (1999).
68. B. Honig and A. Nicholls, Classical electrostatics in biology and chemistry, Science **268**:1144 (1995).
69. F. J. Luque, M. Bachs, C. Alemán, and M. Orozco, Extension of the MST/SCRF method to organic solvents. Ab initio and semiempirical parametrization for neutral solutes in CCl_4, J. Comput. Chem. **17**:806 (1996).
70. V. Barone, M. Cossi, and J. Tomasi, A new definition of cavities for the computation of solvation free energies by the polarizable continuum model, J. Chem. Phys. **107**:3210 (1997).
71. P. Debye, Polar Molecules, Chemical Catalog Co., New York (1929), p. 109.
72. C. J. F. Bottcher, O. C. van Belle, P. Bordewijk, and A. Rip, Theory of Electric Polarization, 2nd. ed., Elsevier, Amsterdam (1973), p. 289.
73. H. Hellmann, Einführing in die Quantenchemie, Franz Deuticke, Vienna (1937).
74. R. P. Feynman, Forces in molecules, Phys. Rev. **56**:340 (1939).
75. J. C. Slater, Quantum Theory of Matter, McGraw-Hill, New York (1968), pp. 381, 410-411.
76. A. Ben-Naim, Hydrophobic interactions in biological systems, Top. Mol. Pharmacol. **2**:1 (1983).
77. C. Hansch, J. P. Björkroth, and A. Leo, Hydrophobicity and central nervous system agents: On the principle of minimal hydrophobicity in drug design, J. Pharm. Sci. **76**:663 (1987).
78. Lipophilicity in Drug Action and Toxicology, V. Pliska, B. Testa, and H. van de Waterbeemd, eds., VCH, New York (1996).
79. A. D. J. Haymet, K. A. T. Silverstein, and K. A. Dill, Hydrophobicity reinterpreted as "minimization of the entropy penalty of solvation", Faraday Discuss. **103**:117 (1996).
80. A. M. Davis and S. J. Teague, Hydrogen bonding, hydrophobic interactions, and failure of the rigid receptor hypothesis, Angew. Chem. Int. Ed. **38**:736 (1999).
81. P. Jolliet-Riant and J.-P. Tillement, Drug and nutrient transfers through the blood-brain barrier and their pharmacological changes, Encephale **25**:135 (1999).
82. G. L. Amidon, S. H. Yalkowski, S. T. Anik, and S. C. Valvani, Solubility of nonelectrolytes in polar solvents. 5. Estimation of solubility of alphatic monofunctional compounds in water using a molecular surface area approach, J. Phys. Chem. **79**:2239 (1975).
83. G. D. Rose, A. R. Geselowitz, G. J. Lesser, R. H. Lee, and M. H. Zehfus, Hydrophobicity of amino acid residues in globular proteins, Science **229**:834 (1985).
84. D. Eisenberg and A. D. McLachlan, Solvation energy in protein folding and binding, Nature **319**:199 (1986).
85. T. Ooi, M. Oobatake, G. Nemethy and H. A. Scheraga, Accessible surface areas as a measure of the thermodynamic parameters of hydration of peptides, Proc. Natl. Acad. Sci. USA **84**:3086 (1987).
86. S. J. Stuart and B. J. Berne, Surface curvature effects in the aqueous ionic solvation of the chloride ion, J. Phys. Chem. A **103**:10300 (1999).

87. G. D. Hawkins, C. J. Cramer, and D. G. Truhlar, Parameterized model for aqueous free energies of solvation using geometry-dependent atomic surface tensions with implicit electrostatics, J. Phys. Chem. B **101**:7147 (1997); erratum: to be published.
88. G. D. Hawkins, D. A. Liotard, C. J. Cramer, and D. G. Truhlar, OMNISOL: Fast prediction of free energies of solvation and partition coefficients, J. Org. Chem. **63**:4305 (1998).
89. M. H. Abraham, New solute descriptors for linear free energy relationships and quantitative structure-activity relationships, in: Quantitative Treatments of Solute/Solvent Interactions, P. Politzer and J. S. Murray, eds., Elsevier, Amsterdam (1994) pp. 83–134.
90. W. Vaes, E. Urrestarazu Ramos, H. J. M. Verhaar, C. J. Cramer, and J. L. M. Hermens, Understanding and estimating membrane/water partition coefficients: Approaches to derive quantitative structure property relationships, Chem. Res. Toxicol. **11**:847 (1998).
91. C. J. Cramer and D. G. Truhlar, Development and biological applications of quantum mechanical continuum solvation models, in: Quantitative Treatments of Solute/Solvent Interactions, P. Politzer and J. S. Murray, eds., Elsevier, Amsterdam (1994), pp. 9–54. [Theor. Comp. Chem. **2**:9 (1994).]
92. M. F. Nicol, Solvent effects on electronic spectra, Appl. Spect. Rev. **8:**183 (1974).
93. B. T. Thole and P. T. van Duijnen, On the quantum mechanical treatment of solvent effects, Theor. Chim. Acta **55**:307 (1980).
94. A. G. Ángyán and G. Jensen, Are direct reaction field methods appropriate for describing dispersion interactions?, Chem. Phys. Lett. **175**:313 (1990).
95. J. Li, C. J. Cramer, and D. G. Truhlar, A two-response-time model based on CM2/INDO/S2 electrostatic potentials for the dielectric polarization component of solvatochromic shifts on vertical excitation energies, Int. J. Quan. Chem. **77**:264 (2000).
96. R. A. Marcus and N. Sutin, Electron transfers in chemistry and biology, Biochim. Biophys. Acta **811**:265 (1985).
97. M. M. Kreevoy and D. G. Truhlar, Transition state theory, in Investigation of Rates and Mechanisms of Reactions, 4th ed., C. F. Bernasconi, ed., John Wiley and Sons, New York (1986), Part 1, pp. 13–95. [Tech. Chem. (N.Y.), 4th ed., **6**/Pt. 1, pp. 13-95 (1986)]
98. M. Maroncelli, V. P. Kumar, and A. Papazyan, A simple interpretation of polar solvation dynamics, J. Phys. Chem. **97**:13 (1993).
99. H. A. Kramers, Brownian motion in a field of force and the diffusion model of chemical reactions, Physica **7**:284 (1940).
100. R. F. Grote and J. T. Hynes, The stable states picture of chemical reactions. II. Rate constants for condensed and gas phase reaction models, J. Chem. Phys. **73**:2715 (1980).
101. S. Lee and J. T. Hynes, Solution reaction path Hamiltonian for reactions in polar solvents. I. Formulation, J. Chem. Phys. **88**:6853 (1988).
102. D. G. Truhlar, G. K. Schenter, and B. C. Garrett, Inclusion of nonequilibrium continuum solvation effects in variational transition state theory, J. Chem. Phys. **98**:5756 (1993).
103. M. V. Basilevsky, G. E. Chudinov, and D. V. Napolov, Calculation of the rate constant for the reaction $CL^- + ClCH_3CL \rightarrow ClCH_3 + Cl^-$ in the framework of the continuum medium model, J. Phys. Chem. **97**:3270 (1993).
104. B. C. Garrett and G. K. Schenter, Nonequilibrium solvation for an aqueous-phase reaction: Kinetic isotope effects for the addition of hydrogen to benzene, in: Structure and Reactivity in Aqueous Solution, ACS Symposium Series, C. J. Cramer and D. G. Truhlar, eds., American Chemical Society, Washington, DC (1994), Vol. 568, pp. 122-142.
105. M. F. Ruiz-López, D. Rinaldi, and J. Bertrán, Nonequilibrium solvent effects on the S_N2 reaction using a self-consistent reaction field continuum model based on multipole expansions, J. Chem. Phys. **103**:9249 (1995).
106. Y.-Y. Chuang and D. G. Truhlar, Nonequilibrium solvation effects for a polyatomic reaction in solution, J. Am. Chem. Soc. **121**:10157 (1999).
107. D. G. Truhlar and B. C. Garrett, Multidimensional transition state theory and the validity

of Grote-Hynes theory, J. Phys. Chem. B **104**:1069 (2000).

108. G. van der Zwan and J. T. Hynes, Theory of multi-dimensional transition state and Kramers expression, J. Chem. Phys. **78**:4174 (1983).
109. Y. I. Dakhnovskii and A. A. Ovchinnikov, The transition-state theory and generalized Kramers diffusion model of chemical reactions, Khim. fiz. **5**:36 (1986).
110. E. Pollak, Theory of activated rate processes: A new derivation of Kramers' expression, J. Chem. Phys. **85**:865 (1986).
111. S. C. Tucker, Variational transition state theory in condensed phases, in: New Trends in Kramers' Reaction Rate Theory, P. Talkner and P. Hänggi, eds., Kluwer, Dordrecht (1995) pp. 5-46.

Chapter 5

Relative Solvation Free Energies Calculated Using Explicit Solvent

Atul Agarwal,[†] Frank K. Brown,[‡] and M. Rami Reddy[*]
[†]*Wyeth-Ayerst Research/AHP, Pearl River, NY 10965*
[‡]*R.W. Johnson, PRI, Raritan, NJ 08869*
[*]*Metabasis Therapeutics, Inc., San Diego, CA 92121*

1. INTRODUCTION

Solvation free energies for small molecules are important in drug design and in the understanding of the relationship between solvation and ligand binding. Given that, ΔG (binding) = ΔG (affinity) – ΔG (desolvation), the ligand with the best affinity for the binding site may not be the one with the best overall binding free energy if the cost to desolvate the ligand is large. For example, if ligand A has an affinity of 14 kcal/mol for a given enzyme and a desolvation cost of 6 kcal/mol, while ligand B has an affinity of 12 kcal/mol for the same enzyme and a desolvation cost of 3 kcal/mol. Ligand B will exhibit a 1 kcal/mol better binding free energy. This could lead to a misinterpretation of the SAR by medicinal chemists as was the case for collagenase. Collagenase is reported to have water molecule in the binding pocket and exhibits selectivity for Leu side-chains in the P1' site. It was further shown that replacement of Leu in the collagenase peptide inhibitors with Gln resulted in a 50% decrease in potency. This result led to the postulate that the site was relatively hydrophobic. However, if the desolvation cost difference between Leu and Gln (~11 kcal/mol) was considered, one realizes that Gln must have a much greater affinity for the site than Leu to compensate for the much larger desolvation cost of Gln and therefore that the P1' binding site is less hydrophobic than originally postulated.[1]

Desolvation free energies are computed using either explicit solvent or an implicit solvent model. While explicit solvent simulations are usually considered more accurate or at least more representative of the true molecular environment, simulations using implicit solvent are often chosen due to their reduced CPU requirements. Explicit solvent models are required in a simulation in which a detailed picture of solvent structure is important for which there is evidence that a particular structure feature of the solvent is playing a key role, e.g., a specific water-protein hydrogen bond. Relative solvation free energies of small molecules are calculated with reasonable accuracy using implicit solvent models such as GB/SA.[2] In contrast, accurate relative binding affinities of inhibitors to an enzyme is best accomplished using explicit solvent models. Many explicit solvent models have been used to represent solvent effects in small and macromolecular systems. Some examples include: MCY,[3] ST2,[4] TIP2P,[5] TIP3P,[6] TIP4P,[6] TIP5P,[7, 8] SPC,[9] and SPC/E.[10] The TIP5P is the best water model for reproducing experimental water properties. However, it requires more computer time for calculations as compared to simple water models such as TIP3P and SPC/E. Thus, the most extensively used models are TIP3P, SPC and SPC/E, which have shown considerable success in simulations of both small and large molecules. The TIP3P water model reproduces experimental structural properties accurately[6] and the SPC/E water model reproduces experimental water properties such as the dielectric constant and diffusion coefficient accurately.[11] No explicit hydrogen bond term is used in the potential function, because electrostatic forces and the repulsive part of the oxygen – oxygen force, give good description of forces equivalent of hydrogen bonding.

2. METHODOLOGY

Since absolute free energies are difficult to calculate accurately, the thermodynamic cycle perturbation[12-14] is often used to compute the relative changes of free energy for a solvation process by the construction of non-physical paths connecting the desired initial and terminal states. This approach enables calculation of the relative change in solvation free energy difference ($\Delta\Delta G_{sol}$) between two related compounds by computationally simulating the 'mutation' of one to the other. These mutations can be carried out either using single topology or double topology methods. The single topology method entails changing the appropriate reactant atoms to product atoms. These mutations result in geometrical changes as well as changes in partial charges, and van der Waals parameters. In the double topology or thread method[15, 16] a single topology is defined for those atoms which are identical in both molecules in the sense that force constants and equilibrium geometries are the same (partial charges can vary). For the portion of the molecule which must be transformed, both the starting (reactant) and ending (product) topologies are defined using their associated geometries, with one

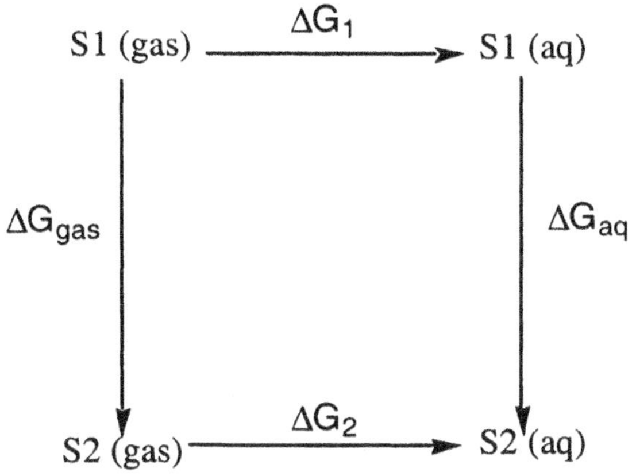

Figure 1. Thermodynamic cycle for computing solvation free energy difference between two solutes S1 and S2.

$$\Delta\Delta G_{sol} = \Delta G_{aq} - \Delta G_{gas} = \Delta G_2 - \Delta G_1 \quad (1)$$

beginning and the other ending the simulation entirely as dummy atoms. Dummy atoms are identical to real atoms except for their Lennard-Jones parameters and charges are set to zero. At intermediate points during the transformation, all atoms in both topologies have fractional Lennard-Jones parameters and charges. Molecules with both topologies interact with the environment, but not with each other. The relative solvation free energy differences for two similar ligands is computed using the thermodynamic cycle shown in Figure 1, as represented in Equation 1.

The free energy change for converting S1 into S2 is computed by perturbing the Hamiltonian of reactant (initial) state S1 into that of the product (final) state S2. This transformation is accomplished through a parameterization of terms comprising the interaction potentials of the system with a change of state variable that maps onto reactant and product states when that variable is 0 and 1, respectively. The total free energy change for the mutation from the initial to the final state is computed by summing 'incremental' free energy changes over several windows visited by the state variable changing from 0 to 1. Alternatively, S1 can be transformed to S2 using the Slow Growth method. This method makes a small perturbation on each time step and sums all the difference for the full run. This method is smoother at the introduction of large changes, but there is a much larger hysteresis.[17, 18]

This chapter summarizes most of the studies reported in the literature to date that use the free energy methodology to calculate relative solvation free energies with explicit solvent. In addition, the results are used to define the current scope and limitations of the methodology.

3. CALCULATED SOLVATION FREE ENERGIES

3.1 Convergence of Results

While calculations involving macromolecules that use explicit solvent are more realistic, they are also associated with significantly longer simulation times in order to achieve convergence. Since the early 1980s, over 100 papers have been published using MD or MC simulations in conjunction with the thermodynamic cycle perturbation approach,[12] for the calculation of relative solvation using explicit solvent. In general, the results from these calculations showed satisfactory agreement with experimental results. In some respects this is remarkable since most of the studies used relatively short simulation times making it likely that the calculations were not fully converged. The likelihood that full convergence was not achieved in these initial calculations was highlighted by the work of McCammon[19] and Kollman[20] who examined the behavior of the calculated free energy with the length of the simulation and concluded that simulation times significantly longer than those typically used (200 ps or greater) were required to yield reliable and accurate results. In addition, others [21-24] have studied the convergence problems in the simulations, particularly, Chipot[24] et al. who investigated the convergence behavior in free energies using several different methods and concluded that proper convergence of the free energy calculations requires simulation times much longer than previously estimated.

Reddy and Erion[25] evaluated for seven systems, the dependence of relative solvation free energy convergence on molecular dynamics simulation length and starting configuration using two studies. In the first study, each simulation started from the same well-equilibrated configuration and the length was varied from 153 ps to 1530 ps. In the second study, the relative solvation free energy differences were calculated starting from three different configurations and using 510 ps of molecular dynamics simulation for each mutation. All the details of calculations and complete relative solvation free energy results for all the systems are described elsewhere,[12] The relative solvation free energies were calculated using simulation lengths that ranged from 153 ps to 1530 ps from the same well-equibrilated configurations for acetone to acetaldehyde, acetone to pyruvic acid, acetone to 1,1,1-trifluoroacetone, acetone to 1,1,1-trichlroacetone, acetone to 2,3-butanedione, acetone to cyclopropanone and formaldehyde hydrate to formaldehyde. The calculated free energies indicated (Table 1) that even for molecules with limited conformational flexibility a simulation length of 510 ps or greater is required to obtain satisfactory convergence. Mutations that represent a large structural changes such as cyclopropanone to acetone require much longer simulation lengths to achieve satisfactory convergence. The other calculations[25] demonstrate that performing one long simulation is better than averaging results from three shorter simulations of the same length using different starting conformations.

Table 1. Calculated relative solvation free energies (kcal/mol)

SYSTEM	Simulation Length	$\Delta\Delta G^{Calc}$
$CH_3COCH_3 - HCOCH_3$[a]	153 ps	-0.56 ± 0.37
	306 ps	-0.52 ± 0.29
	510 ps	-0.44 ± 0.22
	714 ps	-0.45 ± 0.17
	1530 ps	-0.40 ± 0.11
$CH_3COCH_3 - CH_3COCOOH$	153 ps	10.31 ± 1.04
	306 ps	10.09 ± 0.82
	510 ps	10.01 ± 0.72
	714 ps	9.96 ± 0.56
	1530 ps	9.84 ± 0.48
$CH_3COCH_3 - CF_3COCH_3$	153 ps	0.69± 0.50
	306 ps	0.54 ± 0.39
	510 ps	0.58 ± 0.25
	714 ps	0.60 ± 0.18
	1530 ps	0.63 ± 0.13
$CH_3COCH_3 - CCl_3COCH_3$	153ps	3.95 ± 0.59
	306 ps	3.90 ± 0.46
	510 ps	3.79 ± 0.33
	714 ps	3.75 ± 0.26
	1530 ps	3.67 ± 0.19
$CH_3COCH_3 - CH_3COCOCH_3$	153 ps	4.05 ± 0.84
	306 ps	4.24 ± 0.68
	510 ps	3.78 ± 0.57
	714 ps	3.63 ± 0.44
	1530 ps	3.57 ± 0.36
CH_3COCH_3 – Cyclopropanone	153 ps	-0.91 ± 0.70
	306 ps	-0.79 ± 0.61
	510 ps	-0.68 ± 0.51
	714 ps	-0.51 ± 0.36
	1530 ps	-0.46 ± 0.20
$HCOH - HC(OH)_2H$	153 ps	8.10 ± 0.97
	306 ps	7.81 ± 0.82
	510 ps	7.46 ± 0.67
	714 ps	7.30 ± 0.56
	1530 ps	7.20 ± 0.40

[a] $\Delta\Delta G^{Expt} = -0.3$

3.2 Validation of Methodology

Initial reports on solvation free energy calculations focused on modeling of a cavity in water.[26, 27] Simulations of simple solutes and their impact on the structure of bulk water followed the basic study. These studies were fundamental for their contribution to the deconvolution of the systems into several components. Jorgensen and Ravimohan,[28] were among the first to calculate relative solvation free energies (SFEs). They used the free energy perturbation (FEP) method and MC simulations to calculate free energies of hydration for inter-conversion of methanol and ethane using OPLS parameters and TIP4P water. The inter-conversion was carried out in 6 windows using double wide sampling in the NPT ensemble with a spherical cutoff of 7.5 Å. The equilibration phase consisted of 0.5×10^6 configurations followed by averaging for properties over an additional 1.5×10^6 configurations. The calculated relative solvation free energy for the mutation of methanol to ethane was 6.75±0.38 kcal/mol, which was in good agreement with the experimental relative solvation free energy of 6.93 kcal/mol. In addition to this work, the Jorgensen group also demonstrated that the desolvation free energy of water itself could be accurately calculated using the TIP4P model. The calculation gave 6.1 kcal/mol while the experimental value is 6.3 kcal/mol.[29] This was an important contribution because it showed that the water model was capable of reproducing the most basic system for desolvation studies, namely its own model of water was self-consistent. These results also proved that the hydrophobic – hydrophilic difference in small molecules could be accurately computed. These first studies gave way to an avalanche of solvation free energy studies on numerous topics as discussed below.

3.3 Solvation Free Energies between Neutral Systems

Kollman and coworkers[30] calculated the relative solvation free energies of several fluorocarbons and demonstrated the necessity of the use of bond-PMF correction in order to obtain good agreement with experiment (Table 2). Intramolecular interaction energy was not included in the solvation free energy calculation and a non-bonded cut off of 8 Å was used. The relative solvation free energies were calculated using AMBER and TIP3P water model in 40 windows where each window consisted of an equilibration of 0.5 ps and data collection of 0.5 ps with a time step of 1 fs. ESP fitted charges were calculated with 6-31G* basis set for all the molecules. The calculated relative solvation free energies for the mutation of CF_4 to CH_4, CFH_3 to CHF_3, CFH_3 to CH_4, CHF_3 to CH_2F_2 and CHF_3 to CF_4 without and with bond-PMF corrections were 0.77±0.06 kcal/mol, -0.8±0.3 kcal/mol (expt: -1.2 kcal/mol), 0.3±0.3 kcal/mol, 0.7±0.1 kcal/mol (expt: 1.0

Table 2. Calculated and experimental relative solvation free energies

Mutations	Force Field	NB	Simulation Length	$\Delta\Delta G_{sol}^{calc}$ (kcal/mol)	$\Delta\Delta G_{sol}^{exp}$ (kcal/mol)	Ref.
Arginine→Arginine$^{(+)}$	AMBER	8	50 ps	-32.28±0.71		17
Alanine→Arginine	AMBER	8	50 ps	-14.40±1.49	-12.86	17
Alanine→Cysteine	AMBER	8	50 ps	-3.04±0.37	-3.19	17
Histidine→Histidine$^{(+)}$	AMBER	8	50 ps	-35.49±0.38		17
Alanine→Histidine	AMBER	8	50 ps	-13.32±0.20	-12.19	17
Alanine→Leucine	AMBER	8	50 ps	0.65±0.20		17
Lysine→Lysine$^{(+)}$	AMBER	8	50 ps	-58.85±0.60		17
Alanine→Lysine	AMBER	8	50 ps	-6.75±0.10	-6.33	17
Alanine→Phenylalanine	AMBER	8	50 ps	-2.87±0.45	-2.71	17
Alanine→Serine	AMBER	8	50 ps	-7.04±0.35	-7.02	17
Alanine→Tryptophan	AMBER	8	50 ps	-7.29±0.35		17
Nothing→Methane	AMBER	8	50 ps	2.12±0.90	1.95	17
Methane→3-Methylindole	AMBER	8	50 ps	-6.68±0.93	-7.85	17
Methane→Isobutane	AMBER	8	50 ps	1.08±0.35	0.34	17
Benzene→Phenol	AMBER	8	50 ps	-4.57±0.19	-5.36	17
Methane→Acetamide	AMBER	8	50 ps	-10.30±0.21	-11.66	17
Acetamide→Acetate$^{(-)}$	AMBER	8	50 ps	-71.86±0.88		17
Methane→Acetic acid	AMBER	8	50-ps	-9.03 ±0.45	-8.64	17
Methane→Adenine	AMBER	8	50-ps	-14.73±0.76		17
Methane→Thymine	AMBER	8	50 ps	-9.61±1.15		17
Thymine→Cytosine	AMBER	8	50 ps	5.24±0.33		17
Adenine→Guanine	AMBER	8	50 ps	6.95±0.54		17
Methane→Acetamide	AMBER	8	50 ps	-10.30±0.21	-11.7±0.21	17
Methane→N-Methylacetamide	AMBER	8	50 ps	-10.29±0.76	-12.2±0.20	17
Methane→N,N-Dimethylacetamide	AMBER	8	50 ps	-9.16±0.10	-10.6±0.30	17
Methanol→Ethane	AMBER	8	20 ps	6.86±0.12	6.9	18
Ammonium→Oxonium	AMBER	8	20 ps	-21.2±0.4		18
Alanine→Glycine	AMBER	8	60 ps	0.98±0.18		18
Alanine→Phenylalanine	AMBER	8	63 ps	-3.01±0.50	-2.71	18
Acetone→Acetaldehyde	AMBER	11	1530 ps	-0.40±0.11	-0.30	25
Acetone→Pyruvic acid	AMBER	11	1530 ps	9.84±0.48		25
Acetone→1,1,1-Trifluoroacetone	AMBER	11	1530 ps	0.63±0.13		25
Acetone→1,1,1-Trichloroacetone	AMBER	11	1530 ps	3.67±0.19		25
Acetone→2,3-Butanedione	AMBER	11	1530 ps	3.57±0.36		25
Acetone→Cyclopropanone	AMBER	11	1530 ps	-0.46±0.20		25
Acetone→Formaldehyde hydrate	AMBER	11	1530 ps	7.20±0.40		25
Methanol→Ethane	BOSS^a	7.5	$^b1.5X10^6$	6.75±0.38	6.93	28
Water→Ethane	BOSS^a	8.5	$^b2X10^6$	-8.41	-8.33	29
Chloride ion→Methane	BOSS^a	8.5	$^b2X10^6$	-79.3	-77	29
CF_4→CH_4	AMBER	8	40 ps	-0.8±0.3	-1.2	30
CFH_3→CHF_3	AMBER	8	40 ps	0.7±0.1	1.0	30
CFH_3→CH_4	AMBER	8	40 ps	2.3±0.2	2.2	30
CHF_3→$CH2F_2$	AMBER	8	40 ps	-1.0±0.1		30
CHF_3→CF_4	AMBER	8	40 ps	2.19±0.05	2.4	30
Methane→Ethane	AMBER	8	102 ps	-0.42	0.17	31
Methane→Ethane (polarization)	AMBER	8	102 ps	-0.18	0.17	31
Methane →Neopentane	AMBER	8	102 ps	-0.86	-0.50	31
Methane→Neopentane (polarization)	AMBER	8	102 ps	0.91	-0.50	31
Ammonium ion→ Tetramethyl Ammonium ion	AMBER	8	102 ps	-29.82	-31.70	31
Ammonium ion→ (polar) Tetramethyl Ammonium ion	AMBER	8	102 ps	-23.56	-31.70	31

Mutations	Force Field	NB	Simulation Length	$\Delta\Delta G^{calc}_{sol}$ (kcal/mol)	$\Delta\Delta G^{exp}_{sol}$ (kcal/mol)	Ref.
Chloride ion→Bromide ion	AMBER	8	102 ps	-3.23	-3.33	31
Chloride ion→ Bromide ion (polarization)	AMBER	8	102 ps	-1.95	-3.33	31
Sodium ion→Potassium ion	AMBER	8	102 ps	-20.09	-17.05	31
Sodium ion→Potassium ion (polarization)	AMBER	8	102 ps	-24.66	-17.05	31
Ethane→Methane	AMBER	8	62 ps	-0.42±0.03	0.17	32
Propane→Ethane	AMBER	8	62 ps	0.03±0.01	-0.12	32
Butane→Ethane	AMBER	8	62 ps	0.06±0.01	-0.12	32
$Me_4C \rightarrow CH_4$	AMBER	8	62 ps	-0.86±0.08	-0.50	32
$Et_4C \rightarrow Me_4C$	AMBER	8	62 ps	0.40±0.01	0.28	32
$Pr_4C \rightarrow Et_4C$	AMBER	8	62 ps	-0.59±0.02		32
$Bu_4C \rightarrow Pr_4C$	AMBER	8	62 ps	-0.51±0.03		32
$MeNH_2 \rightarrow NH_3$	AMBER	8	62 ps	0.07±0.13	0.27	32
$Me_2NH \rightarrow MeNH_2$	AMBER	8	62 ps	-1.93±0.08	-0.27	32
$Me_3N \rightarrow Me_2NH$	AMBER	8	62 ps	-1.17±0.06	-1.07	32
$MeNH_3^+ \rightarrow NH_4^+$	AMBER	8	62 ps	-9.08±0.12	-7.30	32
$Me_2NH_2^+ \rightarrow MeNH_3^+$	AMBER	8	62 ps	-6.35±0.13	-6.40	32
$Me_3NH^+ \rightarrow Me_2NH_2^+$	AMBER	8	62 ps	-5.87±0.05	-7.00	32
$Me_4N^+ \rightarrow NH_4^+$	AMBER	8	62 ps	-29.82±0.52	-31.70	32
$Et_4N^+ \rightarrow Me_4N^+$	AMBER	8	62 ps	-6.74±0.26	-7.00	32
$Pr_4N^+ \rightarrow Et_4N^+$	AMBER	8	62 ps	-5.59±0.11		32
$Bu_4N^+ \rightarrow Pr_4N^+$	AMBER	8	62 ps	-2.75±0.08		32
$C_6H_5CH_3 \rightarrow C_6H_6$	AMBER	8	62 ps	-0.42±0.03	-0.86	32
$C_6H_5OH \rightarrow C_6H_6$	AMBER	8	62 ps	3.52±0.01	2.20	32
$C_6H_5NH_2 \rightarrow C_6H_6$	AMBER	8	62 ps	2.21±0.02	1.68	32
$Cl^- \rightarrow Br^-$ (H_2O)	AMBER	8	41 ps	3.23±0.05	3.3	33
$Cl^- \rightarrow Br^-$ (MeOH)	AMBER	18	41 ps	3.03±0.05	3.8	33
$Cl^- \rightarrow Br^-$ (DMSO)	AMBER	18	41 ps	1.26±0.04	0.2	33
$Na^+ \rightarrow K^+$ (H_2O)	AMBER	8	41 ps	20.09±0.07	17.5	33
$NA^+ \rightarrow K^+$ (MeOH)	AMBER	18	41 ps	20.13±0.08	17.8	33
$NA^+ \rightarrow K^+$ (DMSO)	AMBER	18	41 ps	20.17±0.09	17.6	33
$Me_4N^+ \rightarrow NH_4^+$ (H_2O)	AMBER	8	41 ps	-29.82±0.52	-31.7	33
$Me_4N^+ \rightarrow NH_4^+$ (MeOH)	AMBER	12	41 ps	-29.64±0.28	-31.9	33
$Me_4N^+ \rightarrow NH_4^+$ (DMSO)	AMBER	18	41 ps	-29.74±0.63		33
$Et_4N^+ \rightarrow Me_4N^+$ (H_2O)	AMBER	8	41 ps	-6.74±0.26	-7.0	33
$Et_4N^+ \rightarrow Me_4N^+$ (MeOH)	AMBER	12	41 ps	-5.22±0.34	-5.8	33
$Et_4N^+ \rightarrow Me_4N^+$ (DMSO)	AMBER	18	41 ps	-5.03±0.32	-5.3	33
$C_2H_6 \rightarrow CH_4$ (H_2O)	AMBER	8	62 ps	-0.42±0.03	0.2	33
$C_2H_6 \rightarrow CH_4$ (MeOH)	AMBER	12	62 ps	0.31±0.01	0.9	33
$C_2H_6 \rightarrow CH_4$ (DMSO)	AMBER	18	62 ps	0.15±0.06	1.1	33
$Me_4C \rightarrow CH_4$ (H_2O)	AMBER	8	62 ps	-0.86±0.08	-0.5	33
$Me_4C \rightarrow CH_4$ (MeOH)	AMBER	12	62 ps	1.90±0.02	1.9	33
$Me_4C \rightarrow CH_4$ (DMSO)	AMBER	18	62 ps	1.29±0.04		33
$Et_4C \rightarrow Me_4C$ (H_2O)	AMBER	8	62 ps	0.40±0.01	0.2	33
$Et_4C \rightarrow Me_4C$ (MeOH)	AMBER	12	62 ps	1.71±0.04	2.5	33
$Et_4C \rightarrow Me_4C$ (DMSO)	AMBER	18	62 ps	1.85±0.02		33
$Cl^- \rightarrow Br^-$ (hydrazine)	AMBER	14	41 ps	2.65±0.37	2.5	34
$Cl^- \rightarrow Br^-$ (CCl_4)	AMBER	18	41 ps	0.71±0.04		34
$Na^+ \rightarrow K^+$ (hydrazine)	AMBER	14	41ps	23.32±0.11	17.0	34
$Na^+ \rightarrow K^+$ (CCl_4)	AMBER	18	41 ps	1.46±0.03		34
$Me_4N^+ \rightarrow NH_4^+$ (hydrazine)	AMBER	12	41 ps	-25.54±0.35		34
$Et_4N^+ \rightarrow Me_4N^+$ (hydrazine)	AMBER	12	41 ps	-4.16±0.28		34
$C_2H_6 \rightarrow CH_4$ (hydrazine)	AMBER	14	62 ps	0.67±0.05		34
$C_2H_6 \rightarrow CH_4$ (CCl_4)	AMBER	18	62 ps	1.08±0.05	1.2	34
$Me_4C \rightarrow CH_4$ (hydrazine)	AMBER	12	62 ps	3.64±0.07		34
$Me_4C \rightarrow CH_4$ (CCl_4)	AMBER	18	62 ps	3.53±0.17		34
Acetamide→N-Methylacetamide	AMBER	8	404 ps	2.09±0.11	-0.40	35
N-Methylacetamide→ N,N-Dimethylacetamide	AMBER	8	404 ps	1.05±0.02	1.53	35
$NH_3 \rightarrow MeNH_2$	AMBER	8	404 ps	0.62 ± 0.05	-0.3	35

Mutations	Force Field	NB	Simulation Length	$\Delta\Delta G^{calc}_{sol}$ (kcal/mol)	$\Delta\Delta G^{exp}_{sol}$ (kcal/mol)	Ref.
$MeNH_2 \rightarrow Me_2NH$	AMBER	8	404 ps	1.62±0.01	0.3	35
$Me_2NH \rightarrow Me_3N$	AMBER	8	404 ps	2.34±0.02	1.1	35
$NH_3 \rightarrow Me_3N$	AMBER	8	606 ps	4.36±0.05	1.07	35
$NH_3 \rightarrow H_2O$	AMBER	8	404 ps	-2.17±0.00	-2.01	35
$NH_3 \rightarrow NH_2CH_3$	BOSS	8	400 ps	0.3±0.5	-0.3	36
$NH_2CH_3 \rightarrow NH(CH_3)_2$	BOSS	8	400 ps	2.5±0.6	0.3	36
$NH(CH_3)_2 \rightarrow N(CH_3)$	BOSS	8	400 ps	0.6±0.6	1.1	36
Propane→Ethane	CHARMM	8.5	60 ps	0.87±0.3	-0.13	38
Propane→Ethane	OPLS	8.5	60 ps	0.56±0.1	-0.13	38
Butane→Propane	OPLS	8.5	60 ps	-1.0±0.2	-0.13	38
Methanol→Ethane	OPLS	8.5	60 ps	7.5±0.2	6.9	38
Ethanol→Propane	OPLS	8.5	60 ps	7.2±0.1	7.0	38
Toluene→Benzene	BOSS[a]	10.5	[b]$4X10^6$	0.13±0.1	0.10±0.1	39
Cyanobenzene→Toluene	BOSS[a]	10.5	[b]$4X10^6$	3.8±0.1	2.6± 0.5	39
p-Xylene→Toluene	BOSS[a]	10.5	[b]$4X10^6$	-0.12±0.1	0.0± 0.1	39
p-Xylene→Hydroquinone	BOSS[a]	10.5	[b]$4X10^6$	-10.5±0.2	-9.3±0.1	39
p-Xylene→p-Dicyanobenzene	BOSS[a]	10.5	[b]$4X10^6$	-7.7±0.3		39
p-Xylene→p-Cresol	BOSS[a]	10.5	[b]$4X10^6$	-5.6±0.2	-5.3±0.1	39
p-Cresol→Phenol	BOSS[a]	10.5	[b]$4X10^6$	-0.5 ± 0.1	-0.4±0.1	39
Anisole→Phenol	BOSS[a]	10.5	[b]$4X10^6$	-5.1 ± 0.2	-4.7±0.8	39
Phenol→Benzene	BOSS[a]	10.5	[b]$4X10^6$	5.2 ± 0.2	5.8± 0.1	39
Anisole→Benzene	AMBER	8	72 ps	0.90	1.1-1.6	40
Trimethoxybenzene→Benzene	AMBER	8	72 ps	4.31	4.1-4.5	40
Methane→Ethane	AMBER	8	300 ps	0.16± 0.04	-0.16	41
Ethane→Propane	AMBER	8	300 ps	0.20± 0.04	0.21	41
N-Acetylalanine N-Methylamide →Valine dipeptide	AMBER	8	300 ps	1.1± 0.1		41
Adenine→Guanine	AMBER[a]	9	189 ps	-10.3		43
Uracil→Thymine	AMBER[a]	9	189 ps	-0.3		43
Thymine→Cytosine	AMBER[a]	9	189 ps	-6.9		43
Adenine→Uracil	AMBER[a]	9	210 ps	-1.0		43
Guanine→Cytosine	AMBER[a]	9	210 ps	1.4		43
9-Methyladenine→Methane	AMBER	8	1000 ps	13.90± 0.14		44
9-Methylguanine→Methane	AMBER	8	1000 ps	24.34± 0.19		44
1-Methylcytosine→Methane	AMBER	8	1000 ps	20.30± 0.24		44
1-Methyluracil→Methane	AMBER	8	1000 ps	15.92± 0.06		44
1-Methylthymine→Methane	AMBER	8	1000 ps	14.34± 0.14		44
9-Methyladenine→9-Methylguanine	AMBER	8	1000 ps	-11.41± 0.16		44
1-Methylcytosine→1-Methyluracil	AMBER	8	1000 ps	3.87± 0.07		44
1-Methyluracil→1-Methylthymine	AMBER	8	1000 ps	1.80± 0.11		44
1-Methylcytosine→1-Methylthymine	AMBER	8	1000 ps	5.73± 0.18		44
Methanol→Ethane	BOSS	8	[b]$1.4X10^6$	8.3	6.9	45
Thr→Ala (tri-peptide)	BOSS	8	[b]$1.4X10^6$	7.05	6.9	45
Thr→Ala (penta-peptide)	BOSS	8	[b]$1.4X10^6$	6.61	6.9	45

[a]OPLS force field; [b]Monte Carlo simulation; NB = Non-bonded cutoff distance in Å

kcal/mol), 2.4±0.3 kcal/mol, 2.3±0.2 kcal/mol (expt: 2.2 kcal/mol), -0.71± 0.07 kcal/mol, -1.0±0.1 kcal/mol and 1.5±0.1 kcal/mol, 2.19±0.05 kcal/mol (expt: 2.4 kcal/mol), respectively. The authors concluded that agreement with the experiment is better when bond-PMF correction is used because it allows an accurate estimation of the length of bonds involved in perturbation and the effect of the size of the solute.

Singh and coworkers[31] have computed the relative solvation free energies for small molecules to examine the effect of the polarization energy using AMBER 3.3. Each solute was solvated in a rectangular box with 216 TIP3P water molecules. During the equilibration and subsequent perturbations the periodic boundary conditions were applied only for solute-solvent and solvent-solvent non-bonded interactions. Also, all solute-solute non-bonded interactions were included. Each perturbation was achieved in two stages: charge mutation followed by the mutation of the van der Waals and the polarizability parameters. The electrostatic run (charge mutation) was carried out in 21 windows where the system was equilibrated for 1 ps and data was collected for the next 1 ps in each window. The van der Waals parameters were mutated by using a 201 windows simulation with 0.4 ps of equilibration and 0.4 ps of data collection. The calculated solvation free energies with and without polarization and experimental values for the mutations of methane to ethane and methane to neopentane were -0.18 kcal/mol, -0.42 kcal/mol, 0.17 kcal/mol, 0.91 kcal/mol, -0.86 kcal/mol, and -0.50 kcal/mol, respectively. Therefore, polarization energy added a constant positive value to the free energy change for all the transformations.

Jorgensen and coworkers[29] computed free energies of hydration of methane using BOSS and OPLS force field. Two series of MC simulations were performed with a single solute plus 216 TIP4P water molecules in a cubic cell with periodic boundary conditions. In the first series, a single Lennard-Jones particle corresponding to methane was made to vanish. In the second series, a TIP4P water molecule was converted to the methane particle. The intermolecular interactions were truncated smoothly between 8.0 Å and 8.5 Å. Each simulation included 5×10^5 configurations of equilibration followed by 1.5×10^6 or 2.0×10^6 configurations for the averaging. The calculated and experimental free energy difference for the mutation of water molecule to methane were -8.41 kcal/mol, and -8.33 kcal/mol, respectively. The study showed the utility of statistical perturbation theory for computing absolute free energies of solution and the quality of the underlying potential functions.

Rao and Singh[32] calculated relative solvation free energies for normal alkanes, tetra-alkylmethanes, amines and aromatic compounds using AMBER 3.1. Each system was solvated with 216 TIP3P water molecules. The atomic charges were uniformly scaled down by a factor of 0.87 to correct the overestimation of dipole moment by 6-31G* basis set. During the perturbation runs, the periodic boundary conditions were applied only for solute-solvent and solvent-solvent interactions with a non-bonded interaction cutoff of 8.5 Å. All solute-solute non-bonded interactions were included. Electrostatic decoupling was applied where electrostatic run was completed in 21 windows. Each window included 1 ps of equilibration and 1 ps of data collection. The van der Waals parameters were mutated with and without the coordinate coupling. Simulations without the coordinate coupling were divided into 101 windows with equilibration of 0.4 ps and data collection of 0.4 ps in each window. With the coordinate coupling, the simulations were divided into 201 windows with 0.2 ps of equilibration and 0.2 ps of data collection. The calculated relative solvation free energies are in reasonable agreement with the experimental values. The values obtained with coordinate coupling were in better agreement with experimental results as compared to those obtained without coordinate coupling. The solute-solvent interaction energy seems to be the major component in hydration.

In another study, Rao and Singh[33] calculated differences in free energies of solvation for alkanes in water, methanol and dimethyl sulfoxide using AMBER 3.1. The authors created MeOH and DMSO solvent boxes starting with 216 TIP3P water molecules. During minimization, equilibration and perturbation runs, periodic boundary conditions were applied only for solute-solvent and solvent-solvent interactions. For solute-solvent and solvent-solvent non-bonded interactions, cutoff distances of 10 Å and 12 Å were employed for simulations in MeOH and DMSO, respectively. All solute-solute non-bonded interactions were included. The mutations involving molecular solutes were calculated using free energy decoupling method. The partial charges were mutated in 21 windows, each with 1 ps of equilibration and 1 ps of data collection. The van der Waals parameters were mutated with coordinate coupling over 201 windows, each with 0.2 ps of equilibration and 0.2 ps of data collection. A time step of 2 fs was used in all of the simulations. The ΔG varies differently in different solutions due to varying solvation processes. The characteristic feature observed in the variation patterns of ΔG for alkanes in water is not observed in MeOH or DMSO. Rao and Singh calculated relative solvation free energies for normal alkanes and tetra-alkylmethane molecules in hydrazine and carbon tetrachloride as well, using AMBER 3.1.[34] ESP fitted atomic partial charges were obtained with 6-31G* basis set. Periodic boundary conditions were applied only for solute-solvent and solvent-solvent interactions. A constant dielectric of 1 was used in all the simulations. Cutoff distances of 12 Å and 14 Å were used in hydrazine and CCl_4, respectively. The molecular solutes were mutated in two stages with the free energy decoupling method. The partial charges were mutated first by using 21 windows with 1 ps of equilibration and 1 ps of data collection in each window. The van der Waals parameters were mutated with coordinate coupling over 201 windows with 0.2 ps of equilibration and 0.2 ps of data collection. A time step of 2 fs was used during equilibration and data collection. The results were in good agreement with the available experimental results. The solvation behavior in hydrazine resembled that in water for many solutes. The results supported the view that the special phenomenon observed in the hydration of apolar solutes is a result of the structural peculiarity of liquid water.

Morgantini and Kollman[35] calculated relative solvation free energies for several amides and amines. The solvation free energies were calculated using restrained ESP fitted charges calculated using the 6-31G* basis set, the AMBER 4.0 package, the TIP3P water model, a constant dielectric of 1, 8 Å of non-bonded cutoff, 101 windows with 2 fs time step, 2 ps of equilibration and 2 ps of data collection in each window. Bond-PMF correction was employed for all perturbations involving bond length changes. In addition to relative solvation free energies, absolute solvation free energies were calculated for H_2O, NH_3 and $MeNH_2$. The calculated and experimental solvation free energies for acetamide to N-methylacetamide, N-methylacetamide to N,N-dimethylacetamide, ammonia to amino methane, amino methane to dimethyl amine, dimethyl amine to trimethyl amine, ammonia to trimethyl amine and ammonia to water were 2.09±0.11 kcal/mol (expt: -0.40 kcal/mol), 1.05±0.02 kcal/mol (expt: 1.53 kcal/mol), 0.62±0.05 kcal/mol (expt: -0.26 kcal/mol), 1.62±0.01 kcal/mol (expt: 0.27 kcal/mol), 2.34±0.02 kcal/mol (expt: 1.06 kcal/mol), 4.36±0.05 kcal/mol (expt: 1.07 kcal/mol) and -2.17±0.00 kcal/mol (expt: -2.01 kcal/mol), respectively. Unlike experimental data the calculated relative solvation free energies increase monotonically as a function of methyl addition even though the calculated values were within 0.5 kcal/mol of the experiment.

Levy and coworkers[36] reported a monotonous increase in relative solvation free energies as a function of methyl addition to amines and amides as well. ESP fitted charges were calculated for all the amides and amines using the 6-31G* basis set. The solvation free energies were calculated using IMPACT, TIP3P water model, a non-bonded cutoff of 8 Å, time step of 2 fs, 100 and 200 double wide sampling windows were used for amide and amine mutations, respectively, each window included an equilibration (1 ps) and data collection (1 ps) and both, forward and backward perturbations were made. The solvation free energies were calculated with and without many-body polarizable potential. The calculated relative solvation free energies using many-body polarization potential for NH_3 to NH_2CH_3, NH_2CH_3 to $NH(CH_3)_2$ and $NH(CH_3)_2$ to $N(CH_3)_3$ mutations gave 0.3±0.5 kcal/mol (expt: -0.3 kcal/mol), 2.5±0.6 kcal/mol (expt: 0.3 kcal/mol) and 0.6±0.6 kcal/mol (expt: 1.1 kcal/mol), respectively. Those solvation free energies are in better agreement with the experimental result. Kollman and coworkers[37] applied explicit polarizable water and solute potential energy functions and calculated relative solvation free energies that agreed well with Kollman's earlier results but did not reproduce the experimental trend of relative solvation free energies in amines and amides.

Brooks[38] calculated relative solvation free energies for propane, ethane, butane, methanol and ethanol using CHARMM and OPLS force fields. The simulations were carried out using TIP3P water and solute at σ = 1g/cc and T = 298°±5 K. The calculations used double-wide sampling with windows at λ = 0.125, 0.5 and 0.875, each with 8 ps of equilibration dynamics and 12–30 ps of production run with a time step of 1 fs. The calculated and experimental solvation free energies for the mutation of propane to ethane, butane to propane, methanol to ethane and ethanol to propane were 0.56 ± 0.1 kcal/mol, -0.13 kcal/mol, -1.0 ± 0.2 kcal/mol, -0.13 kcal/mol, 7.5±0.2 kcal/mol, 6.9 kcal/mol, 7.2±0.1 kcal/mol and 7.0 kcal/mol, respectively. The author concluded that the parameterization of current empirical potential functions is inadequate in representing apolar to apolar thermodynamics in aqueous solution. However, polar to apolar transformations are governed by the loss of large electrostatic components and are adequately represented.

Jorgensen and Nguyen developed intermolecular potential functions and calculated the relative solvation free energies for substituted benzenes using BOSS.[39] The system consisted of 500 TIP4P water molecules plus the solute in a cubic cell 25 Å on a side with periodic boundary conditions. Each solute was mutated in a series of 5 to 10 simulations guided by insisting that the statistical uncertainty for a computed free energy increment not to exceed 0.1 kcal/mol as obtained from the fluctuations in separate averages over 1 x 10^5 or 2 x 10^5 configurations. Each simulation entailed an equilibration period of at least 10^6 configurations followed by averaging over an additional 2 x 10^6 to 4 x10^6 configurations. The intermolecular interactions were truncated at 8.5 Å for water-water interactions and at 10.5 Å for solute-water interactions. The authors determined that phenol has an average of 2.5 hydrogen bonds with water molecules, 1.0 as donor and 1.5 as acceptor, while anisole and benzonitrile accept only 1 hydrogen bond from water. This simulation was also one of the first to use a cyclic design strategy to show that the calculations where in agreement. Initially benzene was mutated into a simple analog. and these simple changes continued until the circle was completed. The calculated relative solvation free energies for these systems were in good agreement with the experimental results (Table 2).

Kollman and coworkers[40] calculated relative solvation free energies of benzene, anisole, and 1,2,3-trimethoxybenzene (TMB) (Figure 2) in water and demonstrated the sensitivity of results to different charge models. The relative solvation free energies were calculated with AMBER 3.0 and TIP3P water model using electrostatic decoupling. The electrostatic energies were evaluated in 21 windows with 1 ps of equilibration and 1 ps of data collection in each window. The van der Waals energies were evaluated

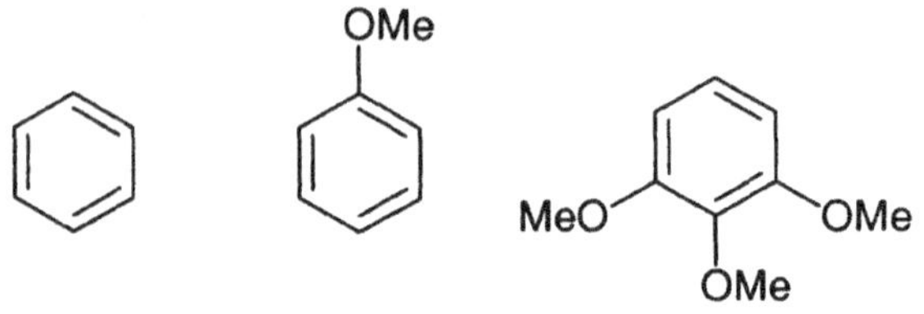

Benzene Anisole 1, 2, 3-Trimethoxy benzene

Figure 2. Molecules considered for solvation free energy calculations.

using slow growth in both forward and backward directions in 30 ps. Only the intermolecular interaction energies were calculated assuming that intramolecular contributions to the free energy are similar in the gas phase and solution. ESP fitted charges were calculated with three different basis sets, namely 6-31G*, 4-31G and STO-3G. The calculated solvation free energies for anisole to benzene and TMB to benzene mutations using ESP fitted charges with 6-31G*, 4-31G and STO-3G basis sets were 0.90 kcal/mol, 3.56 kcal/mol and 1.05 kcal/mol (expt: 1.1–1.6 kcal/mol) and 4.31 kcal/mol, 11.61 kcal/mol and 2.93 kcal/mol (expt: 4.1–4.5 kcal/mol), respectively. Therefore, 6-31G* ESP fitted charges gave the best result. Authors note that van der Waals contribution to the free energy difference was rather small.

Sun et al.[41] implemented bond-PMF correction in calculating relative solvation free energies for methane, ethane, propane, N-acetylalanine N-methylamide and valine dipeptide using AMBER 4.0. The simulations were carried out with explicit solvent (TIP3P water) in a cubic cell at constant pressure of 1 atm and periodic boundary conditions, a time step of 1.5 fs and non-bonded cutoff of 8 Å. The atomic partial charges, Mulliken and ESP fitted, were calculated with 6-31G* basis set. SHAKE constraints[42] on the perturbed bonds were applied and used an explicit potential of mean force (PMF) like calculation to determine the contribution due to bond length changes. This treatment led to relative solvation free energies that were unaffected by the shrinking of the disappearing bond during perturbation. Therefore, a bond PMF correction is applied in the presented calculations. The perturbations involving the shrinking or growing of a methyl group, molecular dynamics simulations were run for 180 ps. For perturbations where only the electrostatic distribution changes (changes in charges), the molecular dynamics simulation times were 90 ps. Two 300 ps simulations were conducted to ensure that appropriate convergence had been obtained. In both cases, each window consisted of 500 steps of data collection. The calculated relative solvation free energies for the mutation of methane to ethane and ethane to propane, were 0.16 kcal/mol (expt: -0.16 ± 0.1 kcal/mol) and 0.20 kcal/mol (expt: 0.21 kcal/mol), respectively. The calculated free energy results for alkanes were within ~0.3 kcal/mol of experimental results and free energy difference for the two dipeptides compared to experimental result was 1.1 ± 0.1 kcal/mol. The authors show that the larger difference in peptides is due to indirect contributions from backbone atoms.

3.4 Solvation Free Energies of Nucleic Acid Bases and Amino Acid Side-Chains

Elcock and Richards calculated relative solvation free energies for the five nucleic acid bases (Figure 3) using AMBER 4.0.[43] The OPLS force filed parameter set was used to represent all non-bonded interactions. In each case, the solute was immersed in a box of 506 TIP3P water molecules and simulations were conducted in the NPT ensemble at 298° K and 1 atm pressure. A time step of 2 fs was used and a 9 Å cutoff was applied to all the non-bonded interactions. The perturbations were performed over a minimum of 21 windows, each consisting of 6 ps equilibration, 9 ps for data collection for the adenine to guanine, uracil to thymine, and thymine to cytosine, mutations and 20 ps equilibration 10 ps data collection for the adenine to uracil and guanine to cytosine mutations. The relative solvation free energies between nucleic acid bases obtained using different methods are in good agreement (Table 2).

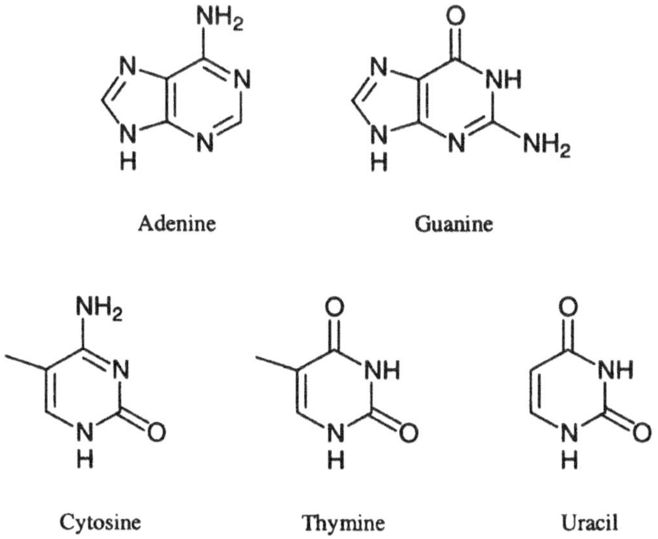

Figure 3. Nucleic acid bases used in solvation free energy calculations.

Miller and Kollman calculated relative solvation free energies of N-methylated nucleic acid bases using AMBER 4.1.[44] All the molecular dynamics and thermodynamic integration calculations were done with a constant dielectric of 1 and an 8 Å cutoff on non-bonded interactions. All simulations were carried out in the NPT ensemble with periodic boundary conditions, a temperature of 300° K, and a pressure of 1 atm. Depending on the perturbation, 217 and 265 TIP3P water molecules were used for solvation. The series of simulation times for most mutations was approximately 200 ps, 400 ps, 800 ps and 1000 ps. The relative solvation free energies for the mutation of 9-methyladenine to methane, 9-methylguanine to methane, 1-methylcytosine to methane, 1-methyluracil to methane, 1-methylthymine to methane, 9-methyladenine to 9-methylguanine, 1-methylcytosine to 1-methyluracil, 1-methyluracil to 1-methylthymine and 1-methylcytosine to 1-methylthymine (same as Figure 3 with methyl group) were 13.90 kcal/mol, 24.34 kcal/mol, 20.30 kcal/mol, 15.92 kcal/mol, 14.34 kcal/mol, -11.41 kcal/mol, 3.87 kcal/mol, 1.80 kcal/mol and 5.73 kcal/mol, respectively. This study includes application of several different methods to compute relative solvation free energies. The absolute solvation free energies calculated for 9-methyladenine and 1-methylthymine are in good agreement with the experimental values (Table 2).

Relative free energies of solvation for amino acid side-chains, four nucleic acid bases and organic molecules were calculated using FEP with AMBER 3.0.[17] The set of compounds studied showed structural dissimilarity and differed in charges. Simulations used 400 TIP3P water molecules at constant temperature and pressure in a box with periodic boundary conditions and an 8.0 Å non-bonded cutoff. The calculations were done in two stages with electrostatic decoupling. The electrostatic changes were simulated in 20 thermodynamic windows, each with 1 ps of equilibration and 1 ps of data collection (time step: 2 fs). Changes in van der Waals parameters were performed with slow growth in 30,000 steps at 0.001 ps time intervals. The results showed good agreement with the experimental values (Table 2). The authors recommended electrostatic decoupling in order to avoid overestimation of electrostatic energies.

In order to answer the questions about the effect of the local environment, Saqi and Goodfellow[45] studied the change of a polar group (OH) to an apolar group (CH3) for a methanol to ethane mutation and a Thr to Val mutation in tri (Ala-Thr-Ala) and penta (Ala-Lys-Thr-Lys-Ala) peptides. Monte Carlo simulations were carried out using BOSS with a double wide sampling in five windows, TIP4P water model, 0.4×10^6 MC steps of equilibration and data collection of 1×10^6 MC steps. The free energy differences were calculated using both NVT and NPT ensembles and better precision was obtained with NPT ensemble. The calculated free energy difference for methanol (OH) to ethane (CH3) mutation was 8.3

kcal/mol as compared to experimental value of 6.9 kcal/mol, whereas relative free energy differences between Thr (OH) to Ala (CH3) in tri and penta peptides are 7.05 kcal/mol, and 6.61 kcal/mol, respectively. The solvation free energy differences between methanol (OH) to ethane (CH3) and Thr (OH) to Ala (CH3) in tri and penta peptides are attributed due to local environmental effects. Kollman and coworkers[18] performed methanol to ethane, glycine to alanine and alanine to phenylalanine inter-conversions using FEP and slow growth methods of AMBER 3.0. Calculated relative solvation free energies for the mutation of methanol to ethane, alanine to glycine and alanine to phenylalanine were 6.86±0.12 kcal/mol, 0.98±0.18 kcal/mol and -3.01±0.50 kcal/mol, respectively. The results were in reasonable agreement (Table 2) with the related experimental data and suggested that the methodology could be used in a wide range of chemical and biochemical situations.

3.5 Solvation Free Energies between Charged Systems

Kollman and coworkers[18] calculated relative solvation free energies between ammonium(NH_4^+) to oxonium (OH_3^+) using both slow growth and windowing procedures of AMBER program. The simulations were run at constant pressure (1 atm) and temperature (300° K), and with 215 TIP3P water molecules. The accumulated change in free energy for the entire simulations lie between -20.6 and -21.8 kcal/mol. Experimentally the enthalphy of solvation has been estimated to be between 75 and 81 kcal/mol for NH_4^+ and approximately 100 kcal/mol for OH_3^+. Thus, the ΔΔHs is roughly 19–25 kcal/mol, and as in the gas-phase studies the ΔΔGs would be expected to be about 2–3 kcal/mol less than the ΔΔHs. The calculated values were within the estimated experimental value. These simulations show that both slow growth and windowing procedures do equally well.

Singh and coworkers[31] examined the effect of polarization energy on computed relative solvation free energies of ions as well. The calculated solvation free energies with and without polarization and experimental values for the mutations of ammonium ion to tetramethyl ammonium ion, chloride ion to bromide ion and sodium ion to potassium ion were -23.56 kcal/mol, -29.82 kcal/mol, -31.70 kcal/mol, -1.95 kcal/mol, -3.23 kcal/mol, -3.33 kcal/mol, -24.66 kcal/mol, -20.09 kcal/mol, and -17.05 kcal/mol, respectively. Therefore, polarization energy added a constant positive value to the free energy change for all the transformations except for the mutation of sodium ion to potassium ion.

Rao and Singh[32] calculated relative solvation free energies for alkyl- and tetra-alkylammonium ions using same conditions as described before for neutral molecules (except, atomic partial charges were not scaled for ions). The values obtained with coordinate coupling were in better agreement with experimental results. Also, using the conditions described before for alkanes, Rao and Singh[33] calculated differences in free energies of solvation for closed shell ions and tetra-alkylammonium ions in water, methanol and dimethyl sulfoxide using AMBER 3.1. All the transformations involving closed shell ions were done in 101 windows with 0.4 ps of equilibration followed by 0.4 ps of data collection in each window with the solute-solvent and the solvent-solvent non-bonded interactions cutoff distance of 14 Å. The nature of solvent seems to have almost negligible influence on the variation of ΔG with the change in the solute size for positively charged alkali ions and smaller tetra-alkylammonium ions. The ΔG of negatively charged halides are affected by the solute size to different extents in different solvents. The characteristic feature observed in the variation patterns of ΔG for large tetra-alkylammonium ions in water is not observed in MeOH or DMSO. Rao and Singh[34] calculated relative solvation free energies for closed shell ions and tetra-alkylammonium ions in hydrazine and carbon tetrachloride using same conditions as described before for alkanes. For the transformations involving closed shell ions, 101 windows were employed with 0.4 ps of equilibration followed by 0.4 ps of data collection at each window. The results were in good agreement with the available experimental results. The solvation behavior in hydrazine resembled that in water for many solutes.

3.6 Solvation Free Energies between Neutral and Charged Systems

Using the aforementioned conditions Jorgensen and coworkers[29] computed free energies of hydration between charged and neutral systems. Two series of MC simulations were performed. A single Lennard-Jones particle corresponding to methane was made to vanish and a chloride ion was converted to the methane particle. The calculated and experimental free energy difference for the mutation of chloride ion to methane were -79.3 kcal/mol and -77 kcal/mol, respectively. The study showed the utility of statistical perturbation theory for computing absolute free energies of solution and the quality of the underlying potential functions.

Kollman and coworkers calculated relative free energies of solvation between neutral and charged amino acids using FEP with AMBER 3.0.[17] The same set of conditions were used as those for the mutations of neutral molecules. The alanine to lysine mutation was calculated using the two stage approach, i.e. perturbation of alanine to neutral lysine followed by perturbation of neutral lysine to protonated lysine. The calculated relative solvation free energies between these molecules were in good agreement with the experimental results (Table 2).

4. CONCLUSIONS

As we mentioned in the beginning of this chapter, solvation plays a very crucial role in the binding of inhibitors to an enzyme. As a result several theoretical methods have been developed and used for calculating relative solvation free energies with explicit solvent water. These calculations clearly indicated that convergence in the calculated solvation free energies is a potential problem. Several papers[19, 20, 25] described the length of the simulation required to achieve satisfactory convergence as well as the effect of size and conformational flexibility on convergence.

5. REFERENCES

1. J. Berman, M. Green, E. Sugg, R. Anderegg, D. S. Millington, D. L. Norwood, J. McGeehan, and J. Wiseman, Rapid optimization of enzyme substrates using defined substrate mixtures, J. Biol. Chem. **267**:1434 (1992).
2. W. C. Still, A. Tempczyk, R. C. Hawley, and T. Hendrickson, Semianalytical treatment of solvation for molecular mechanics and dynamics, J. Am. Chem. Soc. **112**:6127 (1990).
3. G. C. Lie and E. Clementi, Study of the structure of molecular complexes XII: Structure of liquid water obtained by Monte Carlo simulation with the Hartree-Fock potential corrected by inclusion of dispersion forces, J. Chem. Phys. **62**:2195 (1975).
4. F. H. Stillinger and A. Rahman, Improved simulation of liquid water by molecular dynamics, J. Chem. Phys. **60**:1545 (1974).
5. W. L. Jorgensen, Transferable intermolecular potential functions for water, alcohols, and ethers: Application to liquid water, J. Am. Chem. Soc. **103**:335 (1981).
6. W. L. Jorgensen, J. Chandrasekhar, J. D. Madura, R. W. Impey, and M. L. Klein, Comparison of simple potential functions for simulating liquid water, J. Chem. Phys. **79**:926 (1983).
7. M. W. Mahoney and W. L. Jorgensen, A five-site model for liquid water and the reproduction of the density anomaly by ridge, nonpolarizable potential functions, J. Chem. Phys. **112**:8910 (2000).
8. M. W. Mahoney and W. L. Jorgensen, Diffusion constant of the TIP5P model of liquid water, J. Chem. Phys. **114**:363 (2001).
9. H. J. C. Berendsen, J. P. M. Postma, W. F. v. Gunsteren, and J. Hermans, Interaction models for water in relation to protein hydration, in: Intermolecular Forces, B. Pullman, ed., D. Reidel Publishing Co., Boston (1981), pp. 331-342.
10. H. J. C. Berendsen, J. R. Grigera, and T. P. Straatsma, The missing term in effective pair potentials, J. Phys. Chem. **91**:6269 (1987).
11. M. R. Reddy and M. Berkowitz, The dielectric constant of SPC/E water, Chem. Phys. Lett. **155**:173 (1989).
12. M. R. Reddy, M. D. Erion, and A. Agarwal, Free energy calculations: Use and limitations in predicting ligand binding affinities, in: Reviews in Computational Chemistry, vol. 16, K. B. Lipkowitz and D. B. Boyd, eds., VCH Publishers, New York (2000), pp. 217-304.

13. B. L. Tembe and J. A. McCammon, Ligand-receptor interactions, Comput. Chem. **8**:281 (1984).
14. P. A. Bash, U. C. Singh, F. K. Brown, R. Langridge, and P. A. Kollman, Calculation of the relative change in binding free energy of a protein-inhibitor complex, Science **235**:574 (1987).
15. U. C. Singh, Probing the salt bridge in the dihydrofolate reductase-methotrexate complex by using the coordinate-coupled free-energy perturbation method, Proc. Natl. Acad. Sci. USA **85**:4280 (1988).
16. M. R. Reddy, R. J. Bacquet, D. Zichi, D. A. Matthews, K. M. Welsh, T. R. Jones, and S. Freer, Calculation of solvation and binding free energy differences for folate-based inhibitors of the enzyme thymidylate synthase, J. Am. Chem. Soc. **114**:10117 (1992).
17. P. A. Bash, U. C. Singh, R. Langridge, and P. A. Kollman, Free energy calculations by computer simulation, Science **236**:564 (1987).
18. U. C. Singh, F. K. Brown, and P. A. Kollman, An approach to the application of free energy perturbation methods using molecular dynamics: Applications to the transformations of $CH_3OH \rightarrow CH_3CH_3$, $H_3O^+ \rightarrow NH_4^+$, Glycine→Alanine, and Alanine→Phenylalanine in aqueous solution and to $H_3O^+(H_2O_3) \rightarrow NH_4^+(H_2O)_3$ in the gas phase, J. Am. Chem. Soc. **109**:1607 (1987).
19. M. J. Mitchell and J. A. McCammon, Free energy difference calculations by thermodynamic integration: Difficulties in obtaning a precise value, J. Comp. Chem. **12**:271 (1991).
20. D. A. Pearlman and P. A. Kollman, The overlooked-stretching contribution in free energy perturbation calculations, J. Chem. Phys. **94**:4532 (1991).
21. F. Guarnieri and W. C. Still, A rapidly convergent simulation method: Mixed Monte Carlo/stochastic dynamics, J. Comp. Chem. **15**:1302 (1994).
22. D. A. Pearlman, Free energy derivatives: A new method for probing the convergence problem in free energy calculations, J. Comp. Chem. **15**:105 (1994).
23. C. Chipot, C. Millot, B. Maigret, and P. A. Kollman, Molecular dynamics free energy perturbation calculations: Influence of nonbonded parameters on the free energy of hydration of charged and neutral species, J. Phys. Chem. **98**:11362 (1994).
24. C. Chipot, P. A. Kollman, and D. A. Pearlman, Alternative approaches to potential of mean force calculations: Free energy perturbation versus thermodynamic integration: Case study of some representative nonpolar interactions, J. Comp. Chem. **17**:1112 (1996).
25. M. R. Reddy and M. D. Erion, Calculation of relative solvation free energy differences by thermodynamic perturbation method: Dependence of the free energy results on the simulation length, J. Comp. Chem. **20**:1018 (1999).
26. J. P. M. Postma, H. J. C. Berendsen, and J. R. F. Haak, Thermodynamics of cavity formation in water: A molecular dynamics study, Symp. Chem. Soc. **17**:55 (1982).
27. K. Remerie, W. F. van Gunsteren, J. P. M. Postma, H. J. C. Berendsen, and J. B. F. N. Engberts, Molecular dynamics computer simulation of the hydration of two simple organic solutes. Comparison with the simulation of an empty cavity, Mol. Phys. **53**:1517 (1984).
28. W. L. Jorgensen and C. Ravimohan, Monte Carlo simulation of differences in free energies of hydration, J. Chem. Phys. **83**:3050 (1985).
29. W. L. Jorgensen, J. F. Blake, and J. K. Buckner, Free energy of TIP4P water and the free energies of hydration of methane and chloride from statistical perturbation theory, Chem. Phys. **129**:193 (1989).
30. C. A. Gough, D. A. Pearlman, and P. A. Kollman, Calculations of the relative free energies of aqueous solvation of several fluorocarbons: A test of the bond potential of mean force correction, J. Chem. Phys. **99**:9103 (1993).

31. K. Ramnarayan, B. G. Rao, and U. C. Singh, The effect of polarization energy on the free energy perturbation calculations, J. Chem. Phys. **92**:7057 (1990).
32. B. G. Rao and U. C. Singh, Hydrophobic hydration: A free energy perturbation study, J. Am. Chem. Soc. **111**:3125 (1989).
33. B. G. Rao and U. C. Singh, A free energy perturbation study of solvation in methanol and dimethyl sulfoxide, J. Am. Chem. Soc. **112**:3803 (1990).
34. B. G. Rao and U. C. Singh, A free energy perturbation study of solvation in hydrazine and carbon tetrachloride, J. Am. Chem. Soc. **113**:3481 (1991).
35. P. Y. Morgantini and P. A. Kollman, Solvation free energies of amides and amines: Disagreement between free energy calculations and experiment, J. Am. Chem. Soc. **117**:6057 (1995).
36. Y. Ding, D. N. Bernardo, K. Krogh-Jespersen, and R. M. Levy, Solvation free energies of small amides and amines from molecular dynamics/free energy perturbation simulations using pairwise additive and many-body polarizable potentials, J. Phys. Chem. **99**:11575 (1995).
37. E. C. Meng, J. W. Caldwell, and P. A. Kollman, Investigating the anomalous solvation free energies of amines with a polarizable potential, J. Phys. Chem. **100**:2367 (1996).
38. C. L. Brooks III, Thermodynamic calculations on biological molecules, J. Quant. Chem.: Quant. Biol. Symp. **15**:221 (1988).
39. W. L. Jorgensen and T. B. Nguyen, Monte Carlo simulations of the hydration of substituted benzenes with OPLS potential functions, J. Comput. Chem. **14**:195 (1993).
40. L. F. Kuyper, R. N. Hunter, D. Ashton, K. M. Merz Jr., and P. A. Kollman, Energies of benzene, anisole and 1,2,3-trimethoxybenzene: Theoretical and experimental analysis of aromatic methoxy solvation, J. Phys. Chem. **95**:6661 (1991).
41. Y. Sun, D. Spellmeyer, D. A. Pearlman and P. Kollman, Simulation of the solvation free energies for methane, ethane, and propane and corresponding amino acid dipeptides: A critical test of the "Bond-PMF" correction, a new set of hydrocarbon parameters, and the gas phase-water hydrophobicity scale, J. Am. Chem. Soc. **114**:6798 (1992).
42. D. A. Pearlman and P. A. Kollman, The overlooked bond-stretching contribution in free energy perturbation calculations, J. Chem. Phys. **94**:4532 (1991).
43. A. H. Elcock and W. G. Richards, Relative hydration free energies of nucleic acid bases, J. Am. Chem. Soc. **115**:7930 (1993).
44. J. L. Miller and P. A. Kollman, Solvation free energies of the nucleic acid bases, J. Phys. Chem. **100**:8587 (1996).
45. M. A. S. Saqi and J. M. Goodfellow, Free energy changes associated with amino acid substitution in proteins, Protein Eng. **3**:419 (1990).

Chapter 6

Tautomerism and Ionisation Studies Using Free Energy Methods

Graham A. Worth[†] and Paul M. King[‡]
[†]*Dept. of Chemistry, King's College London, UK*
[‡]*School of Biological and Chemical Sciences, Birkbeck College, London, UK*

1. INTRODUCTION

Tautomerism refers to the inter-conversion between chemical structures which differ in the placement of an atom or a group, most commonly a proton (in which case it is termed prototropy). Many different types of organic functionality can undergo tautomerism, common examples being keto–enol, nitroso–oxime and imine–enamine.[1] In all cases the equilibrium is dynamic, with rapid inter–conversion between the various forms. Often the equilibrium lies strongly in favour of one form or the other, with the position of equilibrium depending upon a variety of factors including molecular structure (e.g. stability through enhanced resonance or intra-molecular hydrogen-bonding) and environmental effects (such as solvent polarity, hydrogen-bonding capacity, concentration, pH, temperature etc.).

Many naturally occurring biologically active molecules are able to adopt tautomeric forms, in particular those containing heterocyclic components.[2] Most common are those of the nucleic acid bases: all of the five major bases of DNA/RNA exist predominantly in the amino– and keto-tautomeric forms at physiological pH. However, if the less prevalent, minor tautomer exists at the time of replication it is possible for abnormal base-pairings to occur, i.e. the imino form of cytosine can hydrogen bond to adenine while the hydroxy form of guanine can hydrogen bond to thymine. Such mispairings may well be one of the mechanisms by which mutations occur. The tautomeric form of biologically active molecules also plays a major structural role in the binding of molecules to receptors and active sites within proteins and enzymes, and to the reactions that the molecules undergo within the host. It is thus useful to be able to determine the tautomeric equilibria that potential drug molecules may undergo.[3]

Many compounds of biological importance contain ionizable groups and in particular many drugs tend to be weak acids (e.g. aspirin, pK_a = 3.6) or weak bases (e.g. amphetamine, pK_a = 9.9). Within a biological system a series of reversible equilibria are established amongst the various tissues and fluids which compose that system. The pH of the environment will be the predominant factor in determining the relative concentrations of the compound in its neutral and charged forms and hence the position of the various equilibria governing the distribution of the compound within the system.

In terms of drug action, the ability of the functional groups within a molecule to ionize will affect both its pharmacokinetic and pharmacodynamic properties.[4] Once the drug has been introduced into the body both the site of absorption into the blood and the rate at which it occurs will depend upon the ability of the drug to permeate the membranes within the system, as will the maintenance of equilibrium concentrations within the blood plasma itself. Crossing the membrane to tissue where action is required and specific binding within the active-site will also depend on the nature of the ionizable groups. Eventual elimination from the body (e.g. by way of urine) will similarly depend on the pK_a of the compound, and so the total dosage of the administered drug can be said to be partly determined by its acid/base properties.

Besides the affecting the absorption, transport and eventual elimination of a drug, the pK_a may also have therapeutic utility by way of differentiating between normal and tumour cells. For example, it has been found[5] that pH in both experimental and human tumours is lower than in normal tissues, and that even within a single tumour there is a marked heterogeneity of pH. The transmembrane pH gradient in tumour cells would favour the uptake of weak acids in contrast to the converse preference shown by cells in normal tissue, and thus an agent with pK_a in the required range could be made selective for tumour regions. Thus some a priori method for the determination of the pK_a of a novel compound would prove useful in the prediction of the activity of a proposed biologically active compound.

Whilst this Chapter is primarily concerned with the methods of determining the free energies of tautomeric or ionisation equilibria via computer simulation of free energy differences, many of the issues raised relate also to the determination of other molecular properties upon which behaviour of the molecule within the body may depend, such as the redox potential or the partition coefficient.[6] In the next section, we shall give a brief explanation of the methods used to calculate these free energy differences – namely the use of a thermodynamic cycle in conjunction with ab initio and free energy perturbation (FEP) methods. This enables an explicit representation of the solvent environment to be used. In depth descriptions of the various simulation protocols, or the accuracy limiting factors of the simulations and methods of validation, have not been included. These are described elsewhere.[7, 8] It is worth emphasizing, however, that the use of computer simulation methods incorporating explicit solvent models has the advantage over empirical or continuum approaches of giving a more accurate description of the solvation of molecular solutes. In addition, provided sufficient time is allowed for the simulation, all-important degrees of freedom can be sampled. This may be crucial in determining a free energy where multiple conformers of a molecule are possible.

In Section 3, an overview of studies published in the literature using the methods described is then given. The first study on tautomeric systems using the thermodynamic cycle method was made in 1987 by Cieplak et al.[9] Since then a number of systems have been studied. Many of these molecules are known to be important for their biological activity, others are important for their chemical natures. Far less work has been done on the free energy of ionisation. One area that has received interest is the calculation of the acid-base equilibrium constant, given as pK_a for weak organic acids in aqueous solution. The final section highlights the problems of the methodology using illustrations from the published studies. In this way a modus operandi is given for studies on systems of interest.

2. METHODS

2.1 Tautomerism

When determining the tautomeric equilibrium constant for a process $T1(aq) \rightleftharpoons T2(aq)$, where T1 and T2 are the two tautomeric forms of the compound, one requires the free energy change, $\Delta G(aq) = -RT \ln K_T$, associated with this reaction. Standard simulation methods could be used to calculate the free energy difference, using for example the perturbation approach,

$$\Delta G(aq) = -k_B T \ln\langle \exp[-\Delta H / k_B T] \rangle \tag{1}$$

incorporating a suitable windowing protocol, or alternatively a method such as thermodynamic integration,

$$\Delta G(aq) = \int_{\lambda 1}^{\lambda 2} d\lambda \left\langle \frac{\partial H(\lambda)}{\partial \lambda} \right\rangle_{\lambda} \tag{2}$$

In either case, the Hamiltonian of the system, $H(\lambda)$, is made to depend on a coupling parameter, λ, which smoothly defines a pathway between the two tautomers, defined by λ_1 and λ_2. Issues relating to the basic simulation methods and the choice of protocols are discussed in Chapter 2 and also by a number of other authors.[7, 10]

Using Equations 1 and 2 involves evaluating contributions from all terms which arise in the force field. In practice intra-molecular terms are not evaluated, as their parameterization is often not accurate and their values are often large and similar for both tautomers. This leads to an overall inaccurate value for $\Delta G(aq)$ as the evaluation may involve taking the difference between two large numbers. This approximation also facilitates the use of bond constraints via SHAKE[11] for performing the simulation, without the overhead required for the evaluation of the constraint contribution to the free energy difference.[7]

In omitting the intra-molecular free energy change during the simulation one is making the assumption that the free energy difference can be split into identifiable contributions. This is generally not a valid assumption,[7, 12] i.e. while the overall free energy is a state function its components are not, and are path dependent. The assumption is valid where the force-field terms are evaluated over independent regions of phase space. In the case of tautomerism this condition is not rigorously met, i.e. changes in intra-molecular terms will clearly affect solute–solvent interactions (especially rotation about torsion angles). However, where changes in intra-molecular structure are relatively small the approximation is deemed valid.

In neglecting intra-molecular terms during the free energy simulation one is effectively calculating the differences in hydration free energy, ΔG_{hyd} between the tautomers. Figure 1 shows how the free energy is partitioned, with

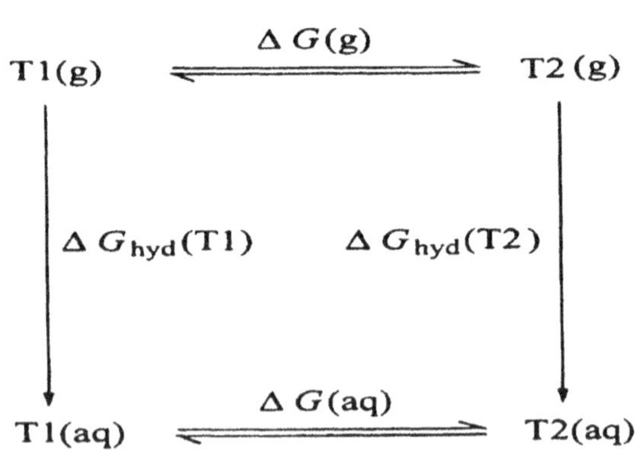

Figure 1: Cycle for the calculation of K_T

$$\Delta G(aq) = \Delta G(g) - \Delta G_{hyd}(T1) + \Delta G_{hyd}(T2) \tag{3}$$

$$= \Delta G(g) + \Delta\Delta G_{hyd} \tag{4}$$

$\Delta\Delta G_{hyd}$ is evaluated by calculating the free energy difference between the two tautomers in aqueous solution, omitting the contributions from intra-molecular terms. This involves changing the Hamiltonian of the system from T1 to T2 in an explicit solvent box using a suitable simulation protocol.[9, 13, 14] Naturally there are alternative methods for determining the difference in hydration free energy between two compounds: if the compounds are distinct chemical species (which is *not* the case with tautomers) one may be able to take the experimental value. Alternatively the hydration free energies may be calculated using a variety of methods of differing complexity and accuracy, e.g. empirical approaches based upon the composition of the molecule, or reaction field methods (see Chapter 4). However, traditional simulation methods using an explicit solvent model (see Chapter 5) are necessary where solute–solvent bonding, in particular hydrogen-bonding, is critical in correctly describing the solvation process.

The gas phase free energy difference, ΔG(g), can be calculated quantum mechanically. The absolute energy of each molecule can be calculated using ab initio methods. To obtain sufficient accuracy the inclusion of polarization and diffuse basis functions is often necessary, while the method should incorporate electron correlation (which is most commonly performed at the MP2 level).[15] The standard energy reported from an ab initio calculation is an energy (or enthalpy) at 0° K. To correct this to a free energy at ambient temperature one needs to determine the partition function of the molecule, from which one may calculate the necessary correction terms.[16] Contributions from rotation and translation have approximately their classical values at room temperature, and so simply depend on the geometry of the molecule and are straightforward to calculate. The zero-point energy also requires calculation, while the vibrational contribution should be calculated explicitly as it may be non-classical. This involves determining the frequencies of the normal modes of the molecule through the calculation of the molecular Hessian, which, depending upon the computational power available, may be calculated either using semi-empirical or ab initio quantum mechanics. Methods for performing these calculations are available in standard packages, e.g. MOPAC[17] and Gaussian.[18]

2.2 Ionisation

Given the task of calculating the pK_a of an acid AH one needs to calculate the free energy change, ΔG(aq), associated with the reaction

$$AH(aq) \Leftrightarrow A^-(aq) + H^+(aq) \tag{5}$$

with $pK_a = \Delta G(aq)/2.303RT$ (a similar approach can be used to calculate the basicity of a compound by considering the acidity of its conjugate acid). This could be determined by simulating the changes $AH \rightarrow A^-$ and nothing $\rightarrow H^+$ in aqueous solution, but as explained above this would result in poor accuracy if intra-molecular terms are included. The required free energy can be obtained from the cycle shown in Figure 2 where

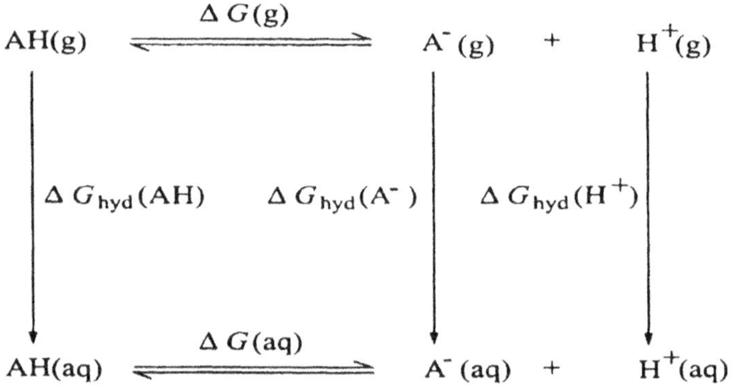

Figure 2: Cycle for the calculation of absolute pK_a values

$$\Delta G(aq) = \Delta G(g) - \Delta G_{hyd}(AH) + \Delta G_{hyd}(A^-) + \Delta G_{hyd}(H^+) \tag{6}$$

As in the case of tautomerism, the difference in gas phase free energy ΔG(g) can be calculated relatively accurately using high level ab initio quantum mechanics, while the required hydration terms could in principle be obtained from simulation of the changes $AH \rightarrow A^-$ and nothing $\rightarrow H^+$ in aqueous solution, where only inter-molecular interactions are evaluated. Both of these changes, however, would be subject to substantial error as a result of truncating the long-range Coulombic interactions (a common procedure in the vast majority of simulations), since there would be a disproportionate loss in long range interaction between the neutral species and the formally charged ion. In terms of simple electrostatics, the contribution of a given sphere of solvent can be calculated using the continuum-based Born equation.[19] Using a typical non-bonded cutoff of 8 Å one can calculate that the hydration free energy would be underestimated by approximately 86 kJ/mol, which would thus be the order of the error involved in performing a change where a formal charge is created or destroyed (the only way to avoid this is to use simulation methods such as the Ewald summation, a very large non-bonded cutoff or various augmented dielectric continuum approaches,[20] all of which would significantly add to the time required to perform the simulation). Free energies of this magnitude cannot be neglected if reasonable differences in hydration terms are required. The term for the free energy of hydration for a proton could be taken from the literature,[21] but the error involved in the remaining simulation is sufficient to make the calculation of absolute pK_a's infeasible.

The calculation of relative pK_a's is more promising. The difference in strength between two acids AH and BH is given by

$$pK_a(BH) - pK_a(AH) = \Delta G(aq)/2.303RT \tag{7}$$

where ΔG(aq) is now defined in Figure 3, with

$$\Delta G(aq) = \Delta G(g) - \Delta G_{hyd}(AH) + \Delta G_{hyd}(BH) + \Delta G_{hyd}(A^-) - \Delta G_{hyd}(B-) \tag{8}$$

$$= \Delta G(g) + \Delta G(BH \rightarrow AH) + \Delta G(A^- \rightarrow B^-) \tag{9}$$

Once again a gas phase free energy difference is required and this can be calculated using ab initio quantum mechanics. The required difference in free energies of hydration can be determined from two simulations in aqueous solution, one between charged species ($A^- \rightarrow B^-$) and the other between the neutral forms (BH→AH). In neither of these simulations is a formal charge created or destroyed, and thus the errors introduced through truncation of the non-bonded interactions can be expected to be very similar at each end of the calculation, and therefore cancel out to a large degree.

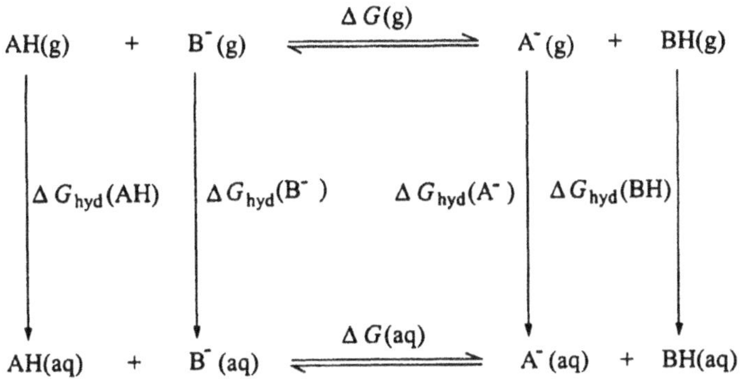

Figure 3. Cycle for the calculation of relative pK_a values

3. OVERVIEW OF PUBLISHED STUDIES

3.1 Tautomerism in Simple Heterocycles

Histamine is an important hormone, responsible for many different physiological responses. Its interaction with the H_2 receptor is responsible for acid secretion in the stomach, and blocking this action is central to the treatment of various stomach disorders. Tautomerism has been implicated in this interaction, and this was an important factor in the discovery of the major drug cimetidine (Tagamet). As shown in Figure 4, the main component is an imidazole ring. Tautomerism occurs by a proton moving between the two ring nitrogen atoms. These are referred to here as the N1H and N3H tautomers. The putative mechanism for binding to the H_2 receptor site is that the N3H tautomer is favoured in solution, whereas the N1H tautomer is favoured in a non-aqueous environment (modeled by gas phase calculations). This change in tautomer on entering the enzyme activates the receptor.

Worth et al[13] used the thermodynamic cycle method to study the simpler variant 4-(5-) methylimidazole (Figure 4). Here the side-chain is a methyl group, and the calculated equilibrium constant for this system is in good agreement with the known experimental data. It is found that the free energy difference in both phases is small, slightly favouring the N3H tautomer.

Compared to this system, the histamine flexible side-chain introduces the complication of multiple conformers. The histamine base and monocation have been studied by Worth et al,[14, 22] and by Nagy et al.[23] The latter study

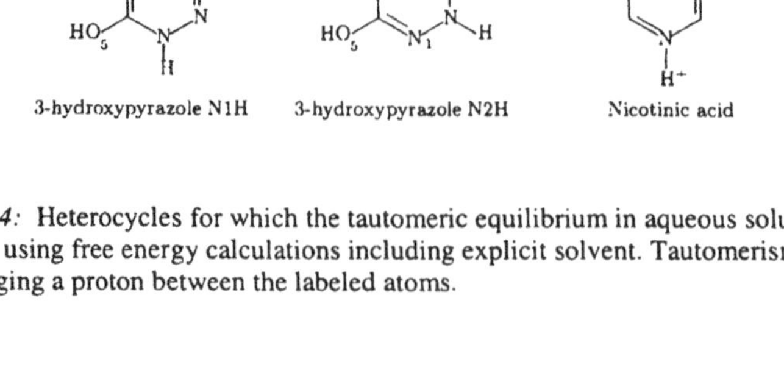

Figure 4: Heterocycles for which the tautomeric equilibrium in aqueous solution has been studied using free energy calculations including explicit solvent. Tautomerism occurs by exchanging a proton between the labeled atoms.

also looked at the H_2 receptor agonist (αR, βS) -α, β-dimethylhistamine. In the most complete study, Nagy et al used Monte-Carlo (MC)-FEP to calculate the free energy difference not only for the tautomerism, but also between the various conformers. In both phases, mixtures of all species occur. The details of these mixtures however are very different for the base and monocation. For histamine base in the gas phase, the gauche-N1H tautomer dominates, due to the intra-molecular hydrogen bond formed between the side-chain amino group and the N1H proton. In solution this interaction is outweighed by interaction with the solvent, and the trans-N3H species accounts for 83% of the population. Only 5% of the population is of N1H tautomers.

For the monocation in the gas phase, the gauche-N3H species dominates, again due to intra-molecular hydrogen bonding between the side-chain amino group and the ring N_1 atom. In this case, however, this bonding is strong enough to be retained in solution, and this species provides 64% of the mixture in the aqueous phase. The N1H tautomer then provides 34% of the mixture, all in the trans conformer. The theoretical studies show that the N1H tautomer is unfavoured under all conditions. This is in contradiction to the Weinstein model for interaction with the H_2 receptor. Nagy et al discuss possible alternative mechanisms in the light of the calculations.

The tautomeric equilibria of 1,2,3- and 1,2,4-triazoles have also been studied using the thermodynamic cycle method.[24] See Figure 4 for the numbering used for these molecules. For 1,2,3-triazole in the gas phase only the N2H tautomer is experimentally found, while in aqueous solution both tautomers are present, with the ratio of N2H:N1H being around 2:1. For 1,2,4-triazole, the N1H tautomer is overwhelmingly the more stable in both phases, with less than 1% N4H tautomer being present. Both these results are obtained in the calculations, although the ratio of tautomers in the 1,2,3-triazole case is found to be much larger than in experiment, with a value of 10:1. In the case of 3-hydroxypyrazole there is however a qualitative disagreement between experiment and theory.[25] Four tautomeric species are possible. As shown in Figure 4, these can be thought of as a keto-enol tautomerisation with N1H, and a keto-imino tautomerism with N2H.

Although the tautomeric ratios of the 4 species have not been measured directly, it is known that in aqueous solution the keto-N2H form dominates, while the keto-N1H form is only detectable in non-polar solvents. An analysis of experimental data concluded that in aqueous solution the stability (lowest free energy) is in the order keto-N2H > imino-N2H > enol-N1H > keto-N1H. In the gas phase, calculations predict that the keto-N2H form is the least stable. While solvation is found to favour this species, which is the most polar, this stabilisation is not enough to reverse the order of stability. It is thus clearly predicted that the keto-N1H tautomer is the most stable in both the gas phase and aqueous solution. The reason for this discrepancy is still to be resolved.

Tautomerism is also found to play a role in the partitioning equilibria between aqueous and organic solvents for ampholytes. Nicotinic acid, for example, can exist in a neutral or a zwitterionic form. The latter is shown in Figure 4. It is often assumed that only the neutral form is the partitioning tautomer in such systems, even when the zwitterionic form dominates in aqueous solution. Nagy and Takacs-Novak[26] have used the thermodynamic cycle method to study this hypothesis, calculating the equilibrium constant for the neutral and zwitterionic forms of nicotinic acid and isonicotinic acid. To relate the results to partitioning a range of solvents were used: water, methanol, THF, and water/methanol and water / THF mixtures. It is found that in the gas phase the neutral form dominates by around 147 kJ/mol. Solvation then favours the zwitterionic form in all media. The size of this stabilization however varies a lot with the polarity of the solvent, and the zwitterionic form dominates when the medium contains more than 60 % water. In THF only the neutral form is found. Thus in this case only the neutral form partitions from water to a pure organic solvent. The results in general agree with experimental values, but the population of zwitterionic forms are underestimated in the solvent mixtures.

3.2 Tautomerism in DNA Bases and Related Compounds

As mentioned in the Introduction, tautomerism plays a major role in the chemistry of DNA and RNA bases. Colominas et al[27] have made an extensive study of the tautomeric equilibria of guanine and cytosine (see Figure 5). The importance of protonation on the equilibria was also studied. To begin with all possible tautomers were considered. The least stable species were then screened out using increasingly accurate quantum chemical calculations with a SCRF solvent term. FEP calculations were then used for the most important equilibria. In the neutral cytosine molecule, four important tautomers are based on the amine form shown in Figure 5 as cytosine I. Another 2 tautomers are based on the imino form shown in the Figure 5 as cytosine II. In the gas phase, an admixture of forms is predicted, with the depicted cytosine I structure the most stable. In solution, all tautomers are destabilised with respect to this tautomer, which then dominates by about 20 kJ/mol.

For neutral guanine two stable tautomers are based on the keto form, labeled guanine I in Figure 5. The remaining three stable structures are based on the hydroxy form, guanine II. In the gas phase, guanine is found in a mixture dominated by the keto forms. In solution, the relative stability of the

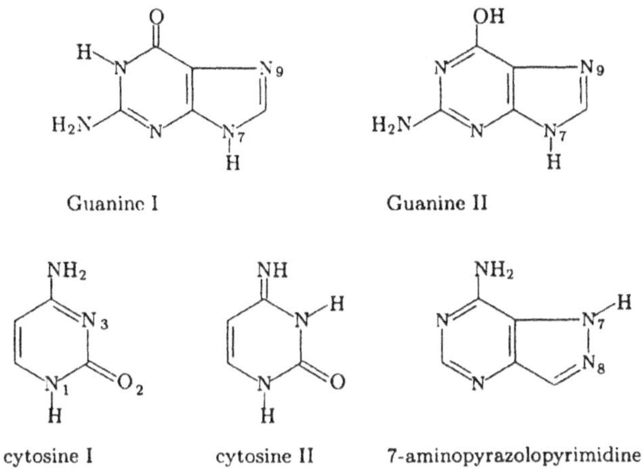

Figure 5. DNA bases and related molecules for which the tautomeric equilibrium in aqueous solution has been studied using free energy calculations including explicit solvent. Tautomerism occurs by exchanging a proton between the numbered atoms.

keto forms increases, with the N9H tautomer dominating. Experiments have determined that only the keto form is present, in agreement with these findings. Which of the forms is the more stable has however not been measured.

The equilibrium between the amino and imino forms of cytosine has also been studied by Cieplak et al.[9] Their results agree with those of Colominas et al. In this study, the related 2-oxopyridine and 2-oxopyrimidine molecules were also treated. In both these molecules the amine group of cytosine is not present, and in oxopyridine only one ring nitrogen is present. This enabled the keto-imino tautomerism to be studied in isolation. In both cases the imino form dominates in the gas phase, but the keto form is stabilised by solvation, and dominates in solution, in agreement with experiment.

The molecule 7-aminopyrazolopyrimidine is related to the DNA base adenine. It is the base attached to ribose in formycin A, which is believed to have potential therapeutic value. It is also shown in Figure 5. There is a paradox in this system. This molecule is deactivated by the enzyme adenosine deaminase (ADA). In solution the N7H tautomer predominates. This structure however inhibits ADA, and this tautomer of formycin A would not be deactivated by the enzyme.

Orozco and Luque[28] studied the equilibrium between these tautomers. Protonated species were also considered. In the gas phase, it was found that in the neutral molecule the N8H tautomer dominates by around 4 kJ/mol. In solution, however, the N7H tautomer is more populous. This change in population with change in environment is thus able to explain the experimental observations.

3.3 Tautomerism in Other Systems

Reddy et al[29] studied the molecule N_6,N_6-dimethyl-2,6-diaminobenz[cd]-indole, which is shown in Figure 6. The aim was to assess its suitability as an inhibitor of thymidylate synthase, an enzyme which is important in DNA biosynthesis. Three structures were investigated: N1H, which is the amine form depicted, and N2H, the imine form, in anti- and syn-conformations, i.e. with the proton on N_1 away from or towards the N_2 proton, respectively. In the gas phase it was found that the order of stability is N1H > anti-N2H > syn-N2H, in a ratio of 73.1: 20.3: 6.6. Solvation then further favours the N1H form, and the ratios in solution are calculated to be 98.5: 0.5: 1.0. Unfortunately, the syn-N2H form is likely to be the conformation that binds most strongly to the enzyme, and so the calculations indicate that this molecule is unlikely to be a suitable inhibitor.

Another polycyclic molecule that has been studied is pterin.[30] This is the major component of the folate molecule, and is shown in Figure 6. One of the reasons for interest in this molecule is that the enzyme dihydrofolate reductase (DHFR) is a target of several important drugs, including the anti-cancer drug methotrexate. The N1H and N3H tautomers of pterin are likely to bind in different orientations, due to differing hydrogen-bonding capabilities. Calculations show that in the gas phase, the N3H tautomer predominates. In solution, however, solvation strongly favours the N1H form, and one molecule in seven could be this tautomer. Engineering this

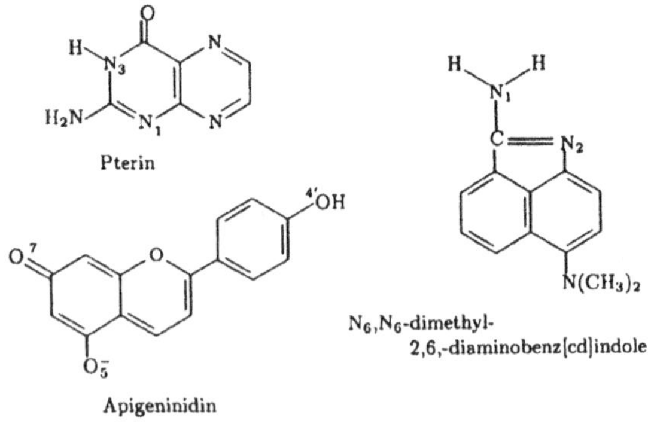

Figure 6. Polycyclic molecules for which the tautomeric equilibrium in aqueous solution has been studied using free energy calculations including explicit solvent. Tautomerism occurs by exchanging a proton between the labeled atoms.

ring could thus lead to either tautomer being dominant, with consequences for its activity in the enzyme.

Apegininidin is one of the most common anthacyanidins, which are widely found in nature, and are responsible for the colouring in many flowers. This molecule is also thought to have pharmacological activity. As shown in Figure 6, it has three acidic oxygen atoms, and can therefore be found in a range of tautomers. Rastelli et al[31] looked at the mono-anion, which has three tautomers. In the gas phase, the tautomer with the proton on O_7 is found to be the most stable. In solution, however, the tautomer with the proton on $O_{4'}$ is strongly stabilised, by over 85 kJ/mol solvation free energy with respect to the other tautomers. As a result this tautomer, which is highly unstable in the gas phase, is found to dominate the equilibrium in an aqueous environment.

Finally, Nagaoka et al have made a very interesting study applying MC-FEP techniques to the vinyl alcohol – acetaldehyde tautomerism.[32] Using a cluster of the solute with three water molecules as a "solute", the free energy for the tautomerism was calculated along different reaction pathways, which had been previously found by ab initio calculations including an SCRF solvation term. They were able to deduce that a two-step mechanism is favoured over a concerted one for the transfer of the proton.

3.4 Very Weak Acid - Base Ionisation

The concept of acid and base is fundamental to the understanding of reactivity in organic chemistry. For this reason there is great interest in the study of extremely weak acids in aqueous solution. Unfortunately it is impossible to directly measure the acid-base equilibrium constant in water if $pK_a > 17$. In two papers Jorgensen et al[33, 34] applied the theoretical methodology described above in Section 2.2, using MC-FEP, to the weak acids CH_3SH, CH_3OH, CH_3CN, CH_3NH_2, and CH_3CH_3. In the gas phase, the order of acidity is as listed: methanethiol is the most acidic, and ethane the least, all with free energy values of a few hundred kJ/mol. The final pK_a values, calculated using experimental values for CH_3SH and CH_3OH, agree well with the experimental values. The order of decreasing acidity in solution is $CH_3OH > CH_3SH > CH_3CN > CH_3NH_2 > CH_3CH_3$. Of particular interest is the value of $pK_a = 52 \pm 2$ for ethane. This lies in the middle of the measured values, which span a range of 42 – 60. The simulations were then analysed to provide information about the solvent structure around these rare species.

In a different study on acidity, King et al[35] studied the amine series NH_3, CH_3NH_2, $(CH_3)_2NH$, and $(CH_3)_3N$. The experimental values order the amines $CH_3NH_2 > (CH_3)_2NH > NH_3 > (CH_3)_3N$, with trimethylamine known to be a very weak acid. The accuracy of the calculations is found to be reasonable between homologues, i.e. calculating the pK_a for a molecule relative to its nearest neighbour in the series. Calculating the values relative to the NH_3 / NH_4^+ system, while able to produce pK_a values to within 2.5 units (14.3 kJ/mol) of the experimental values, failed to reproduce the correct order. It was concluded that the method is not sensitive enough for the fine differences in this series.

4. MODUS OPERANDI

4.1 Gas Phase Free Energy

The energy at 0° K is obtained by gas phase ab initio, or semi-empirical calculations. It is the difference between two large numbers and it is thus important to check that the basis set used is sufficient. If it is not large enough, qualitatively wrong values may result. The importance of electron correlation should also be checked. For example, in the case of 4-(5-) methylimidazole,[13] the ΔE(g) results vary from +4.22 kJ/mol to -0.92 kJ/mol on moving from an RHF/STO-3G method to RHF/6-31G**. Electron correlation is also found to play a significant role: MP2/6-31G** further changing the value to -1.37 kJ/mol. Thus the small STO-3G basis set is found to incorrectly predict that the N1H tautomer is more stable than the N3H tautomer. This pattern is generally found in the tautomeric systems studied. The delocalised nature of the systems, and the fine balance of energetics in the systems means that large basis sets are needed, and electron correlation is often important. In general at least a 6-31G* basis set is required.

Unfortunately, semi-empirical methods often produce poor results. While they may be qualitatively correct, they are often quantitatively inaccurate. Density functional methods have rarely been used in these calculations. Colominas et al in their study of DNA base tautomerism concluded, however, that the "B3LYP-DFT formalism seems less than adequate for the study of tautomerism in heterocycles."[27] The energy difference must then be corrected to a free energy at 298° K, ΔG (g, 298° K). Calculating the partition functions using a harmonic analysis is a convenient way to do this. As semi-empirical methods are good at calculating frequencies for many systems, it is generally found that either semi-empirical or full ab initio calculations produce similar values for the correction. In most cases this is of the order of 1 kJ/mol. Despite being small, the correction may be important. This is especially likely when the species under study differ in the number of bonds, or vary markedly in rigidity. For example, in the histamine system the intra-molecular interaction in the gauche forms makes these conformers more rigid than the trans forms. This is reflected in the partition functions, which add terms to the free energy differences very favourable to the trans conformers.[14] The N_6,N_6-dimethyl-2,6-diaminobenz[cd]indole system provide an example of how many small terms must be added together. The free energy difference between the N1H and anti-N2H forms has the following components (in kJ/mol)[29]:

$\Delta E(0)$	HF/6-31G**	6.195
$\Delta E(0)$	MP2 correction	-7.196
$\Delta E_{ZPE}(0)$	Zero-point energy	-1.787
$\Delta E_{vib}(0)$	vibrational enthalpy	0.745
$T\Delta S$	Entropy at 298° K	-1.134
Total ΔG(g, 298° K)		-3.177

Note that here the electron correlation is critical. Such a re-ordering of stabilities on the addition of electron correlation is also found in the studies on guanine and cytosine by Colominas et al, and in the study on 7-aminopyrazolopyrimidine.

4.2 Force-Field Parameters

To make an accurate FEP calculation, a good description of the system is required. This means that the parameters for the chosen force field must reproduce the dynamic behaviour of both species correctly. A realistic description of the environment, e.g. size of water box, and the treatment of the solute–solvent interaction energy is also required. The majority of the parameters can usually be taken from the standard atom types of a force field. The electrostatic description of the species at both ends of the perturbation is, however, the key to a good simulation of many systems. This is also the part that usually requires tailoring to the system of interest. Most force fields require atom centered charges obtained by fitting to the molecular electrostatic potential (MEP), usually over the van der Waals surface. Most authors in the studies discussed above used RHF/6-31G or higher methods to obtain the MEP.

For consistency, however, the same method should be used to derive the charges that was used to obtain the original force field parameters. For example, the standard charges in many force- fields (e.g. AMBER)[36, 37] were obtained using a STO-3G basis set. The use of a larger basis set often leads to a higher dipole moment, and more polar charges are obtained. These charges are then not consistent with the other non-bonded parameters. Studies on quinone molecules[38] and GDP / GTP,[39] support this fact, and show that incorrect dynamics can result from using 6-31G* derived charges. Simulations on DNA segments using 6-31G* charges[40] are also found to be questionable, with B-DNA transforming into the less stable A-DNA.[41] This does not happen using the original STO-3G derived charges. To make use of the better quality description afforded by a large basis set, and yet retain the relevant magnitude of the charges required by the force fields, use can be made of linear scaling factors that have been derived.[38, 42] For example 6-31G* derived charges are on average a factor of 1.15 larger than the comparable STO-3G derived charges.

To the best of our knowledge the sensitivity of FEP calculations to this overestimation of electrostatic interaction has received little attention. Cox et al[24] did use scaled 6-31G* charges in their successful treatment of the 1,2,3- and 1,2,4-triazole tautomer equilibria, but no comparison was made with unscaled charges. This factor could however be of crucial importance in calculation of tautomeric equilibria, where fine differences are important.

4.3 Long-Range Forces

The treatment of non-bonded forces is a standard problem in MD and MC simulations. A non-bonded cut-off is used, treating all interactions within a certain distance, and assuming that the ignored contributions are unimportant both for the energetics and dynamics of the system. The problems that this introduces are well documented in the literature, and will not be discussed here. Usually it is assumed that the problems of using a cutoff are small for FEP simulations. The reason for this is that one is calculating the difference between two similar molecules. The long forces should thus cancel, and the ignored contribution is not important. Worth and Richards[14] tested the effect of the long-range forces on the calculated $\Delta\Delta G_{hyd}$ for the histamine monocation. First, the cut-off was systematically changed from 8 Å - 11 Å. Over this range the calculated energies varied insignificantly.

The long-range coulombic forces were then treated with a reaction field correction, calculating the reaction field using the image point charge model of Friedman.[43] Here, the reaction field at a charge is modeled by an image charge, the strength and position of which are given by a simple relationship to the charge in question. This scheme is remarkably simple to implement, and the long-range interactions are reduced to a few coulombic pair terms.

In the case of histamine, inclusion of this reaction field correction was found to change the result by almost 4 kJ/mol. Unfortunately, the highly sensitive nature of the histamine monocation results to protocol and parameters used meant that no definite conclusion could be drawn about the importance of long-range forces in these calculations. It does however appear that they do play a role. The correct treatment of long-range forces is more crucial in the study of nicotinic acid tautomers, were the equilibrium is between a neutral molecule and a zwitterion. Nagy and Takacs-Novak[26] used an Onsager reaction field to correct for the solvation term from solvent outside the cutoff range. This correction was found to be of the order of 1–4 kJ/mol. The results for this system were also found to be very sensitive to the charges used.

4.4 Correct and Sufficient Sampling

One of the major practical problems in an FEP simulation is to know whether the sampling is sufficient to represent the relevant ensemble of solute and solvent. Even if the parameters are reasonable, and the long-range forces are insignificant, incorrect or insufficient sampling may lead to spurious results. In all the studies outlined here, the windowing FEP methodology was used. At the start of each window it is necessary to equilibrate the system to the relevant system parameters. If this is not done, the sampling in the window will be dominated by the ensemble for the system in the previous window. This will be termed the problem of correct sampling. After equilibration, enough data collection steps must then be made to sample all the important regions of phase space for the ensemble. This will be termed the problem of sufficient sampling.

The original tautomeric study of Cieplak et al[9] used a protocol in which 1 ps of equilibration followed by 1 ps of data collection at each value of λ. It is now generally accepted that this is not long enough for good results. A systematic study of the effects of the windowing protocol used is given in the study on histamine monocation by Worth and Richards.[14] They found that for this system, at least 4 ps of equilibration, and 4 ps of data collection are required for stable, converged results. The amount of equilibration required between windows depends on the time required for solvent relaxation to the new configuration. Thus if the change in perturbation parameter used between windows, $\Delta\lambda$, results in a large perturbation to the system more equilibration is required than if a small change is made.

This study also shows that care has to be taken in treating a conformationally flexible system with MD. The simulation is best made so that a transformation takes place between two distinct species (conformer or tautomer). If a mixture of conformers must be sampled, very long simulations will be needed for good results. The simulations on histamine were always started in the trans conformation. For the histamine base this gave good results as the trans conformers dominate. In the case of histamine monocation, however, the gauche N3H conformer is the most important and this species was never sampled during the simulations, resulting in poor agreement with experiment.

Differences between the calculations on organic amines in aqueous solution by King et al[35] and Rao and Singh[44] show the importance for the sampling of coordinate coupling. In coordinate coupling the internal coordinates of the evolving atoms are also changed at each window, i.e. they are coupled to the perturbation parameter, λ. This prevents what are known as "end point catastrophies": the production of spurious energy terms due to the sudden appearance or disappearance of a charged group from a solvent. The results of Rao and Singh for the series of interest (see Sec. 3.4) have an r.m.s deviation from experimental results of about 5 kJ/mol. Those of King et al, which did not use coordinate coupling, have an r.m.s. deviation of 10 kJ/mol.

4.5 Need for Explicit Solvent

As a final point we shall briefly examine the need for an explicit representation of the solvent in the calculation of $\Delta\Delta G_{hyd}$. In cases where the differences in solvation free energy is dominated by the differences in dipole moment, a continuum model solvent should be able to give a good estimate of this value. In cases where the explicit bonding to the environment, or solvent entropy effects play a role, explicit solvent is required. For example, the free energy of solvation of N_6N_6-dimethyl-2,6-diaminobenz[cd]indole[29] is found to be dominated by the hydrogen-bonding enthalpy. It is unlikely that this situation can be treated correctly by a continuum solvent model.

The most common approach to solvation studies using an implicit solvent is to add a self-consistent reaction field (SCRF) term to an ab initio (or semi-empirical) calculation. One of the problems with SCRF methods is the number of different possible approaches. Orozco and Luque[28] and Colominas et al[27] found that 6-31G* ab initio calculations with the polarizable continuum model (PCM) method of Miertius, Scrocco, and Tomasi (referred to in these papers as the MST method)[45] gave results in reasonable agreement with the MD-FEP results, but the AM1-AMSOL method differed by a number of kJ/mol, and sometimes gave qualitatively wrong results.

Parchment et al[25] also found reasonable agreement between the PCM and MD-FEP methods. Simpler SCRF approaches however differ widely. For example spherical cavity ab initio SCRF calculations predict a solvation free energy of the keto-N2H tautomer of 3-hydroxypyrole (see Sec. 3.1) of -93.5 kJ/mol in comparison to the PCM and FEP values of -9.0 kJ/mol and -12.5 kJ/mol respectively.

5. CONCLUSIONS

For an understanding of many systems involved in biochemistry it is important to know details of their tautomeric and ionic equilibria. For example, moving a molecule from aqueous solution to a polar environment inside a receptor may result in a different tautomer dominating the equilibrium, with consequences for the activity. In this chapter we have outlined how theoretical calculations can be used to study these systems, with the all important solvent environment treated explicitly.

In recent years the FEP method has fallen into disuse. However, as the studies outlined above show, in many cases the results obtained are in good agreement with experimental measurements. In these cases new information may be obtained, which may be difficult or even impossible to measure. Examples of this are the relative ratios of conformers in the histamine system, a detailed breakdown of the tautomers present in the guanine or cystine systems, or the acidity strengths of organic molecules such as ethane in water. In addition to this thermodynamic data, the simulations then also provide detailed information on the solvation of the species of interest.

Care must however be taken with the method. The final value is a sum of many, often small, contributions. Errors in these values can quickly lead to qualitatively incorrect results. The gas phase energy is furthermore the difference of two large numbers, and the ab initio calculations must therefore be of sufficient accuracy. The importance of zero-point energy and other thermodynamic properties must also be checked.

Sources of errors in the solution phase dynamics include the usual sources of errors in simulations using empirical force fields. Correct parametrisation is of course essential, and, as always, the description of the electrostatic forces is a particular problem. In addition to these standard problems, FEP requires carefully converged simulations, i.e. correct and sufficient sampling of the relevant phase space must be made. Present computational resources are such that these calculations are no longer a difficult task. It is perhaps time that some of these old problems be reevaluated, and new systems examined.

6. REFERENCES

1. J. March, Advanced Organic Chemistry: Reactions, Mechanisms, and Structure, McGraw-Hill, 4th ed. (1992).
2. A. R. Katritzky, M. Karelson, and P.A. Harris, Prototropic tautomerism of heterocyclic compounds, Heterocycles **32**:329 (1991).
3. V. I. Minkin, L.P. Olekhnovich, and Y. A. Zhadonov, Molecular Design of Tautomeric Compounds, Reidel, Dordrecht–Boston (1988).
4. R. W. Foster, Basic Pharmacology, Butterworth–Heinemann, 4th ed. (1996).

5. W. A. Denny and W. R. Wilson, Considerations for the design of nitrophenyl mustards as agents with selective toxicity for hypoxic tumour cells, J. Med. Chem., **29**:879 (1986).
6. P. M. King, C. A. Reynolds, J. W. Essex, G. A. Worth, and W. G. Richards, Calculation of pharmaceutically important properties, Mol. Sim. **5**:265 (1990).
7. W. F. van Gunsteren, T. C. Beutler, F. Fraternali, P. M. King, A. E. Mark, and P. E. Smith, Computation of free energy in practice: Choice of approximations and accuracy limiting factors, in: Computer Simulation of Biomolecular Systems, vol. 2, W. F. van Gunsteren, P. K. Weiner, and A. J. Wilkinson, eds., Escom, Leiden (1993), pp. 315–348..
8. W. F. van Gunsteren and A. E. Mark, Validation of molecular dynamics simulation, J. Chem. Phys. **108**:6109 (1998).
9. P. Cieplak, P. Bash, U. C. Singh, and P. A. Kollman, A theoretical study of tautomerism in the gas-phase and aqueous-solution: A combined use of "state-of-the-art" ab initio quantum mechanics and free energy perturbation, J. Am. Chem. Soc. **109**:6283 (1987).
10. P. M. King, Free energy via molecular simulation: A primer, in: Computer Simulation of Biomolecular Systems, vol. 2, W. F. van Gunsteren, P. K. Weiner, and A. J. Wilkinson, eds., Escom, Leiden (1993), pp. 267–314.
11. J. P. Ryckaert, G. Ciccotti, and H. J. C. Berendsen, Numerical integration of the cartesian equations of motion of a system with constraints: Molecular dynamics of n-alkanes, J. Comp. Phys. **23**:327 (1977).
12. A. Hodel, T. Simonson, R. O. Fox, and A. Brunger, Conformational substates and uncertainty in macromolecular free energy calculations, J. Phys. Chem. **97**:3409 (1993).
13. G. A. Worth, P. M. King, and W. G. Richards, Theoretical calculation of tautomer equilibria in solution. (5-)-Methyl imidazole, Biochim. Biophys. Acta **993**:134 (1989).
14. G. A. Worth and W. G. Richards, Calculation of the tautomer ratio of histamine in aqueous solution using free energy perturbation methods: An in depth study, J. Am. Chem. Soc. **116**:239 (1994).
15. W. J. Hehre, L. Radom, P. v. R. Schleyer, and J. A. Pople, Ab Initio Molecular Orbital Theory, J. Wiley, New York (1986).
16. D. A. McQuarrie, Satistical Mechanics, Harper and Row, New York (1976).
17. J. J. P. Stewart, Mopac 7.0. Technical Report, QCPE, University of Indiana, Bloomington, IN, USA (1993).
18. M. J. Frisch, G. W. Trucks, H. B. Schlegel, G. E. Scuseria, M. A. Robb, J .R. Cheeseman, V. G. Zakrzewski, J. A. Montgomery Jr, R. E. Stratmann, J. C. Burant, S. Dapprich, J. M. Millam, A. D. Daniels, K. N. Kudin, M. C. Strain, O Farkas, J. Tomasi, V. Barone, M. Cossi, R. Cammi, B. Mennucci, C. Pomelli, C. Adamo, S. Clifford, J. Ochterski, G. A. Petersson, P. Y. Ayala, Q. Cui, K. Morokuma, D. K. Malick, A. D. Rabuck, K. Raghavachari, J. B. Foresman, J. Cioslowski, J. V. Ortiz, B. B. Stefanov, G. Liu, A. Liashenko, P. Piskorz, I. Komaromi, R. Gomperts, R. L. Martin, D. J. Fox, T. Keith, M. A. Al-Laham, C. Y. Peng, A. Nanayakkara, C. Gonzalez, M. Challacombe, P. M.W. Gill, B. Johnson, W. Chen, M. W. Wond, J. L. Andres, C. Gonzalez, M. Head-Gordon, E. S. Replogle, and J. A. Pople, Gaussian 98. Gaussian Inc., Pittsburgh (1998).
19. M. Born, Volumen und Hydratationswärme der ionen, Z. Physik **1**:45 (1920).
20. P. E. Smith and W. F. van Gunsteren, Methods for the evaluation of long range electrostatic forces in computer simulation of molecular systems, in: Computer Simulation of Biomolecular Systems, vol. 2, W. F. van Gunsteren, P. K. Weiner, and A. J. Wilkinson, eds., Escom, Leiden (1993), pp. 315.
21. R. M. Noyes, Thermodynamics of ion hydration as a measure of effective dielectric properties of water, J. Am. Chem. Soc. **84**:513 (1962).
22. G. A. Worth, P. M. King, and W. G. Richards, Histamine tautomerism and its mode of action, Biochim. Biophys. Acta **1036**:158 (1990).

23. P. I. Nagy, G. J. Durant, W. P. Hoss, and D. A. Smith, Theoretical analysis of the tautomeric and conformational equilibria of histamine and (αR,βS)-dimethylhistamine in the gas phase and aqueous solution, J. Am. Chem. Soc. **116**:4898 (1994).
24. J. R. Cox, S. Woodcock, I. H. Hillier, and M. A. Vincent, Tautomerism of 1,2,3- and 1,2,4-triazole in the gas phase and in aqueous solution: A combined ab initio quantum mechanics and free energy perturbation study, J. Phys. Chem. **94**:5499 (1990).
25. O. G. Parchment, D. V. S. Green, P. J. Taylor, and I. H. Hillier, The prediction of tautomer equilibria in hydrated 3-hydroxypyrazole: A challenge to theory, J. Am. Chem. Soc. **115**:2352 (1993).
26. P. I. Nagy and K. Takacs-Novak, Theoretical and experimental studies of the zwitterions neutral form equilibrium of ampholytes in pure solvents and mixtures, J. Am. Chem. Soc. **119**:4999 (1997).
27. C. Colominas, F. J. Luque, and M. Orozco, Tautomerism and protonation of guanine and cytosine. Implications in the formation of hydrogen-bonded complexes, J. Am. Chem. Soc. **118**:6811 (1996).
28. M. Orozco and F. J. Luque, Theoretical study of the tautomerism and protonation of 7-aminopyrazolopyrimidine in the gas phase and in aqueous solution, J. Am. Chem. Soc. **117**:1378 (1995).
29. M. R. Reddy, R. J. Bacquet, and M. Varney, Tautomerization of N6,N6-dimethyl-2,6-diaminobenz[cd]indole in the gas phase and in aqueous solution: A combined quantum mechanical and free energy perturbation study, J. Chim. Phys. **88**:2605 (1991).
30. C. H. Schwalbe, D. R. Lowis, and W. G. Richards, Pterin 1H – 3H tautomerism and its possible relevance to the binding of folates to dihydrofolate reductase, J. Chem. Soc. Chem. Commun. 1199 (1993).
31. G. Rastelli, L. Costanino, and A. Albasini, Solvent effects on the tautomerism of apigenidin, Tetrahedron. Lett. **52**:9751 (1994).
32. M. Nagaoka, K. Suenobu, and T. Yamabe, On the hydronium ion cataylsed mechanism in vinyl alcohol-aldehyde isomerization: Ab initio molecular orbital theory and Monte Carlo simulation, J. Am. Chem. Soc. **119**:8023 (1997).
33. W. L. Jorgensen, J. M. Briggs, and J. Gao, A priori calculations of pKas for organic compounds in water. The pKa of ethane, J. Am. Chem. Soc. **109**:6857 (1987).
34. W. L. Jorgensen and J. M. Briggs, A priori pKa calculations and the hydration of organic anions, J. Am. Chem. Soc. **111**:4190 (1989).
35. P. M. King, C. A. Reynolds, and W. G. Richard, The theoretical calculation of basicities: An homologous amine series, J. Mol. Struct. (THEOCHEM) **208**:205 (1990).
36. S. J. Weiner, P. A.Kollman, D. A. Case, U. C. Singh, C. Ghio, G. Alagona, S. Profeta, Jr., and P. Weiner, A new force field for molecular mechanical simulation of nucleic acids and proteins, J. Am. Chem. Soc. **106**:765 (1984).
37. S. J. Weiner, P. A. Kollman, D. T. Nguyen, and D. A. Case, An all atom force field for simulations of proteins and nucleic acids, J. Comput. Chem. **7**:230 (1986)
38. S. G. Lister, C. A. Reynolds, and W. G. Richards, Theoretical calculation of electrostatic potentials: Electron withdrawing compounds, Int. J. Quant. Chem. **41**:293 (1992).
39. G. A. Worth, C. Edge, and W. G. Richards, Molecular dynamics of the Ha-ras protein: Nucleotide atom-centred charges within the amber force field, J. Mol. Mod. **1**:115 (1995).
40. F. H. Hausheer, U. C. Singh, T. C. Palmer, and J. D. Saxe, Dynamic properties and electrostatic potential surface of neutral DNA heteropolymers, J. Am. Chem. Soc. **112**:9468 (1990).
41. I. Haworth and A. Elcock, Private communication, (1991).
42. K. M. Merz, Analysis of a large database of electrostatic potential derived charges, J. Comp. Chem. **13**:749 (1992).
43. H. L. Friedman, Image approximation to the reaction field, Mol. Phys. **29**:1533 (1975).

44. B. G. Rao and U. C. Singh, Hydrophobic hydration: A free energy perturbation study, J. Am. Chem. Soc. **111**:3125 (1989).
45. S. Miertius, E. Scrocco, and J. Tomasi, Electrostatic interaction of a solute with a continuum. A direct utilization of ab initio molecular potentials for the prevision of solvent effects, Chem. Phys. **55**:117 (1981).

Section Three

Ligand Binding

Chapter 7

Free Energy Calculations on Enzyme-Inhibitor Complexes: Studies of Thermolysin and Rhizopus Pepsin

B. Govinda Rao
Vertex Pharmaceuticals, Inc., Cambridge, MA 02139

1. INTRODUCTION

Thermolysin and rhizopus pepsin were used as model targets for designing drugs against hypertension. Thermolysin is a zinc endopeptidase and rhizopus pepsin is an aspartic proteinase. These two enzymes are not involved in any human physiological pathway, but are homologous to angiotensin-converting enzyme (ACE) and renin of the renin-angiotensin cascade implicated in hypertension, respectively. As crystal structure of ACE is not available, the extensive structural information on inhibitor complexes of thermolysin was utilized[1] to design ACE inhibitors, many of which are popular high blood pressure drugs on the market now. The crystal structure of renin[2] was reported in late 1980s, but the coordinates were not available for some time. Therefore, the crystal structures of three homologous enzymes, rhizopus pepsin, penicillopepsin and endothiapepsin or homology models based on these three crystal structures were utilized for designing many renin inhibitors.[3] Because of this importance, these enzymes were among the first few to be extensively investigated both structurally and enzymatically during 1980s. The resulting information served as basis for many successful applications of modern computational methods.[4] This chapter will discuss the first few free energy simulation studies on these two enzymes.

2. THERMOLYSIN

Several phosphorous containing substrate analogs of thermolysin act as potent inhibitors. For example, Cbz-Gly^P-(NH)-Leu-Leu, a tripeptide analog that has phosphoramidate (-PO_2NH-) group in place of the scissile peptide bond (-CONH-) between Gly and Leu, binds[5] to thermolysin with Ki of 9.1 nM. When the NH group of the phosphoramidate is replaced with O, the Ki increases[5] by about 1000-fold to 900 nM. It is very interesting that such a small change leads to a very large difference in binding. The crystal structure of the phosphonamide inhibitor in complex with thermolysin was available,[6] but not the structure of phosphonate ester analog.

2.1 Mutation of Phosphoramidate Inhibitor

Kollman's group[7] found it to be a perfect test case for applying free energy perturbation method to understand the binding difference between these two thermolysin inhibitors. A thermodynamic cycle approach (Chapter 2) was used to calculate binding free energy difference between the two inhibitors which requires perturbation of the first inhibitor to the second in solution phase and in complex with the enzyme to obtain ΔG_{aq} and ΔG_{com}, respectively. The binding free energy difference of the two inhibitors is calculated as $\Delta\Delta G_{bind} = \Delta G_{com} - \Delta G_{aq}$. In this study, AMBER program was utilized to perturb the NH group of the phosphoramidate inhibitor into O to

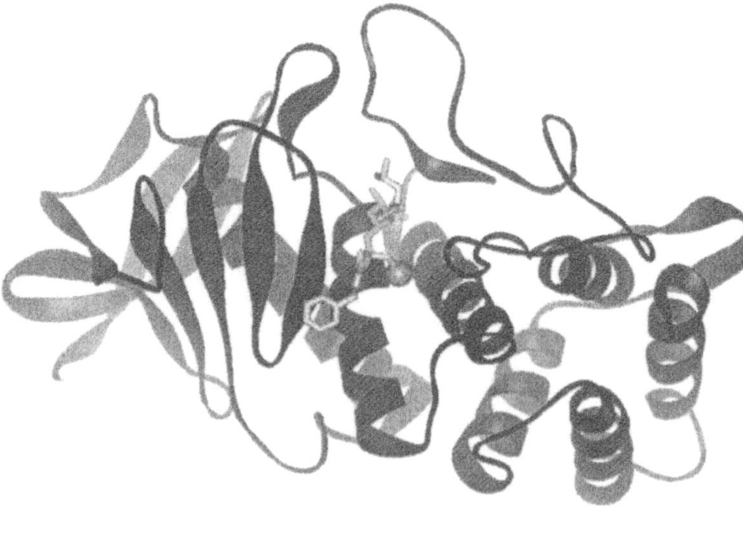

Figure 1. Ribbon diagram of thermolysin complexed the inhibitor Cbz-GlyP-NH-Leu-Leu (stick) and Zn (sphere) in the active site (PDB code:5TMN).

form a phosphonate ester analog using 20 windows with 1 ps of equilibration and 1 ps of data collection at each window. They calculated $\Delta\Delta G_{bind}$ to be 4.1 kcal/mol from ΔG_{com} = 7.64 kcal/mol and ΔG_{aq} = 3.44 kcal/mol. The calculated result is in excellent agreement with the experimental binding difference of 4.2 kcal/mol. Based on the X-ray structure[6] and the results of the calculations, it was concluded that the phosphoramidate exhibited higher affinity in the complex than the phosphonate ester because the NH of the phosphoramidate formed a hydrogen bond to the carbonyl of Ala113 whereas the phosphonate ester oxygen produced an unfavorable electrostatic interaction.

Subsequently, the same group extended these calculations on thermolysin inhibitors to test the effect of different force field parameters on calculated results and to predict the binding constant of an untested inhibitor.[8] The new inhibitor, yet to be synthesized, had a CH_2 group instead of the NH of the phosphonamide inhibitor. The NH to O perturbation was re-examined by using a new set of atomic charges and a larger C-P-X bond angle (92° to 96°) derived from higher level geometry optimizations (4-31G). In addition, the ionic states of the two active-site residues were modified, namely the Glu143 side-chain was treated as neutral and the His231 side-chain had positively charged imidazolium ion. In this case, the perturbation was conducted over 10 windows for a total simulation time of 17.6 ps. A short simulation time was chosen since longer times resulted in more distortion of the active-site residues. These changes led to a significant increase in $\Delta\Delta G_{bind}$ (5.9 vs. 4.2 kcal/mol). The difference was attributed to the new charge sets, which was supported by conducting a thermodynamic cycle in which the new charges were converted to the old charges. A second calculation using a modified force field parameter for the torsion representing the neutral Glu143 side chain (X-C-O-H) led to a value of 3.3 kcal/mol.

The perturbation converting the NH to a CH2 was calculated in a similar manner and in advance of inhibitor synthesis. In this case, the ΔG_{aq} and ΔG_{com} were -2.4 ± 0.28 and -2.72 ± 0.84 kcal/mol, respectively, which predicted a $\Delta\Delta G_{bind}$ to be -0.3 kcal/mol. This result is in close agreement with the experimental result of. -0.1 kcal/mol, determined after the prediction. Since, like the phosphonate ester, the phosphinate lacks the hydrogen bond to the carbonyl of Ala113, it should be less potent than the phosphanamide inhibitor. However, it is more potent than the phosphonate ester due to more favorable desolvation (ΔG_{sol} = -2.4 ± 0.3) and reduced electrostatic repulsion.

These studies showed, for the first time, the usefulness of free energy simulation methods to understand enzyme-inhibitor interactions and demonstrated the predictive power of this method. Furthermore, these studies showed that the use of less accurate force field parameters and partial charges could have a significant effect on results.

3. RHIZOPUS PEPSIN

Rhizopus pepsin is an aspartic proteinase, belonging to an important class of proteolytic enzymes.[9] Several of aspartic proteinases are therapeutic targets for drug design. In particular, human renin, HIV-1 proteinase, and most recently BACE (Beta-site APP-Cleaving Enzyme) are targeted for developing drugs against hypertension, AIDS, and Alzheimer's disease, respectively. Structures of these enzymes and their inhibitor complexes have been determined at high resolution.[2, 10-12] A large number of inhibitors have been designed which bind to these enzymes at very low concentrations and their mode of inhibition has also been studied. However, at the time of first free energy perturbation study on aspartic proteinase by Rao and Singh,[13] several questions remained unanswered regarding (1) the nature of the acidity of the active site Asp dyad of aspartic proteinases and (2) the nature of the enzyme inhibition by tight binding inhibitors like pepstatin. Rao and Singh[13] investigated rhizopus pepsin - pepstatin system to address some of these questions.

3.1 Acidity of the Active Site

Aspartic proteinases have two catalytically important aspartic acid residues (Asp35 and Asp218 in rhizopus pepsin) at the center of the active site cleft. The pH activity profiles of pepsin catalyzed hydrolysis suggest pKa values of 1.2 and 4.7 for the two Asp residues. In rhizopus pepsin the low pKa is usually attributed to Asp35 and the high pKa is assigned to Asp218 based on the reactions of these groups with DAN (diazo-DL-norleucine methyl ester) and EPNP (1,2-epoxy-3-(p-nitrophenoxy)propane). Structural studies[10] show that the center of the enzyme's active site is fairly rigid and that the two-carboxylate groups of the Asp dyad at this center are held in almost a plane due to intricate hydrogen bonding from the neighboring residues. In rhizopus pepsin,[14, 15] the two-carboxyl oxygens of Asp35 make hydrogen bonds with HN of Gly37 and OG of Ser38. Similarly, the two-carboxyl oxygens of Asp218 make hydrogen bonds with HN of Gly220 and OG1 of Thr221. A tightly bound water molecule found close to the Asp dyad makes hydrogen bonds with all the four-carboxyl oxygens of the two Asp residues. Further, two additional water molecules are found to be making hydrogen bonds with the two outer oxygens of the Asp dyad, respectively. Similar arrangements around the two Asp residues are also observed in the crystal structures of penicillinopepsin,[16] endothiapepsin[17]

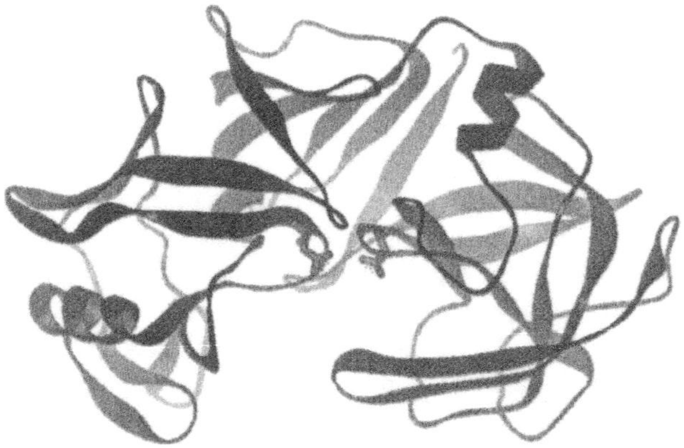

Figure 2. A ribbon diagram of rhizopus pepsin (PDB code: 5APR). The catalytically important Asp dyad (Asp218 and Asp35) side-chains are shown in stick diagrams. The β-hair pin flap that covers the active site cleft is located in the bottom of the diagram.

and human renin.[2] Even though the lower pKa is assigned to Asp35 based on chemical studies, the structural data was initially interpreted[16] to suggest that the negative charge is associated with Asp218. On the other hand, Pearl and Blundell[17] pointed out that the environments of the two Asp residues in endothiapepsin are related by a local two fold axis. Therefore, the hydrogen is expected to occupy positions closer to the two Asp residues for equal times like in a dicarboxylic acid. The abnormal pKa values of the Asp dyad resemble the values reported for dicarboxylic acids. Experimental determination of actual location of hydrogen atoms, which is possible by neutron diffraction, has not been reported for any of the aspartic proteinases. Free energy simulations were used[13] to get important insight into the energetics of the active site, in addition to locating the energetically favorable hydrogen atom positions.

3.2 Mutation of the Proton Position

For calculating the free energy difference between the two states of the enzyme with protonation of Asp35 or Asp218, the hydrogen position from one Asp to the other Asp is perturbed through an intermediate state by the free energy perturbation method. All calculations were carried out by the AMBER and QUEST programs. The intermediate state was obtained from the gas phase calculations on acetic acid – acetate complex. The partial charges for all atom models of the two Asp residues at different proton positions were obtained from ab initio calculations on the optimized structures of acetic acid – acetate complex. The electrostatic potential method was used for calculation of charges at the level of 6-31G* basis set. For mutation of the proton position from one Asp to the intermediate position between the two Asp residues, 11 or 25 windows were used with 1.0 ps of equilibration followed by 1.0 ps of data collection at each window. In the next run, the proton was transferred from the intermediate position to the inner oxygen of the second Asp residue using the same number of windows. This process was repeated for the reverse direction as well. In a second set of simulations, an ammonium ion was placed near the Asp dyad in the active site and the proton position was mutated in the forward and backward directions. The same process was repeated for a third set of simulations, wherein a pepstatin molecule was included in the active site instead of an ammonium ion.

The results of the geometry optimization studies on the model systems in the gas phase showed that the geometry observed for the carboxylates of the two active site Asp residues was not the most stable configuration. The two inner oxygens preferred to be at a distance of about 2.5 to 2.6 Å in the absence of the enzyme environment rather than 2.9 Å observed in the crystal structure. Moreover, the barrier to proton transfer in this model system was calculated to be about 0.8 kcal/mol with inter oxygen distance at about 2.5 Å and it is about 15 kcal/mol if the inter oxygen distance is fixed at 2.9 Å. The energy of the acetic acid - acetate complex was not only sensitive to the change in the distance between the inner oxygens but also to changes in the relative orientations of the two carboxylates. It was also found that the water molecule near the complex, which makes strong hydrogen bond with the acetate ion did not affect the relative energies of the complex with different proton positions. The calculated ab initio energies for the Asp dyad inside the enzyme are higher than that in the gas phase. However, the contributions of the surroundings lower the energy substantially in each case. It is also interesting to note that the total energies of the dyad inside the enzyme complexed with pepstatin are lower when the distance between the two inner oxygens is smaller. These results, therefore, suggest that the larger distance between the two inner oxygens observed in the crystal structure is not necessarily the lowest energy configuration of the enzyme active site. It may be of interest here to note that this distance is 2.6 Å in HIV-1 proteinase.[18]

The change in the free energy with the mutation of the proton position is about 7.0 kcal/mol, in the presence of an ammonium ion, suggesting that the initial state is stable compared to the final state. In the presence of pepstatin the results suggest a barrier of about 1.0 kcal/mol for the transfer of proton from one site to the other site. This low energy barrier should allow the proton to shuttle between the two sites. The same barrier of about 1.0

kcal/mol is observed for the gas phase optimization calculations of the acetic acid - acetate complex in planar configuration.

From this study, it was concluded that: (1) there is little preference for one Asp to be ionized over the other inside the active site complexed with a neutral ligand; (2) water molecule in the active site does not alter the relative energies of the two states of the dyad in the gas phase; (3) the neighboring residues affect the energies of the two states of the dyad inside the enzyme; and (4) the energies of the two states are very sensitive to changes in their inter oxygen distances. In addition to these important conclusions, the study showed that the barrier to proton transfer could be higher if the distance between the two oxygens is close to 2.9 Å and is much less (about 1.0 kcal/mol) if the distance is close to 2.5 Å. It is possible that the carboxylates of the two Asp residues in the active site assume the configuration found in the gas phase to facilitate proton transfer due to the dynamic nature of the protein. It is also possible that proton shuttles between the two oxygens by quantum mechanical tunneling,[19] which can not be addressed by free energy simulation studies.

3.3 Pepstatin Binding

Pepstatin (Iva-Val-Val-Sta-Ala-Sta) containing the unusual amino acid, statine [4(S)-amino-3(S)-hydroxy-6-methyl heptanoic acid] is a tight-binding inhibitor of aspartic proteinases.[20] The structure of pepstatin is shown in Figure 3. It binds to the active site of porcine pepsin with unusually low dissociation constant of 45.7 pM.[21] Since pepstatin is a tight binding inhibitor of most aspartic proteinases and the 3(S)-hydroxyl group on the central statine is structurally related to a hydroxyl group in the tetrahedral intermediate formed during amide hydrolysis, it was considered a transition state analog inhibitor. It has been established from kinetic studies that the hydroxyl group of the central statine residue is important for the tight binding of pepstatin.[22]

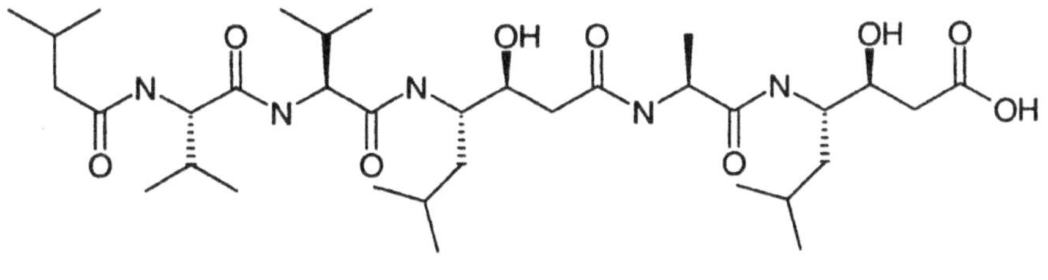

Figure 3. Rhizopus pepsin inhibitor pepstatin

The derivatives of pepstatin, containing a modified central statine residue lacking the hydroxyl group in the S configuration, are weaker inhibitors of these enzymes. In fact, deletion of this hydroxyl group on the central statine residue leads to a decrease of about 4000-fold in its binding to pepsin and when the configuration of the hydroxyl group is changed from S to R, the resulting inhibitor is about 1000-fold weaker than pepstatin.[23] Two different roles had been attributed to the hydroxyl group of the central statine residue based on the kinetic and structural studies. The structural data[10] on the complexes of aspartic proteinases with pepstatin and other inhibitors containing central statine residue show that the hydroxyl oxygen of the central statine is placed in the middle of the carboxylates of the two active site Asp residues and the hydroxyl group makes multiple hydrogen bonds to the carboxylate oxygens. Therefore, it is possible that the binding of pepstatin is aided by these multiple hydrogen bond interactions in addition to its role as a transition state analog. The derivatives of pepstatin, lacking the hydroxyl group in S configuration on central statine are weaker inhibitors due to the absence of these hydrogen bond interactions with the active site Asp residues. Alternatively, the tighter binding of pepstatin may result from the positive entropy change (of about 10-16 eu or 3-5 kcal/mol) associated with displacement of the active site bound water molecule into the bulk solvent, as suggested by Rich.[23] It has been his contention that pepstatin is a bisubstrate inhibitor, in addition to being a transition state analog inhibitor. Though both the roles suggested for the hydroxyl group of statine appear very reasonable, it is not clear which one of them has major contribution to the tight binding of pepstatin. In spite of the impressive amount of structural information and kinetic data, this question is not resolved satisfactorily.

3.4 Binding of Pepstatin Analogs

Free energy perturbation calculations were employed to understand the importance of the interactions between the hydroxyl group of pepstatin and the enzyme active site. The central statine residue of pepstatin was mutated into several of its derivatives. The most important mutation simulated is that of the central statine to dehydroxystatine, which determines the difference in the free energies of binding between the two inhibitors. It is assumed in these calculations that pepstatin analog, containing a dehydroxystatine in the place of central statine, binds in the active site of the enzyme in the same way as pepstatin does. These calculations, therefore, do not include the entropic contribution to binding arising due to the displacement of the active site bound water molecule. If the later contribution is significant as postulated by Rich,[23] the calculated free energy difference should be smaller than the experimental value of about 5 kcal/mol. If the calculated value is closer to the experimental value, then it would suggest that the major contribution to the free energy of binding is enthalpic and the entropic contribution caused by the displacement of water is not significant.

The free energy perturbation calculations on mutation of the central statine residue of pepstatin to its dehydroxy and other derivatives were carried out using the window method. The crystal structure reported by Suguna et al.[14-15] was used for these calculations. In most simulations, the mutations were achieved either in 101 or 51 windows with 0.4 ps of equilibration and 0.4 ps of data collection at each window. The calculation for each mutation was repeated in water to determine the difference in the free energies of solvation and to complete the thermodynamic cycle.

The free energy change for the mutation of the central statine to dehydroxystatine in water, ΔG_{aq} is equal to 3.01 kcal/mol. For the same mutation inside the active site of the enzyme, the free energy change, ΔG_{com} was 8.18 kcal/mol. These values lead to $\Delta\Delta G_{bind}$ of -5.17 kcal/mol, which is in good agreement with the experimental value of 4.9 kcal/mol. These results, therefore, suggest that the difference in binding between pepstatin and dehydroxypepstatin is solely due to the interaction of the hydroxyl group with the enzyme active site. It appears that the favorable situation for the strong interaction of statine hydroxyl group with the negatively charged Asp dyad is provided by the enzyme structure that keeps the two carboxylates of the Asp residues in a somewhat rigid configuration. These results, therefore, clearly establish that the major contribution to the difference in binding of pepstatin and dehydroxypepstatin is due to the strong interaction between the Sta hydroxyl group and the active site Asp residues. Hence, it is not necessary to presume that the tighter binding of pepstatin over its analog lacking the OH group is due to the displacement of the active site water molecule by pepstatin. The later explanation was advanced based on the kinetic data that the active site bound water molecule (which is displaced when pepstatin binds to the enzyme) is not displaced when dehydroxypepstatin binds to the enzyme. This is not ascertained by direct-structural studies. No structural results are reported for any aspartic proteinase complexed with the inhibitors lacking the OH group. On the contrary, the structure of rhizopus pepsin complexed with a reduced peptide inhibitor shows that the water molecule bound between the two Asp residues is displaced by the inhibitor, though no group of the inhibitor occupies the position of the water molecule.[14] The free energy results suggest that dehydroxypepstatin also displaces the active site water molecule.

The mutation of the hydroxyl group positioned in R-configuration at the C(3) atom of the central statine (rSta) residue of the inhibitor gives rise to $\Delta\Delta G_{bind}$ of -0.51 kcal/mol, which is very close to the experimental value of -0.8 kcal/mol. It may be noted here that the starting configuration of the inhibitor in the enzyme-inhibitor complex is the same as that of pepstatin. The crystal structure of rhizopus pepsin or any other aspartic proteinase complexed with the pepstatin analog containing rSta in place of the central Sta residue is not yet available. The agreement between the experimental and calculated values suggests that the modeled configuration of the inhibitor is correct. In this configuration the hydroxyl group of the central rSta residue makes only one hydrogen bond interaction with the outer oxygen of Asp35.

Rich and coworkers[24] studied the binding of methylated analogs of pepstatin to pepsin and obtained some very surprising results. They found that when the hydrogen at the C(3) atom of the central Sta is substituted by a methyl group, its inhibition goes down by a factor of about 12000. On the other hand, if the positions of the hydroxyl group and the methyl group are interchanged, the new inhibitor (with the hydroxyl group in pro-R position) binds to pepsin almost as strongly as pepstatin. These workers attributed the strong binding affinity of the second inhibitor to the ability of the methyl group (in the pro-S position) to displace the active site bound water, but provided no explanation for the highly reduced potency of the first inhibitor with the hydroxyl group on the pro-S side. In fact, this inhibitor has the ability to displace the water molecule as well as interact with the active site Asp dyad and it should, therefore, bind as strongly as pepstatin.

The present simulation studies show that the inhibitor with the methyl group in pro-R position binds to rhizopus pepsin better by about 1.5 kcal/mol than pepstatin, which is in disagreement with the experimental value of -5.6 kcal/mol. The experimental results point to the possibility that the pro-R methyl group of the inhibitor may interfere with the optimal binding of the inhibitor due to steric reasons. Such steric interference between the inhibitor and the active site of rhizopus pepsin is not observed in the model. Since the structures of the two enzymes in the active site region are expected to be similar, it is not clear why pro-R methyl-statine containing inhibitor is such a weaker inhibitor of pepsin. It is possible that the inhibitor binds in a conformation which is different from the model. When the methyl group is present in pro-S configuration, it is positioned between the two carboxylates of the Asp dyad in the model. A mutation of the methyl group in the hydrophilic site gives rises $\Delta\Delta G_{bind}$ equal to -1.27 kcal/mol. Though the calculated result appears reasonable, it is in disagreement with the experimental result. The calculated result suggests that pepstatin with central rSta is better than the methylated rSta analog, while the experimental results suggest the opposite. As in the earlier case, the disagreement causes a suspicion that the actual binding modes of the two inhibitors are different from the binding modes modeled for this simulation. These calculations, however, find it hard to explain that the pepstatin analog with a methyl group in *pro-R* configuration is so much weaker (by about -5.6 kcal/mol) than pepstatin.

4. CONCLUSIONS

The coplanar configuration of the dyad is found to be crucial for optimal binding of pepstatin. In the absence of improved force fields to mimic the configuration of the Asp dyad, the use of constraints on the dyad gave results that are in good agreement with the experiments. The results pointed out that the contribution of about 5 kcal/mol to binding from the hydroxyl group of the central statine residue is mainly due to the strong interaction of this group with the negatively charged Asp dyad. It implies, therefore, that entropic contribution to the binding due to the displacement of the active site bound water molecule may not be significant. These results pointed out the possibility that water molecule is displaced even when pepstatin analog lacking the hydroxyl group in S-configuration binds to the enzyme. Following this study, several calculations on related HIV-1 protease inhibitors (Chapter 16) have been carried out and affirmed the conclusions of this study regarding the hydroxyethylene based inhibitors. These calculations have been helpful in designing novel HIV-1 protease inhibitors that are approved drugs or undergoing clinical trials.[25]

5. REFERENCES

1. D. G. Hangauer, Computer-aided design and evaluation of angiotensin converting enzyme inhibitor, in: Computer-Aided Drug Design, T. J. Perun and C. L. Propst, eds., Marcel Dekker, New York (1989) pp. 253-326.
2. A. R. Sielecki, K. Hayakawa, M. Fujinaga, M. E. Murphy, M. Fraser, A. K. Muir, C. T. Carilli, J. A. Lewicki, J. D. Baxter, and M. N. James, Structure of recombinant human renin, a target for cardiovascular-active drugs, at 2.5 A resolution, Science **243**:1346 (1989).
3. C. Hutchins, and J. Greer, Comparative modeling of proteins in the design of novel renin inhibitors, Crit. Rev. Biochem. Mol. Biol. **26**:77 (1991).
4. M. R. Reddy, M. D. Erion, and A. Agarwal, Free energy calculations: use and limitations in predicting ligand binding affinities, in: Reviews in Computational Chemistry, K. B. Lipkowitz and D. B. Boyd, eds. VCH Publishers, New York, vol. 16 (2000), pp. 217-304.
5. P. A. Bartlett, and C. K. Marlowe, Evaluation of intrinsic binding energy from a hydrogen bonding group in an enzyme inhibitor, Science **235**:569 (1987).
6. D. E. Tronrud, H. M. Holden, and B. W. Matthews, Structures of two thermolysin-inhibitor complexes that differ by a single hydrogen bond, Science **235**:571 (1987).
7. P. A. Bash, U. C. Singh, F. K. Brown, R. Langridge, and P. A. Kollman, Calculation of the relative change in binding free energy of a protein-inhibitor complex, Science **235**:574 (1987).
8. K. M. Merz, and P. A. Kollman, Free energy perturbation simulations of the inhibition of thermolysin: Prediction of the free energy of binding of a new inhibitors, J. Am. Chem. Soc. **111**:5649 (1989).
9. B. M. Dunn, Structure and Function of the Aspartic Proteinases: Genetics, Structures, and Mechanisms, Plenum Press, New York (1991).

10. D. Davies, The structure and function of the aspartic proteinases, Annu. Rev. Biophys. Biophys. Chem. **19**:189 (1990).
11. A. Wlodawer, and A. Gustchina, Structural and biochemical studies of retroviral proteases, Biochim. Biophys. Acta. **1477**:16 (2000).
12. L. Hong, G. Koelsch, X. Lin, S. Wu, S. Terzyan, A. K. Ghosh, X. C. Zhang, and J. Tang, Structure of the protease domain of memapsin 2-(beta-secretase) complexed with inhibitor, Science **290**:150 (2000).
13. B. G. Rao, and U. C. Singh, Studies on the binding of pepstatin and its derivatives to Rhizopus pepsin by quantum methanics, molecular mechanics, and free energy perturbation methods, J. Am. Chem. Soc. **113**:6735 (1991).
14. K. Suguna, E. A. Padlan, C. W. Smith, W. D. Carlson, and D. R. Davies, Binding of a reduced peptide inhibitor to the aspartic proteinase from Rhizopus Chinensis: Implications for a mechanism of action, Proc. Natl. Acad. Sci. USA **84**:7009 (1987).
15. K. Suguna, E. A. Padlan, R. Bott, J. Boger, K. D. Parris, and D. R. Davies, Structures of complexes of rhizopus pepsin with pepstatin and other statine-containing inhibitors, Proteins **13**:195 (1992).
16. M. N. James, and A. R. Sielecki, Stereochemical analysis of peptide bond hydrolysis catalyzed by the aspartic proteinase penicillopepsin, Biochemistry **24**:3701 (1985).
17. L. Pearl, and T. Blundell, The active site of aspartic proteinases, FEBS Lett. **174**:96 (1984).
18. M. Miller, J. Schneider, B. K. Sathyanarayana, M. V. Toth, G. R. Marshall, L. Clawson, L. Selk, S. B. Kent, and A. Wlodawer, Structure of complex of synthetic HIV-1 protease with a substrate-based inhibitor at 2.3 A resolution, Science **246**:1149 (1989).
19. D. DeVault, Quantum Mechanical Tunneling in Biological Systems, 2nd Edn, Cambridge University Press, (1984).
20. H. Umezawa, T. Aoyagi, H. Morishima, M. Matsuzaki, and M. Hamada, Pepstatin, a new pepsin inhibitor produced by actinomycetes, J. Antibiot. (Tokyo) **23**:259 (1970).
21. S. Kunimoto, T. Aoyagi, R. Nishizawa, T. Komai, and T. Takeuchi, Mechanism of inhibition of pepsin by pepstatin. II, J. Antibiot. (Tokyo) **27**:413 (1974).
22. D. H. Rich, and E. T. Sun, Mechanism of inhibition of pepsin by pepstatin. Effect of inhibitor structure on dissociation constant and time-dependent inhibition, Biochem. Pharmacol. **29**:2205 (1980).
23. D. H. Rich, E. T. Sun, and E. Ulm, Synthesis of analogues of the carboxyl protease inhibitor pepstatin. Effects of structure on inhibition of pepsin and renin, J. Med. Chem. **23**:27 (1980).
24. D. H. Rich, M. S. Bernatowicz, N. S. Agarwal, M. Kawai, F. G. Salituro, and P. G. Schmidt, Inhibition of aspartic proteases by pepstatin and 3-methylstatine derivatives of pepstatin. Evidence for collected-substrate enzyme inhibition, Biochemistry **24**:3165 (1985).
25. J. A. Tavel, On-going trials in HIV protease inhibitors, Expert Opin. Invest. Drugs **9**:917 (2000).

Chapter 8

Free Energy Calculations on DNA:Ligand Complexes

Suresh B. Singh[†] and Peter A. Kollman[‡]
[†]*Merck Research Laboratories, Rahway, NJ 07065*
[‡]*Dept. of Pharmaceutical Chemistry, University of California, San Francisco, CA 94143*

1. INTRODUCTION

Ligands that interact physically with DNA have been extensively studied both by experimental techniques and by a variety of theoretical approaches. A diverse set of compounds have been studied, including compounds that intercalate between DNA sequences or bind in the minor groove.[1-7] These studies have identified various factors that influence the stability of DNA:ligand complexes in solution.[6, 8, 9]

Free energy calculations on DNA-drug complexes provided insight into the interactions of daunomycin, acridine,[10] and netropsin[11, 12] with DNA and their role in base specificity. Calculations using continuum electrostatics have been performed to estimate free energies of binding of positively charged ligands such as DAPI, Hoechst 33258, and netropsin to $d(CGCGAATTCGCG)_2$, ethidium to a DNA dodecamer,[13] and binding of five anthracycline antibiotics to DNA.[14]

Some of the best characterized ligands that bind in the minor groove of DNA are distamycin and netropsin. Both these molecules are long and flat and are sterically and electrostatically complementary to the characteristics of the minor groove of DNA. Distamycin and netropsin are known to have specific affinity towards the minor groove of AT rich regions of B-DNA.

The electrostatic potential calculations have shown that the minor groove of AT rich B-DNA sequence has the lowest negative potential.[15] This led to the implication that the cationic drugs will exhibit binding specificity to the minor groove regions of AT rich sequences. The crystal structure of a DNA:distamycin complex showed that there is only one molecule bound to the minor groove of an AATT DNA site.[2] However, experimental studies showed that the introduction of a GC base pair in an AT rich stretch widens the minor groove and leads to the accommodation of two distamycin molecules.[7] According to the experimental studies presented by Dwyer et al.[7] and Geierstanger et al.[16] both distamycin and its imidazole variant are found to bind tightly to the five base pair AAGTT:AACTT site of an 11mer duplex .

Thermodynamic[6] and crystallographic data[3] have characterized the binding of netropsin to the minor groove of AT rich sequences of DNA. Netropsin has a very tight affinity (K_a ~10^9) for DNA at low salt concentrations[6] and binds in the minor groove, causing minor change in the DNA structure. Its association constant with DNA is inversely related to the salt concentration. Thus, there is a significant electrostatic component to DNA-netropsin association. However, it was not evident how much the complementary electrostatics contributes to the absolute free energy of association of netropsin to DNA.

A similar issue was addressed for two protein-ligand interactions.[17] The free energy calculations were able to reproduce the experimental findings for the biotin-streptavidin complex, which has a free energy of association of ~20 kcal/mol and that of N-acetyl tryptophan amide- α-chymotrypsin complex which has a free energy of association of ~5-10 kcal/mol. The decomposition of these free energies into the van der Waals and electrostatic contributions suggested that the biotin-streptavidin binding is predominantly van der Waals in nature and the N-acetyl tryptophan amide-α-chymotrypsin free energy is predominantly electrostatic. These component preferences are quite contrary to expectation, since biotin has a negative charge and a very polar ureido group capable of extensive H-bonding with protein groups that interact with these functionalities. On the other hand, N-acetyl tryptophan is neutral and α-chymotrypsin is known to preferentially accommodate non-polar groups in its P1 pocket.

It is interesting to note that similar observations were made in the calculation of the electrostatic free energy of binding of ethidium bromide to DNA.[13] It was observed that a relatively small electrostatic free energy of binding calculated with a non linear Poisson-Boltzmann model for the association of positively charged ethidium to DNA indicates that nonpolar interactions are more likely to be responsible for the free energy of binding of ethidium to DNA.

The focus of this chapter is on the application of free energy perturbation methodology to evaluate binding strengths of ligands to oligomeric DNA. We present here three case studies related to minor groove binding ligands. The first case is on the calculation of the absolute free energy of binding of netropsin to $d(CGGAATTCGCG)_2$;[18] the second case entails the calculation of the change in free energy of binding of netropsin to $d(IC)_6 \bullet d(IC)_6$ relative to that of $d(GC)_6 \bullet d(GC)_6$;[11] the third case is on the change in free energy of binding of distamycin relative to that of 2-imidazole-distamycin (2-ImD) to $[d(CGCAATTGGC)]_2$.[19]

2. FREE ENERGY PERTURBATION CALCULATION

2.1 Theory

The free energy calculations with complexes of oligomeric DNA with distamycin and netropsin presented in this chapter were carried out using the AMBER force field[20] and the AMBER software.[21] The associated potential energy function is:

$$V_{total} = \sum_{bonds} K_r(r-r_{eq})^2 + \sum_{angles} K_r(\theta-\theta_{eq})^2 + \sum_{dihedrals} \frac{V_n}{2}[1+Cos(n\phi-\gamma)]$$
$$+ \sum_{i<j}\left\{\left[\frac{A_{ij}}{R_{ij}^{12}} - \frac{B_{ij}}{R_{ij}^{6}}\right] + \frac{q_i q_j}{\varepsilon r_{ij}}\right\} + \sum_{Hbonds}\left[\frac{C_{ij}}{R_{ij}^{12}} - \frac{D_{ij}}{R_{ij}^{10}}\right]$$
$$+ \sum_{restraints} K_r(r-r_o)^2 \qquad (1)$$

The change in free energy of a system between two states (A,B) (ΔG_i) is calculated after each step of the transformation. Since the free energy is a state function the total free energy change, ΔG, is the sum of the intermediate ΔG_is. An intermediate state[22] of the system is defined as,

$$G_\lambda = (1-\lambda)G_A + \lambda G_B \qquad (2)$$

where λ is the perturbation parameter, G_A is the free energy of the initial state A, and G_B is the free energy of the final state B. The free energy change is obtained by employing the perturbation method,

$$\Delta G_i = -k_B T \ln \left\langle \exp\left(\frac{-\Delta H}{k_B T}\right) \right\rangle_\lambda \qquad (3)$$

where $\langle\ \rangle_\lambda$ is the ensemble average defined by the parameter λ. λ' is a neighboring state of λ. The total free energy is given by:

$$\Delta G_i = \sum_{i=1}^{n} G_i \tag{4}$$

where n is the total number of windows in a free energy simulation.

2.2 Methodology

Free energy perturbation calculations were performed using molecular dynamics simulations. The electrostatic charges for distamycin, inosine, netropsin, and 2-imidazole-distamycin were calculated by using STO-3G basis set and fitting the point charges that will reproduce the quantum mechanical electrostatic potential. These calculations were carried out using the Singh and Kollman methodology.[23] The force field parameters for the ligands were assigned based on the extrapolations made from AMBER force field parameters.[11, 18, 19] The starting structure for the DNA:netropsin complex was taken from the X-ray structure[3] (Figure 3) and that for the DNA:2-ImD:2-ImD complex was taken from the NMR structure[16] (Figure 5). The phosphates on the oligomeric DNA complexes were neutralized with sodium ions placed at 3.6 Å from the phosphate oxygen bisector and then solvated with TIP3P.H20[24] molecules placed up to 10-12 Å away from the solute (complex + ions) atoms to generate solvated boxes in which to carry out simulations with periodic boundary conditions. The ligands were solvated with TIP3P.H2O molecules placed up to 12-13 Å away from the solute atoms. In the free energy simulations of $d(IC)_6{\bullet}d(IC)_6 \rightarrow d(GC)_6{\bullet}d(GC)_6$ the sodium ions were placed at 3.0 Å from the phosphate oxygen bisector and the resultant neutralized complex was solvated by placing TIP3P.H2O molecules up to 5.0 Å from the solute atoms.

The MD simulations were carried out under standard temperature and pressure. A 1 fs time step was used with SHAKE[25] applied to bonds. A 2 fs time step with SHAKE was used in the $d(IC)_6{\bullet}d(IC)_6 \rightarrow d(GC)_6{\bullet}d(GC)_6$ calculations. The non-bonded interactions for DNA complexes were subject to 10 -12 Å spherical cutoff whereas no cutoff was applied to solute-solute interactions to avoid cutoff artifacts on coulombic interactions between sodium ions with phosphates. In the case of $d(IC)_6{\bullet}d(IC)_6 \rightarrow d(GC)_6{\bullet}d(GC)_6$ calculations an 8 Å spherical cutoff was applied to non-bonded interactions. A weak harmonic restraint of 5.0 kcal/mol was imposed to avoid the disruption of terminal base pairs during FEP simulations of netropsin $\rightarrow$ 0 and 2-imidazole-distamycin $\rightarrow$ distamycin calculations.

3. APPLICATIONS

3.1 Absolute Free Energy of Binding of Netropsin to $[d(CGCGAATTGCGC)]_2$[18]

The free energy method to calculate absolute free energy of binding of two interacting species was first proposed by Jorgensen.[26] This methodology deals with the annihilation of the ligand in complex and in isolation based on the following cycle:

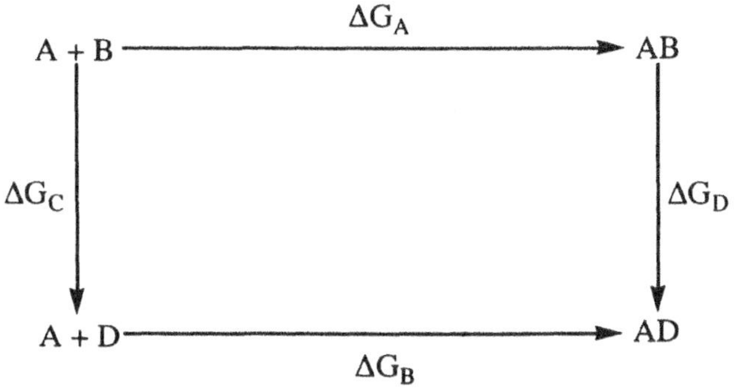

Figure 1. Thermodynamic cycle for an association process

where A and B are association molecules and D is a dummy molecule with no interaction with its surroundings. Since $\Delta G_B = 0$, free energy of association is given by $\Delta G_A = \Delta G_C - \Delta G_D$. For example, this can be accomplished by mutating molecule B both when bound to the macromolecule (ΔG_D) and free in solution (ΔG_C) to nothing. In the free energy perturbation calculations the thermodynamic windows method was used for the electrostatic part and the slow growth method[13] for the van der Waals part. The free energy perturbation was carried out by mutating netropsin $\rightarrow$ 0 in the $[d(CGCGAATTGCGC)]_2$:netropsin complex in water in the forward direction only ($\lambda = 1 \rightarrow 0$) and the free netropsin in solution was mutated to nothing in both forward ($\lambda = 1 \rightarrow 0$) and backward ($\lambda = 0 \rightarrow 1$) directions by changing the force field parameters for netropsin smoothly and uniformly. The structure of netropsin is shown in Figure 2.

The free energy calculations were calculated with electrostatic and van der Waals contributions evaluated separately for both the complex and the isolated ligand. The total free energy change of -10.3 kcal/mol agreed with experimental measurements of -11.5 kcal/mol[6] for netropsin

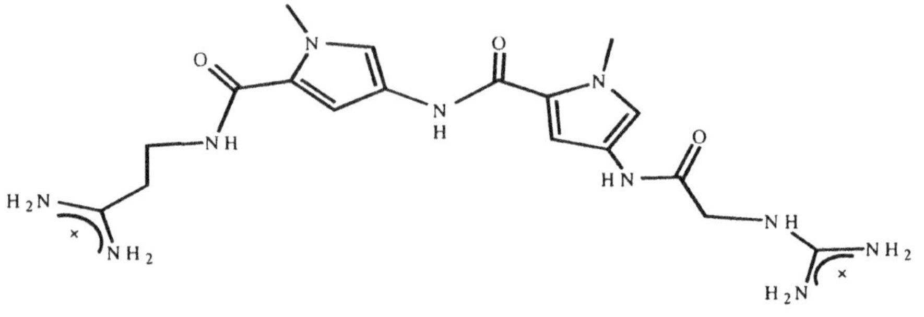

Figure 2. Structure of Netropsin

binding to $[d(GCGAATTCGC)]_2$ and -10.2 kcal/mol for netropsin binding to $[d(CGCAATTGGC)]_2$. The X-ray structure of $[d(GCGAATTCGC)]_2$ complex with netropsin is shown in Figure 3.

The individual free energy components show the relative contributions from the electrostatic and the van der Waals interactions. The free energy change for the annihilation of the charges on the ligand in the complex and in solution is almost identical within the limits of the estimation. However, the van der Waals free energy component for the complex is more than that of the isolated ligand (Table 1).

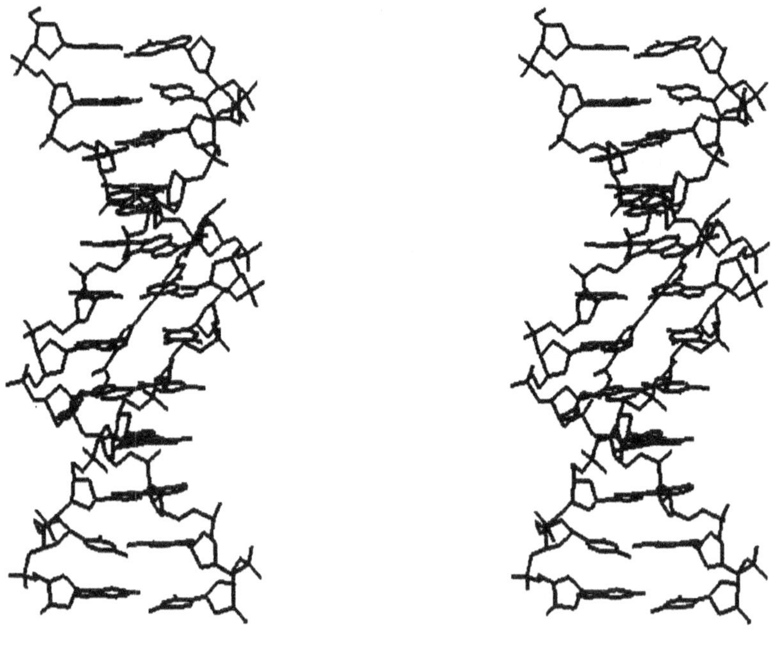

Figure 3. X-ray structure[3] of $[d(CGCGAATTGCGC)]_2$:netropsin complex shown in stereo.

Table 1. Absolute free energy of netropsin binding to [d(CGCGAATTGCGC)]$_2$

Netropsin in Solution ΔG_C (kcal/mol)		Netropsin in Solvated Complex ΔG_D (kcal/mol)	
Electrostatic	van der Waals	Electrostatic	van der Waals
235.2 ± 2.23#	17.5 ± 1.7#	236.7 ± 2.0*	26.3 ± 0.2*
ΔG_I (kcal/mol)			
Electrostatic#	van der Waals #	Total	Experimental*6
-1.5	-8.8	-10.3	-11.5

$\Delta G_I = \Delta G_c - \Delta G_D$; Total $\Delta G_I = \Delta G_{Elc} + \Delta G_{vdW}$
*Free energy of netropsin binding to [d(GCGAATTCGC)]$_2$

These calculations indicate that the major stabilization force for the binding of netropsin to DNA is primarily van der Waals in nature which is contrary to previous postulates suggesting that the electrostatic component contributes significantly towards the binding of the charged ligands. It is quite conceivable that the electrostatic forces play a major role in driving the complexation process by facilitating the rotation and the translation of the charged ligand, while the major stabilization of the complex appears to come from the van der Waals forces. Marky and Breslauer[6] also suggested that the electrostatic forces partially contribute to the binding enthalpy of netropsin to DNA. They hypothesized that the rest of the binding enthalpy most likely is derived from non-polar interactions. Mishra and Honig[13] with the aid of continuum electrostatics calculations on the netropsin DNA complex suggested that the electrostatic free energy of binding is almost fully balanced by loss of solvation upon binding and that non polar interactions could possibly contribute towards the binding affinity of charged ligands to DNA.

3.2 Relative Free Energy of Binding of Netropsin to (IC)$_6$•d(IC)$_6$ vs. d(GC)$_6$•d(GC)$_6$

Gago and Richards[11] used free energy perturbation methodology to calculate the change in the free energy of binding of netropsin to d(IC)$_6$•d(IC)$_6$ relative to that of d(GC)$_6$•d(GC)$_6$. The free energy perturbation calculations utilized the thermodynamic windows method in both the forward and backward directions. The inosine residues were mutated to guanosine residues in the free and the complexed solvated states. The calorimetric measurements by Marky and Breslauer[6] showed that the free energy of binding of netropsin to poly[d(GC)•d(GC)] is -7.1 kcal/mol whereas its binding to poly[d(IC)•d(IC)] is -11.1 kcal/mol (Table 2).

Table 2. Free energy difference (kcal/mol) in binding of netropsin to $d(GC)_6 \bullet d(GC)_6$ versus that to $d(IC)_6 \bullet d(IC)_6$.

	Complex	Free Energy of Binding	Free Energy Difference
Experimental[6]	Poly[d(GC)•d(GC)]:netropsin	-7.1	4.0
	Poly[d(IC)•d(IC)]:netropsin	-11.1	
Free Energy Perturbation	$d(GC)_6 \bullet d(GC)_6$:netropsin	—	4.3
	$d(IC)_6 \bullet d(IC)_6$: netropsin		

These experiments also showed that the electrostatic component of the binding of netropsin to poly[d(GC)•d(GC)] and poly[d(AT)•d(AT)] is very similar indicating that the difference in free energy of binding of netropsin to these two polymers is primarily due to the non-polar interactions between the ligand and the respective polymers. Calculations by Gago and Richards reproduced the experimental results showing that the free energy difference for netropsin binding to $d(IC)_6 \bullet d(IC)_6$ relative to that of $d(GC)_6 \bullet d(GC)_6$ is 4.3 kcal/mol versus that of 4.0 kcal/mol from calorimetric studies. They observed that this free energy change is due to the movement of the drug towards the solvent in the $d(GC)_6 \bullet d(GC)_6$ complex relative to that in the $d(IC)_6 \bullet d(IC)_6$ complex. They also observed that only one out of four hydrogen bonds are preserved between the ligand and $d(GC)_6 \bullet d(GC)_6$ relative to that in $d(IC)_6 \bullet d(IC)_6$. They attribute this to the difference in minor groove widths which results from the presence of the exocylic amino group of guanine in $d(GC)_6 \bullet d(GC)_6$.

3.3 Relative Free Energy of Binding of $[d(CGCAAGTTGGC)]_2$[19] to 2-Imidazole-distamycin versus Distamycin

Distamycin has three pyrrole rings whereas its analog 2-imidazole-distamycin has an imidazole ring substituted for the central pyrrole ring (Figure 4). The pyrrole and the imidazole variant differ only in that the C–H is substituted by an N in the central ring. The solution structure of 2:1 complex 2-ImD bound to $[d(CGCAAGTTGGC)]_2$ was used as the starting structure.[7]

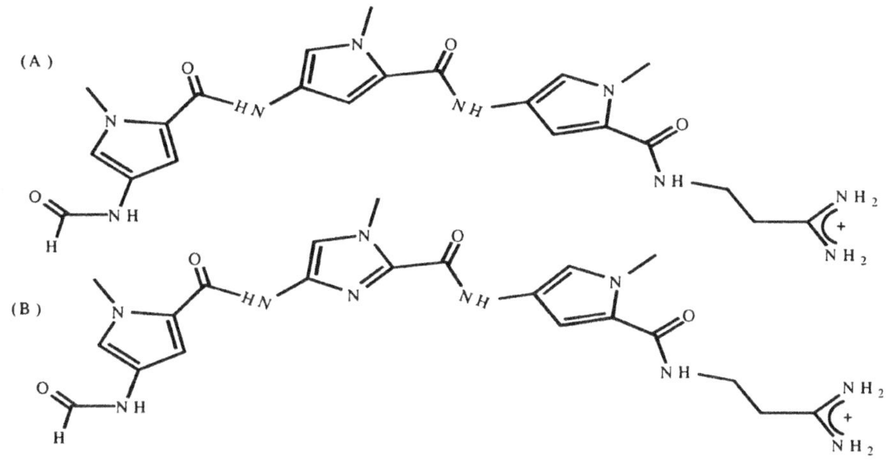

Figure 4. Structures of distamycin (A) and 2-imidazole-distamycin (B).

The first ligand is bound parallel to the AAGTT stretch and the second ligand is bound parallel to the AACTT stretch (Figure 5). The location where the first ligand is bound is labeled site I and the location where the second ligand is bound is labeled site II

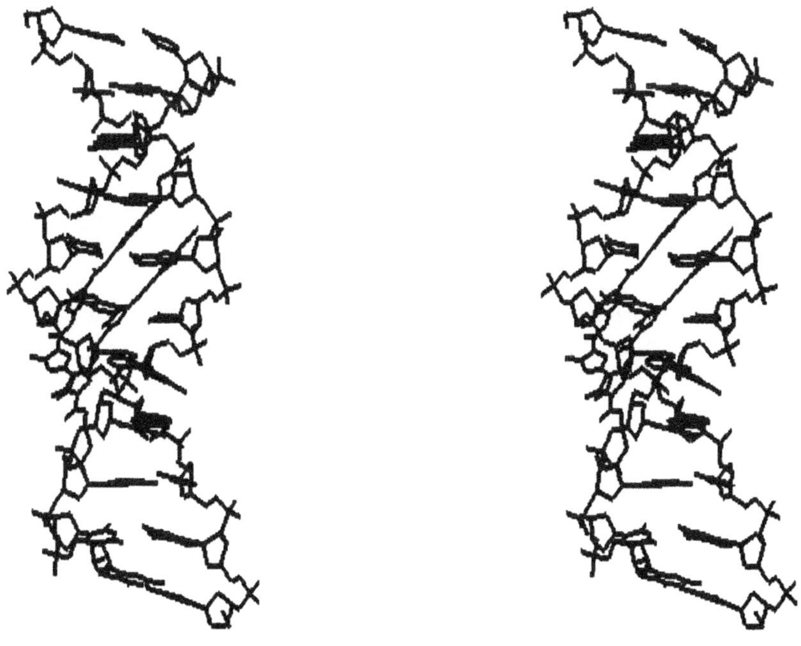

Figure 5. Solution structure[16] of 2:1 complex 2-imidazole-distamycin bound to $[d(CGCAAGTTGGC)]_2$ shown in stereo.

The free energy perturbation calculations were carried out to calculate the solvation free energy difference between distamycin and 2-imidazole-distamycin and the relative binding affinities of distamycin and 2-imidazole-distamycin at sites I and II. The calculation of relative free energies of binding of distamycin and 2-ImD at the two binding sites involved mutating 2-ImD to distamycin at sites I and II independently and then simultaneously mutating both the molecules of 2-ImD to distamycin. These processes are schematically shown in Figure 6. The thermodynamic cycle in Figure 6A shows the independent equilibrium processes involved in the displacement of 2-ImD by distamycin at the two binding sites. The thermodynamic cycle in Figure 6B shows the equilibrium process of the simultaneous displacement of 2-ImD by distamycin at the two binding sites. The processes that are indicated by vertical arrows represent the free energy perturbation calculations and the horizontal arrows represent the experimentally observed processes.

The mutation of 2-ImD to distamycin at site I led to a free energy difference of 0.65 kcal/mol ($\Delta G_4 - \Delta G_3$) (Table 3). This calculation showed that distamycin has weaker affinity for $[d(CGCAAGTTGGC)]_2$ at site I relative to that of 2-ImD. The NMR experiments carried out in Wemmer's

$$
\begin{array}{ccc}
\text{Dst + 2-ImD + DNA} & \xrightarrow{\Delta G_1} & \text{Dst:2-ImD:DNA} \\
\uparrow \Delta G_3 & & \uparrow \Delta G_4 \\
\text{2-ImD + 2-ImD + DNA} & \xrightarrow{\Delta G_2} & \text{2-ImD:2-ImD:DNA} \\
\downarrow \Delta G_3 & & \downarrow \Delta G_6 \\
\text{2-ImD + Dst + DNA} & \xrightarrow{\Delta G_5} & \text{2-ImD:Dst:DNA}
\end{array}
$$

Figure 6A. Thermodynamic cycles for the calculation of the free energy difference of the binding of distamycin relative to 2-imidazole-distamycin at two adjacent sites of $d(CGCAAGTTGGC)_2$.

$$
\begin{array}{ccc}
\text{2-ImD + 2-ImD + DNA} & \xrightarrow{\Delta G_2} & \text{2-ImD:2-ImD:DNA} \\
\downarrow \Delta G_3 \quad \downarrow \Delta G_3 & & \downarrow \Delta G_8 \\
\text{Dst + Dst + DNA} & \xrightarrow{\Delta G_7} & \text{Dst: Dst:DNA}
\end{array}
$$

Figure 6B. Thermodynamic cycles for the calculation of the free energy difference of the binding of distamycin relative to 2-imidazole-distamycin at both sites of $d(CGCAAGTTGGC)_2$.

Table 3. Relative free energy differences between the binding of distamycin and 2-imidazole-distamycin to $[d(CGCAAGTTGGC)]_2$

Free Energy Perturbation	Free Energy Differences (kcal/mol)	
	Calculated	Experimental
2-ImD → Dst	1.82	NA
2-ImD:2-ImD:DNA→Dst:2ImD:DNA	0.65	1.6
2-ImD:2-ImD:DNA → 2-ImD:Dst:DNA	-2.88	NA*
2-ImD:2-ImD:DNA → Dst:Dst:DNA	-1.79	-2.3

*NMR experiments failed to detect 2-ImD:Dst:DNA complex in solution.

lab[18] showed that the relative populations of Dst:Dst:DNA and Dst:2-ImD:DNA in solution are 15:1. This leads to the experimental value of 1.6 kcal/mol ($\Delta G_2 - \Delta G_1$) for the affinity of distamycin relative to that of 2-imidazole-distamycin at site I (Table 3).

The mutation of 2-ImD to distamycin at site II led to a free energy difference of -2.88 kcal/mol ($\Delta G_6 - \Delta G_3$). The NMR experiments could not detect the2-ImD:Dst:DNA complex in solution therefore the calculated number for distamycin's affinity for site II ($\Delta G_5 - \Delta G_2$) cannot be compared with the experiment.

It is interesting that distamycin has greater affinity than 2-imidazole-distamycin at site II whereas it has weaker affinity at site I. This can be understood in terms of the molecular environments present at the two sites and in the intermolecular interactions between distamycin and $[d(CGCAAGTTGGC)]_2$. At site I the exocyclic NH_2 group of guanine is available for the imidazole ring of 2-imidazole-distamycin to establish a hydrogen bond whereas at site II there is no hydrogen bonding partner present hence the weaker affinity for 2-ImD. This was rationalized in terms of the desolvation free energy for 2-ImD relative to that of distamycin, which is 1.8 kcal/mol. At site I the hydrogen bond between N2 of guanine and the imidazole compensates for the desolvation energy of 2-ImD and stabilizes the complex relative to that of distamycin, whereas at site II the lack of an equivalent hydrogen bond between 2-ImD and the DNA cannot compensate for the loss of solvation energy and hence destabilization relative to that of distamycin.

The simultaneous mutation of 2-imidazole-distamycin to distamycin at both the sites I and II led to a free energy change of -1.8 kcal/mol ($\Delta G_8 - 2\Delta G_3$). The NMR experiments showed that the relative populations of Dst:Dst:DNA and 2-ImD:2-ImD:DNA are 50:1 giving an experimental free energy difference of -2.3 kcal/mol ($\Delta G_7 - \Delta G_2$). This indicates that the favorable van der Waals interactions between distamycin and DNA at sites I and II and the stacking interactions between the two distamycin molecules stabilize the 2:1 Dst:DNA complex over the 2:1 2-ImD:DNA complex. The major destabilization factor for the 2:1 2-ImD:DNA complex is the lack of hydrogen bond for the central imidazole ring of 2-ImD at Site II. These calculations show that the order of stability of DNA, distamycin, and 2-imidazole-distamycin complexes are as follows: 2-ImD:Dst:DNA > Dst:Dst:DNA > 2-ImD:2-ImD:DNA > Dst:2-ImD:DNA. This order of affinity is consistent with the solution NMR data and the calculations that 2-ImD has greater affinity for $[d(CGCAAGTTGGC)]_2$ than distamycin at site I and distamycin has greater affinity than 2-ImD at site II.

4. CONCLUSIONS

The chapter summarizes the application of free energy perturbation methods to the study of DNA-ligand interactions. The cases presented here provide a means to evaluate relative and absolute thermodynamic stabilities of DNA:ligand complexes. Given the goals in each case were challenging a careful evaluation of methodologies led to a set of conditions that facilitated the calculation of absolute and relative binding affinities of DNA-ligand complexes in reasonable agreement with the experiment.

5. REFERENCES

1. V. Fratini, M. L. Kopka, H. R. Drew, and R. E. Dickerson, Reversible bending and helix geometry in a B-DNA dodecamer: CGCGAATTBrCGCG, J. Biol. Chem. **257**:14686 (1982).
2. M. Coll, C. A. Frederick, A. H. J. Wang, and A. Rich, A bifurcated hydrogen-bonded conformation in the d(AT) base pairs of the DNA dodecamer d(CGCAAATTTGCG) and its complex with distamycin, Proc. Natl. Acad. Sci. USA **84**:8385 (1987).
3. M. Coll, J. Aymami, G. A. van der Marel, J. H. van Boom, A. Rich, and A. H. J. Wang, Molecular structure of the netropsin-d(CGCGATATCGCG) complex: DNA conformation in an alternating AT segment, Biochem. **28**:310 (1989).
4. K. J. Addess, Sinsheimer, and J. Feigon, Solution structure of a complex between 2 [N-MeCys3,N-MeCys7]tandem and $[d(GATATC)]_2$, Biochem. **32**:2498 (1993).
5. M. Sastry, and D. J. Patel, Solution structure of the mithramycin dimer-DNA complex, Biochem. **32**:6588 (1993).
6. L. A. Marky and K. J. Breslauer, Origins of netropsin binding affinity and specificity: Correlations of thermodynamic and structural data, Proc. Natl. Acad. Sci. USA **84**:4359 (1987).
7. T. J. Dwyer, B. H. Geierstanger, Y. Bathini, J. W. Lown, and D. E. Wemmer, Design and binding of a distamycin A analog to d(CGCAAGTTGGC).d(GCCAACTTGCG): Synthesis, NMR studies, and implications for the design of sequence-specific minor groove binding oligopeptides, J. Am. Chem. Soc. **114**:5911 (1992).
8. K. J. Breslauer, D. P. Remeta, W. Y Chou, R. Ferante, J Curry, D. Zauczkowski, J. G. Snyder, and L. A. Marky, Enthalpy-entropy compensations in drug-DNA binding studies, Proc. Natl. Acad. Aci. USA **84**:892 (1987).

9. D. Rentzeperis, L. A. Marky, T. J. Dwyer, B. H. Geirstanger, J. G. Pelton, and D. E. Wemmer, Interaction of minor groove ligands to an AAATT/AATTT site: Correlation of thermodynamic characterization and solution structure, Biochem. **34**:2937 (1995).
10. P. Cieplak, S. N. Rao, P. D. J. Grootenhuis, and P. A. Kollman, Free energy calculation on base specificity of drug-DNA interactions: Application to daunomycin and acridine intercalation into DNA, Biopol. **29**:717 (1990).
11. F. Gago and W. G. Richards, Netropsin binding to poly[d(IC)].poly[IC)] and poly[d(GC].poly[d(GC)]: A computer simulation, Mol. Pharm. **37**:341 (1990).
12. T. Hard and L. Nilsson, Free energy calculations predict sequence specificity in DNA-drug complexes, Nucleosides Nucleotides **10**:701 (1991).
13. V. K. Mishra and B. Honig, On the magnitude of the electrostatic contribution to ligand-DNA interactions, Proc. Natl. Acad. Sci. USA **92**:4691 (1995).
14. M. Baginski, F. Fogolari, and J. M. Briggs, Electrostatic and non-electrostatic contributions to the binding free energies of anthracycline antiobiotics to DNA, J. Mol. Biol. **28**:253 (1997).
15. R. Lavery and B. Pullman, The molecular electrostatic potential: steric accessibility and hydration of Dickerson's B-DNA dodecamer d(CpGpCpGpApApTpTpCpGpCpG), Nuc. Acid Res. **9**:3765 (1981).
16. H. Geierstanger, T. Dwyer, Y. Bathini, W. J. Lown, and D. E. Wemmer, NMR characterization of a heterocomplex formed by distamycin and its analog 2-ImD with d(CGCAAGTTGGC):d(GCCAACTTGCG): Preference for the 1:1:1 2-ImD:Dst:DNA complex over the 2:1 2-ImD:DNA and the 2:1 Dst:DNA complexes, J. Am. Chem. Soc. **115**:4474 (1993)
17. S. Miyamoto and P. Kollman, What determines the strength of noncovalent association of ligands to proteins in aqueous solution?, Proc. Natl. Acad. Sci. USA **90**:8402 (1993).
18. S. B. Singh and P. A. Kollman, Calculating the absolute free energy of association of netropsin and DNA, J. Am. Chem. Soc. **121**:3267 (1999).
19. S. B. Singh, Ajay, D. E. Wemmer, and P. A. Kollman, Relative binding affinities of distamycin and its analog to d(CGCAAGTTGGC).d(GCCAACTTGCG): Comparison of simulation results with experiment, Proc. Natl. Acad. Sci. **91**:7673 (1994).
20. S. J. Weiner, P. A. Kollman, D. A. Case, U. C. Singh, C. Ghio, G. Alagona, S. Profeta, and P. Weiner, A new force field for molecular mechanical simulation of nucleic acids and proteins, J. Am. Chem. Soc. **106**:765 (1984).
21. D. A. Pearlman, D. A. Case, J. C. Caldwell, G. L. Seibel, U. C. Singh, P. Weiner, and P. A. Kollman, AMBER (4.0), University of California, San Francisco (1991).
22. U. C. Singh, F. K. Brown, P. Bash, and P. A. Kollman, An approach to the application of free energy perturbation methods using molecular dynamics: Applications to the transformations of $CH_3OH \rightarrow CH_3CH_3$, $H_3O^+ \rightarrow NH_4^+$, glycine → alanine, and alanine → phenylalanine in aqueous solution and to $H_3O+(H_2O)^3 \rightarrow NH_4+(H_2O)^3$ in the gas phase, J. Am. Chem. Soc. **109**:1607 (1987).
23. U. C. Singh and P. A. Kollman, An approach to computing electrostatic charges for molecules, J. Comp. Chem. **5**:129 (1984).
24. W. L. Jorgensen, J. Chandrasekhar, J. D. Madura, R. W. Impey, and M. L. Klein, Comparison of simple potential functions for simulating liquid water, J. Chem. Phys. **79**:926 (1983).
25. W. F. van Gunsteren and H. J. C. Berendsen, Algorithm for macromolecular dynamics and constraint dynamics, Mol. Phys. **34**:131 (1977).
26. W. L. Jorgensen, J. K. Buckner, S. Boudon, and J. Tirado-Rives, Efficient computation of absolute free energies of binding by computer simulations. Application to the methane dimer in water, J. Chem. Phys. **89**:3742 (1988).

Section Four

Ligand Design and Analysis

Chapter 9

The Linear Interaction Energy Method for Computation of Ligand Binding Affinities

Johan Åqvist and John Marelius
Department of Cell and Molecular Biology Uppsala University, Biomedical Center, SE-751 24 Uppsala, Sweden

1. INTRODUCTION

A major goal for computational chemistry in the drug discovery process is to be able to predict the strength of non-covalent association between putative ligands and their target receptors. Significant progress has been made in the field of computer-aided ligand design during the last decade and a fairly wide range of computational tools are now being used as complements to experimental studies. This chapter will deal with the type of situation usually referred to as structure-based design, i.e. when the 3D structure of the receptor target site is known. Moreover, we will mainly focus on the problem of estimating the free energy of binding for a given 'conformation' (albeit, subject to thermal configurational averaging) rather than on the problem of finding the optimal fit between the ligand and the receptor through docking. Of course, the two problems of finding the optimal fit (docking) of a putative complex and of estimating its affinity (scoring) is intimately related to each other. That is, the binding energy could be considered as rather ill defined in cases where the complex between ligand and receptor is non-optimal or unstable. Likewise, any useful docking procedure must provide some way of scoring different trial conformations in order to produce a ranking of them. Although many empirical scoring methods assign a binding free energy to a single conformation of the receptor-ligand complex, it may be appropriate to recall that the measured binding constant actually reflects an average over thermally accessible configurations of both the associated and dissociated states.

Several different types of approaches for predicting ligand-binding affinities have been developed and a number of useful reviews of the field have been recently published.[1-3] A key factor when deciding on a computational strategy for trying to predict the potency of a set of ligands is the required time for calculating the score of a typical ligand. If a high throughput of ligands is crucial, e.g., in screening large virtual libraries or when extensive conformational searching is required, such as for docking, the time it takes to evaluate a single compound must be rather short. On the other hand, if the 3D structure of a "lead" compound in complex with the receptor is available and the objective is to explore a limited number of chemical modifications of this lead, then it may be affordable and desirable to carry out more time-consuming and accurate calculations.

The computationally fastest and simplest binding affinity prediction methods are typically so-called empirical or knowledge-based scoring approaches. The empirical scoring methods are primarily based on very simplified energy functions, design to represent different contributions to the binding free energy such as hydrogen bonds, buried hydrophobic surface area and torsional entropy changes etc.[4-6] These types of scoring functions are usually short-ranged so that only receptor-ligand atom pairs more or less in contact with each other are considered. In the knowledge-based approaches the score is simply derived from the frequency of occurrence of different atom-atom contact pairs in complexes of known structure.[7,8] The speed of these methods is thus due to the simplicity of the scoring function, that few interactions are calculated and that no conformational sampling is performed. On the other hand, it is also probably these features, together with the information content of the database used to calibrate the functions that limit the accuracy of entirely empirical or statistical approaches. Another issue is that water molecules are usually not considered at all, which could be a problem for complexes where they appear as an integral part, e.g., by bridging interactions between the ligand and the receptor.

The most time-consuming and rigorous approach for evaluating binding free energies is the free energy perturbation (FEP) method, which involves slow gradual transformations between the states of interest.[3, 9-11] The FEP method and its variants, such as thermodynamic integration, are typically used together with molecular dynamics (MD) or Monte Carlo (MC) simulations for generating conformational ensembles. In order for the FEP technique to yield convergent results extensive conformational sampling is required and, in contrast to the empirical methods above, a large number of (usually) pair-wise interactions must be calculated at each MD or MC step. These facts, together with the requirement that the transformations must be "gentle", i.e. small perturbations, often make the method very slow. Thus, calculation of absolute binding free energies is usually not tractable, and even relative binding energies can pose major problems if the ligands differ substantially from each other.

2. THE LINEAR INTERACTION ENERGY (LIE) METHOD

In order to try to overcome some of the problems with FEP calculations, we decided to examine whether it would be possible to extract any useful information on the binding energetics from simulations of only the physically relevant states (associated and dissociated) of the ligand.[12] The idea was then to consider the absolute binding free energy of a ligand (l) as the change in 'solvation' free energy when it is transferred from aqueous solution (free state) to its solvated receptor-binding site (bound state), that is

$$\Delta G_{bind}(l) \approx \Delta G^{p}_{sol}(l) - \Delta G^{w}_{sol}(l) \tag{1}$$

where the superscripts p and w denote protein (receptor) and water, respectively. The solvation energy of the ligand in a given environment, $\Delta G^{i}_{sol}(l)$, in turn reflects the process of transferring the molecule from the gas-phase to this environment. Such a process can at least formally be considered as consisting of two separate steps: (1) creation of the molecular van der Waals cavity in the given environment and (2) turning on the electrostatic interactions between the molecule and its surroundings. If the ligand is modeled as having intermolecular electrostatic interactions these can, in principle, be referred to either of the two hypothetical steps above. While the solvation energies in Equation 1 include the contributions from internal entropy changes in the ligand, receptor and water solution, it can be noted that the change in translational entropy caused by confining the ligand to a (usually) smaller volume than that implied by a 1M standard state in aqueous solution also needs to be taken into account. This problem is far from trivial since there, e.g., may be a release of bound water molecules accompanying the binding process with a concomitant entropy increase. Since we anyway will deal with empirical parameterization of the terms in Equation 1 here, the translational entropy contribution to binding will not be explicitly considered.

The first version of the LIE method employed the linear response approximation to estimate the electrostatic part of the solvation/binding free energies. The linear response result for this component of the solvation energies, ΔG^{i}_{el} (where i=p or w), can then be written as

$$\Delta G^{i}_{el} = \tfrac{1}{2}\left\{\left\langle V^{el}_{l-s}\right\rangle_{on} + \left\langle V^{el}_{l-s}\right\rangle_{off}\right\} \tag{2}$$

where the two averages are sampled with the electrostatic interactions between the ligand and the surrounding (*l*-s) turned on and off, respectively. The electrostatic interactions within the ligand V_{l-l}^{el} could, in principle, also be included in the averages, but ΔG_{el}^{i} would then also contain the gas-phase 'charging' free energy and therefore not be as directly related to the solvation energy. Furthermore, because of the covalent constraints within the ligand the average $\langle V_{l-l}^{el} \rangle_{off}$, i.e. the intermolecular electrostatic energy that would have been obtained from the configurations generated with these interactions turned off, is generally not zero. One of the simplifying features of the LIE method is, however, that also the term $\langle V_{l-s}^{el} \rangle_{off}$ is neglected. In principle, one would have to carry out one extra simulation for both the bound and free states of the ligand with its intermolecular electrostatic interactions turned off, and calculate from these configurations what the average electrostatic interaction energy would have been if V_{l-s}^{el} had been turned on. While it has been shown that $\langle V_{l-s}^{el} \rangle_{off} = 0$ is reasonable approximation in water[13] it remains unclear how valid the assumption is for the bound state. One could perhaps argue that since proteins (and often also ligands) are flexible there would be substantial relaxation and conformational rearrangements of the complex if such a simulation was carried for long enough, so that $\langle V_{l-s}^{el} \rangle_{off}$ would eventually tend to zero. But, on the other hand, if the binding site is rigid and substantially "prearranged" for accommodating the ligand the $\langle V_{l-s}^{el} \rangle_{off}$ term would inevitably seem to provide a negative contribution to the binding free energy that is neglected. The original idea with the LIE approach was, of course, to keep it as simple as possible and avoid simulations of unphysical states, but the above point should nevertheless be kept in mind.

Non-polar contributions to the binding affinity, e.g., hydrophobic effects and van der Waals interactions, appear to be less straightforward to quantify. We decided to try the simple idea of measuring the non-electrostatic part of the interaction of the ligand with its surrounding environment in the associated and dissociated states, and then just scale these energies by an empirically derived coefficient. These energies are typically given by a Lennard-Jones potential. The basic idea was *not* that hydrophobic effects somehow reside in these energy terms, but rather based on the following observations. Solvation free energies for typical non-polar compounds are experimentally found to scale linearly with solute size measures such accessible surface area.[14] We also found from MD simulations that the average van der Waals (Lennard-Jones) interaction energies scaled approximately linearly with solute size both in polar and non-polar solvents.[12] Combining these two observations would thus suggest that it might be possible to use average ligand van der Waals energies for estimating non-polar binding contributions, simply because they are correlated with the same variables as "hydrophobic free energies'. The above considerations thus lead us to explore an approximate equation for the binding free energy of the following general type

$$\Delta G_{bind} = \alpha\Delta\langle V_{l-s}^{vdw}\rangle + \beta\Delta\langle V_{l-s}^{el}\rangle + \gamma \tag{3}$$

where $\langle\ \rangle$ denotes MD or MC averages of the non-bonded van der Waals (*vdw*) and electrostatic (*el*) interactions between the ligand and its surrounding environment (*l-s*), i.e. either the solvated receptor binding site (bound state) or just solvent (free state). The Δ in Equation 3 denotes the difference between such averages in the bound and free states. The parameters of this equation are the weight coefficients α and β for the non-polar and polar binding energy contributions, respectively, and possibly an additional constant γ. In all calculations discussed herein we have used the Gromos87 force field,[15] while other groups have employed different force fields together with the LIE approach (see below).

In our study of endothiapepsin inhibitors the value $\beta=½$ predicted by the linear response assumption (see Ref. 12 for details) was used and the parameter α was calibrated to an optimal value of 0.16. Furthermore, the constant term γ was set to zero mainly in order not to overparametrize the model in view of the small initial "training set" (however, inclusion of a $\gamma\neq0$ did actually not either significantly improve the fit). It was, however, pointed out in Ref. 12 that it should be possible to use Equation 3 for solvation energies, i.e., without the Δ's, in which case the extra term $\gamma\neq0$ would be needed to properly account for the positive free energy of creating cavities in the solvent. This approach has also been taken by Jorgensen and coworkers for calculating hydration free energies and partition coefficients between water and chloroform.[16, 17] It can be noted again here that the ability of Equation 3, without the Δ's, to describe solvation energies requires the exclusion of the V_{l-l}^{el} terms from the average.

Our initial parametrization of the LIE equation was subsequently used for HIV protease and trypsin inhibitors as well as in a study of sugar binding to a bacterial receptor protein.[10, 18-20] While it was at first suspected that a specific calibration of Equation 3 would have to be carried out for each new system,[12] it turned out that the original endothiapepsin model was reasonably predictive also for the other systems mentioned above. Regardless of whether one really can find a "universal" parametrization of the LIE equation which, as we shall discuss below, might not be the case, there are some novel qualities of this approach that are quite interesting.

First, and perhaps the most important result, is that the binding free energy could be estimated solely from the intermolecular interactions of the ligand in the bound and free states. If we think of the full energetics involved in the binding process how are, e.g., intramolecular relaxation of the ligand and the receptor, desolvation of the receptor and entropic contributions taken into account when only $V_{l\text{-}s}$ terms are considered? Some of these problems can be better understood by considering a simple system such as the binding of an ion to a crown ether. The electrostatic linear response term in Equation 3 with β=½ yields a binding free energy of about −2 kcal/mol for the K^+-18-crown-6 complex in agreement with experiment[21] and the relationship $\Delta G_{sol} = \frac{1}{2}\langle V_{l-s}^{el}\rangle$ also holds well for ion solvation energies.[13] Clearly, there are large relaxation and entropy effects involved in K^+ binding to 18-crown-6 as the host wraps itself around the ion from a more or less "unfolded" state during the binding process. This will lead to a significant conformational entropy decrease of the host and a concomittant entropy increase in the solvent upon binding. What the linear response approximation asserts in this case is that the free energy contributions associated with these effects reflect a simple linear response to the electrostatic forces exerted by the ion, so that the necessary information about the binding and solvation energies is contained in V_{l-s}^{el}. As pointed out earlier,[13, 21] when extending this argument to ligands with intramolecular degrees of freedom Equation 3 also assumes that intramolecular ligand strain/relaxation, just as that of the receptor in the crown ether case, will exhibit a linear response with respect to V_{l-s}^{el}. This appears as a natural assumption, since there is no a priori reason for why certain conformational degrees of freedom (those of the ligand) should behave differently from others (those of the receptor and solvent). Our definition of what is the ligand and what is the receptor is also in this sense arbitrary as one could in principle switch the two labels, which illustrates the above point.

The second somewhat surprising result is that the quantity V_{l-s}^{vdw} appears very useful for obtaining good estimates of the binding free energy. That the V_{l-s}^{el} terms are of importance is not surprising in view of their relation to electrostatic solvation energies within the linear response approximation. However, as emphasized previously[12, 21] the non-polar term of Equation 3 should *not* be interpreted, in analogy with the electrostatic one, as an assumption of linear response also towards forces associated with the Lennard-Jones potential (as done, e.g., in Ref. 22). The idea behind it, again, is simply that V_{l-s}^{vdw} is correlated with ligand size as are non-polar solvation free energies in different environments. Furthermore, for a given ligand or solute V_{l-s}^{vdw} measures the number of Lennard-Jones interaction centers surrounding it, since V_{l-s}^{vdw} is attractive for all but very short distances and van der Waals well-depths are very similar in most force fields. Hence, V_{l-s}^{vdw} provides a measure of the heavy-atom number density in the surrounding and this is the basic reason for why it always tends to be more negative when the ligand is (partly) surrounded by protein than when it is free in water ($\rho_{wat} = 0.033Å^{-3}$ and $\rho_{prot} \approx 0.058$ $Å^{-3}$).[21] Besides its correlation with non-polar or hydrophobic binding contributions, V_{l-s}^{vdw} will of course to some extent also reflect the local packing or steric fit of the ligand in a given site.

It should be clear from the above discussion that the LIE method is thus fundamentally different from other typical molecular mechanics based scoring procedures that are designed to evaluate all separate contributions to the binding enthalpy or free energy explicitly. That is, such methods, e.g. Refs. 23-25, express the binding energy as something like

$$\Delta E_{bind} = E_{inter}^{lig-prot} + \Delta E_{intra}^{lig} + \Delta E_{intra}^{prot} + \Delta E_{sol}^{lig} + \Delta E_{sol}^{prot} \qquad (4)$$

plus possibly extra entropy related terms etc. One problem with this type of expression is that some of the energy terms (e.g., protein intramolecular energies) converge very slowly during MD or MC sampling and formulas of the above type are usually applied only to single (minimized) conformations of the receptor-ligand complex. The question of statistical significance of such minimized conformations therefore appears to be the weak point of that type of approach, although the more detailed bookkeeping of energy terms is potentially quite informative.

2.1 Improving the Original Model

In order to actually examine the validity of the electrostatic linear response approximation for different types of solutes and solvents, Åqvist and Hansson[13] investigated the relationships between electrostatic solvation free energies and the $\langle V_{l-s}^{el} \rangle$ terms for a number of relevant model systems using FEP calculations. It was then found that deviations from linear resonse do occur, particularly for neutral dipolar solutes and more so for those containing hydroxyl groups when the solvent is water. This is reflected by the coefficient $\beta < ½$. This behavior was found to derive from interactions with the hydrogen-bonding network of the solvent, and protic solvents thus displayed larger deviations from linear response than non-protic ones for a given solute. These results suggest that using an electrostatic scaling coefficient of $\beta=\frac{1}{2}$ might not always be the optimal choice.

In a first attempt to refine the original LIE model we considered a more general form of Equation 3 that also relaxes the constraint of equal scaling factors for the bound and free states.[26] This would, e.g., allow possible differences in electrostatic response properties of protein and solvent to be better modeled. The additional constant free energy term γ was also explored, which could arise as a non-zero difference $\gamma_{prot} - \gamma_{wat}$ between constant terms in corresponding linear expressions for the non-polar solvation energies.[12] Such a constant term in the solvation energy has been discussed by Ben-Naim and coworkers.[27, 28] Several empirical scoring functions for estimating binding free energies from structural data also include a constant term, sometimes proposed to represent translational entropy contributions to the free energy of binding. Jorgensen and coworkers[16, 17] have instead used an accessible surface area dependent term in Equation 1 which is similar to the addition of a constant since $\langle V_{l-s}^{vdw} \rangle$ depends approximately linearly on surface area.[12, 26]

The more general LIE equation examined in Ref. 26 was thus written as

$$\Delta G_{bind} = \alpha_{prot} \langle V_{l-s}^{vdw} \rangle_{bound} - \alpha_{wat} \langle V_{l-s}^{vdw} \rangle_{free} + \beta_{prot} \langle V_{l-s}^{el} \rangle_{bound} - \beta_{wat} \langle V_{l-s}^{el} \rangle_{free} + \gamma \tag{5}$$

where the α and β parameters are van der Waals and electrostatic scaling factors, respectively, that now could have different values in water and protein environments, while γ is a constant term as described above. The parameters of this equation were optimized using calculated and observed data for the 18 receptor-ligand complexes studied in Refs. 10, 12, 18-20 under various combinations of the constraints $\alpha_{prot} = \alpha_{wat}$, $\beta_{prot} = \beta_{wat}$ and $\gamma=0$ (such constraints were used in order to keep a reasonable ratio between the number of parameters and observations).

The results of these optimizations are summarized in Ref. 26 and definitely show that the original equation with the linear response value $\beta=\frac{1}{2}$ needs to be reconsidered in order to achieve more accurate predictions. This is also in agreement with the results from FEP investigations of the approximation.[13] The main conclusion from the work in Ref. 13 is that deviations from electrostatic linear response may be particularly important for uncharged ligands with certain dipolar groups, suggesting that a $\beta<\frac{1}{2}$ should then be used. The equation with $\alpha_{prot} = \alpha_{wat}$, $\beta_{prot} = \beta_{wat}$ and $\gamma=0$ (Equation 3 with a free β) was accordingly found to give a statistically significant improvement over the original model (that had an r.m.s. error of

1.57 kcal/mol for the 18 complexes, which was reduced slightly to 1.45 kcal/mol when all of them were included in the calibration set). The resulting values of α and β are 0.186 and 0.320, with an r.m.s. deviation of 0.97 kcal/mol. Since most of the 18 ligands were uncharged, the value of β obtained for this model is in accordance with the results of Ref. 13, where most electroneutral compounds were shown to be associated with β-factors of around 0.3-0.4.

The possibility of assigning different α and β coefficients for the bound and free states, as in Equation 5, was also investigated. It was then found that this did not offer much statistical improvement of the model and that the coefficients converged to very similar values in the two states. The addition of a nonzero γ in the binding free energy equations was also examined but the results showed no significant improvement of the model for the 18 complexes, in spite of the addition of another parameter.[26] The fact that the α and β parameters converge to approximately the same values in the bound and free states (this was the case even for free optimization of all five parameters in Equation 5) is quite remarkable. This finding clearly supports the use of the basic LIE equation (Equation 3).

The results indicate that the introduction of additional model parameters yields relatively little improvement, after adjustment of β away from ½. This suggests that it is important to account for systematic deviations from electrostatic linear response. The results from Ref. 13 can be directly incorporated into Equation 3 by assigning specific values of β to different classes of solutes (ligands). This led to a model, where the chemical composition dependent deviations from linear response are taken into account. A simple implementation of such a scheme divides compounds into four classes: charged, dipolar with no hydroxyl groups, dipolar with one hydroxyl group and dipolar with two or more hydroxyl groups, and assigns a different β parameter value to each class. The values of β (0.50, 0.43, 0.37 and 0.33, respectively) were taken directly from simulations of typical compounds of the different classes (cf. Table 1 of Ref. 13). Examples of such typical compounds are sodium ion, acetone, ethanol and ethylene glycol, respectively.[13] It should be emphasized here that this model effectively has only one free parameter, namely α. The same value of β is thus used for the bound and free states but depends in a systematic way on the nature of the ligand.

The results obtained with the above FEP-derived model[26] were found to yield an excellent agreement with experimental values for the 18 complexes in the calibration set (Figure 1).

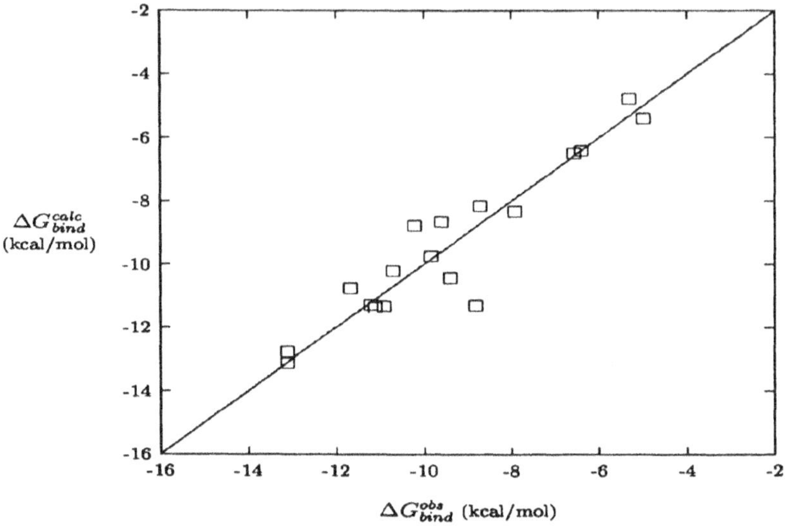

Figure 1. Calculated vs. observed free energies of binding for the 18 receptor-ligand complexes used to derived the LIE model in Ref. 26.

We obtain an r.m.s. deviation of 0.84 kcal/mol with an optimal α of 0.181. One can also note the similarity between the α value of this model and that of the two-parameter model with a free α and β. This suggests that the model is robust in the sense that the actual polar and non-polar free energy contributions are more or less invariant, as long as deviations from linear response are taken into account in a proper way. The FEP-derived model could be considered preferable to the two-parameter model since it contains only one free parameter, *viz.* α. The results of adding a constant γ to the new model was also investigated. Remarkably, the optimal value for such a γ was found to be –0.02 kcal/mol, i.e. virtually zero.

To summarize the attempts to refine the original LIE model, we found that an optimal equation for the binding free energy could be obtained with only one free parameter (α) and with the electrostatic coefficients (β) derived from FEP simulations of some representative compounds in water. For the 18 compound training set that we used this model yielded a mean unsigned error of only 0.58 kcal/mol which seemed very promising.

The above model was used to study the binding of methotrexate, and analogues of it, to wild-type and mutant forms of human dihydrofolate reductase (DHFR).[29] This turned out to be a particularly difficult test because of the three ionized groups of methotrexate. The overall electrostatic interactions of this inhibitor amount to around –500 kcal/mol and MD trajectories of length more than a ns were required in order to get average energies with reasonable error bars. Methotrexate is also very flexible and a wide range of conformations were found to be accessible to the free molecule in solution. Nevertheless, the calculated binding energetics was in fair agreement with experiments. The effects on methotrexate binding of mutations of the active site residue Leu22 to Phe and to Tyr were also reasonably well reproduced by the simulations.[29]

The DHFR calculations are summarized in Figure 2 together with a few other tests on arabinose binding protein, lysine binding protein and fatty acid binding protein (Ref. 30 and Marelius et al., unpublished). One the basis of these cases, one would perhaps be tempted to suggest that the revised LIE model of Ref. 26 now seems to describe the binding energetics very nicely for a reasonably large range of ligand-receptor systems (Figures 1 and 2). However, all of the receptor proteins discussed so far (endothiapepsin, HIV protease, trypsin, DHFR, sugar, lysine and fatty acid binding proteins and even 18-crown-6) have substantially polar binding sites. It turns out that for an entirely hydrophobic binding site, such as that of retinol binding protein (RBP), the affinities of a set of hydrophobic retinoid ligands is completely wrongly predicted by this LIE model. Figure 2 also shows the results of such calculations on RBP, where it can be seen that the binding free energies are offset (underestimated in absolute terms) by about +7 kcal/mol.

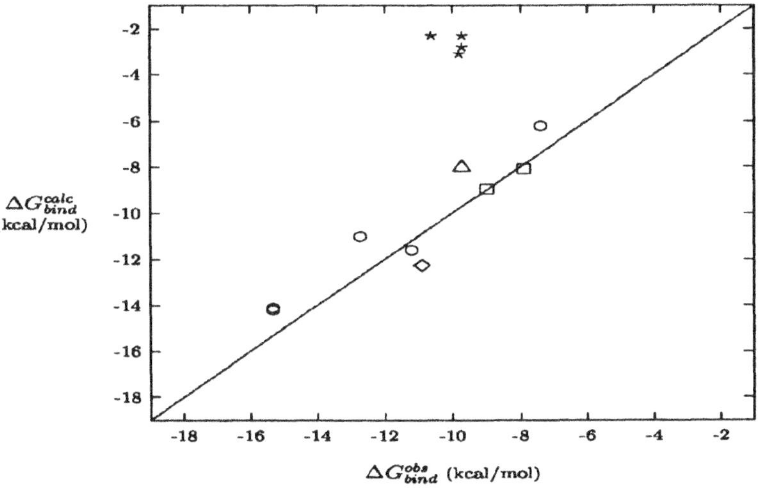

Figure 2. Calculated vs. observed free energies of binding for DHFR complexes[29] (circles), the arabinose complex with arabinose binding protein[50] (triangle), the lysine complex with lysine binding protein[51] (diamond), the complexes of stearate and elaidate with muscular fatty acid binding protein[52] (squares) and the complexes of retinol, retinoic acid, n-ethyl-retinamide and fenretinide with bovine plasma RBP[53] (stars).

In these cases, the electrostatic contribution to binding is essentially zero and the discrepancy therefore illustrates that the hydrophobic binding effect is not reproduced for RBP. From the structural viewpoint, it is of interest to note that both RBP and the muscular fatty acid binding protein belong to a distinct family of protein structures that share the same overall topology as well as an internal "barrel" designed to bind fatty molecules.[31] However, in the case of fatty acid binding proteins a few polar groups have been inserted inside the barrel, in order for it to be able to bind the carboxylate moiety of the acids. Apparently, this feature is enough to give these proteins very different properties from those of RBP, despite their structural similarities.

The examples discussed here show that the new LIE parametrization of Ref. 26, while reliable for a number of systems, could not be the final word in the development of this type of approximate binding free energy calculations. As we will see below there may be more examples of ligand-receptor systems that don't fit the simple picture of Figure 1.

3. OTHER LINEAR RESPONSE AND LIE MODELS

The first attempt to use the electrostatic linear response approach for binding calculations was reported by Lee et al.,[32] who studied the binding of two antigens to an antibody by several different methods. That work was interesting in that it pointed out the possibility of using the approximation for estimating binding free energies, but did not provide evidence that quantitative results could be obtained. That is, extremely short (5 ps) MD trajectories were used to collect the relevant average electrostatic energies together with a rather crude estimate of the non-polar binding contribution (subject to a several kcal/mol errorbar).[32] The resulting binding free energies were also wrong by 7-9 kcal/mol, which really suggested that the method did not work. Another interesting application was reported in Ref. 33, which addressed calculations of intrinsic pK_as in lysozyme, but did not either reach quantitative predictions although the qualitative trends seemed reasonable.

Warshel and coworkers have recently examined the LIE method and different versions of what they call the LRA (linear response approximation) method for the binding of a set of cyclic urea compounds to HIV protease.[34] The key features of their LRA scheme is that both averages of Equation 2 are evaluated, thus requiring two extra simulations of the 'non-polar' states (see above), that the ligand intramolecular electrostatic terms are included in the averages, and that the non-polar contribution is calculated with the PDLD method. Results of similar quality were reported with the different methods.[34] However, it should be noted that the value ½ of the electrostatic coefficient was used in Ref. 34, which, as discussed above, has been shown to not be accurate for the cyclic urea type of compounds. There also appears to be a misunderstanding regarding the thermodynamic cycle used in the LIE method which, contrary to what is said in Ref. 34, has not changed[12, 26, 29] (note that Figure 1 of Ref. 12 does not depict the binding cycle, but just illustrates how the linear reponse approximation can be used to estimate the electrostatic part of the solvation energy in a given environment and it is equivalent to Equation 1 of Ref. 13).

Several other research groups have also investigated the performance and scope of simplified free energy calculation methods of the LIE type. Jorgensen and coworkers started by demonstrating that hydration free energies and partition coefficients between water and a non-polar solvent could be accurately reproduced by Equation 3 with the coefficient γ as a scaling factor for the accessible surface areas.[16, 17] Binding calculations have also been reported by Jorgensen's group for thrombin[35] and FKBP12 inhibitor complexes[36] and in both cases were mean errors of similar magnitude as in Figure 1 obtained, after parametrization of Equation 3. Using the OPLS force field,[37] the thrombin calculations gave an r.m.s. error for seven compounds of 1.02 kcal/mol with α=0.131, β=0.131 and γ=0.014 kcal/(mol Å^2) while the best fits for the twelve FKBP12 complexes was obtained with either α=0.348, β=0.176 and γ=0.008 kcal/(mol Å^2) or α=0.328, β=0.180 and a constant γ of -4.21 kcal/mol (r.m.s. errors of 0.58 and 0.59 kcal/mol).[35, 36]

Paulsen and Ornstein[38] applied the methodology to a series of cytochrome P450-camphor analogue complexes using the CVFF force field[39] and they obtained excellent agreement with experimental binding data using α=1.043, β=½ and γ=0. Another example is provided by Gorse and Gready who used the Amber/OPLS united atom force field in calculations of the binding affinities of charged N5-deazapterins to DHFR.[40] These authors obtain a good fit to experimental data with α=-0.32, β=½ and γ=0, where the negative value of α might appear somewhat unexpected (in fact, a negative α also gave the best fit for thrombin in Ref. 35). In fact, there does not seem to exist any reasonable physical interpretation of a negative α and, in our opinion, such a result most likely reflects some type of problem with the electrostatic interactions (see below). That is, if the polar binding contribution is systematically too negative then this can be compensated in the parametrization by a non-polar scaling coefficient with negative sign.

In a recent study of the avidin-biotin system Wang et al.[41] have used the LIE approach as well as a version of it that includes an extra 'entropy term'. That is, the 'interaction free energy' in the bound and free states was expressed as $\Delta G_{int} = -RT\ln\langle\exp(-E_{int}/RT)\rangle = \langle E_{\rm int}\rangle - T\Delta S_{\rm int}$, where E_{int} denotes the intermolecular interaction energy of the ligand in the given environment (receptor site or aqueous solution). The entropy term was then evaluated as $\Delta S = R\ln \langle\exp(-\Delta E_{int} / RT)\rangle$ which involves the fluctuations of the interaction energy around its average. These formulae actually correspond to the regular FEP equation, i.e. a one-step perturbation calculation. However, in order for the signs to be correct the ensemble averages would be those measured for the non-interacting system, while the sampling was carried out for the interacting system in Ref. 41. One can also note that the calculated entropies associated with the intermolecular interactions are all positive (Table 2 of Ref. 41) which seems counter-intuitive. Moreover, the electrostatic energy and entropy terms in Ref. 41 were scaled by the linear response factor ½, which is inconsistent with the inclusion of an explicit entropy term, since that factor already encompasses entropy effects. An unbiased optimization of the α parameter for both the LIE and the 'interaction free energy' model of Wang et al., utilizing the data in Ref. 41, shows that addition of the entropy term does actually not offer any improvement.[42] However, regardless of which model is used it seems clear that a rather high value of α is required to fit the experimental binding results. Kollman and coworkers have also recently examined what the best value of the LIE α parameter is for seven different proteins using the Amber95 force field.[43] It was then found that while trypsin and HIV protease gave values similar to those reported by us, thrombin, avidin and cytochrome P450cam complexes required higher α's. A method for estimating α on the basis of desolvation surface areas was also presented in Ref. 43 and seems rather promising.

Another interesting type of application of the LIE approach was recently reported by Xu et al.[44] who investigated the catalytic properties of an antibody, where the binding energies of reactants and transition state were evaluated. Adalsteinsson and Bruice[45] have used the LIE method for scanning rapamycin analogues that would bind to the immunophilin FKBP12. Interestingly, that work employed the parametrization of Equation 3 with different β values for the bound and free states, which was found in Ref. 26 yield results of similar quality to that of the FEP-derived model.

4. RECENT CALCULATIONS ON HUMAN THROMBIN INHIBITORS

We have examined the performance of the LIE method for the series of eight thrombin inhibitors shown in Figure 3 (T1–T4 and BZA) using the Gromos87 force field.[46] Table 1 shows the average ligand-surrounding interaction energies from MD simulations of these eight thrombin inhibitors in complex with the enzyme and free in solution, as well as the small electrostatic corrections associated with turning off net charges on distant ionized groups.[19] When applying the earlier LIE calibration[26] with parameter values α=0.181, β=0.5 (all ligands are charged) and γ=0 in Equation 3 one finds that this model does not correctly predict the observed absolute binding free energies, although the relative affinities are well reproduced.[46] In fact, the predicted binding energies turn out to be systematically offset by around +3 kcal/mol. This observation is confirmed by least-squares optimization of the constant γ while leaving α and β fixed at their earlier determined values. Such a calibration yields γ=-2.916 kcal/mol and a mean unsigned error of only 0.68 kcal/mol for the eight complexes. A plot of calculated versus observed free energies of binding using this optimized model is shown in Figure 4, where it is clear that the model rather accurately predicts the experimental data (correlation coefficient=0.88). The above thrombin-specific parameterization of one constant in the LIE model for estimating binding affinities not only yields a very reasonable mean error but also appears robust. A free optimization of all three parameters in Equation 3 gives only a negligible improvement (mean unsigned error = 0.64 kcal/mol), in spite of the addition of two extra parameters. Although such a model is not statistically sound in view of its low ratio of observations to parameters, it is significant that the values of α and β obtained in this case are very close to those determined by us earlier.[26] Hence, one obtains α=0.18, β=0.44 and γ=-3.00 with three free parameters and it is clear that this (statistically unsound) model does not offer any improvement over the one in Ref. 26,

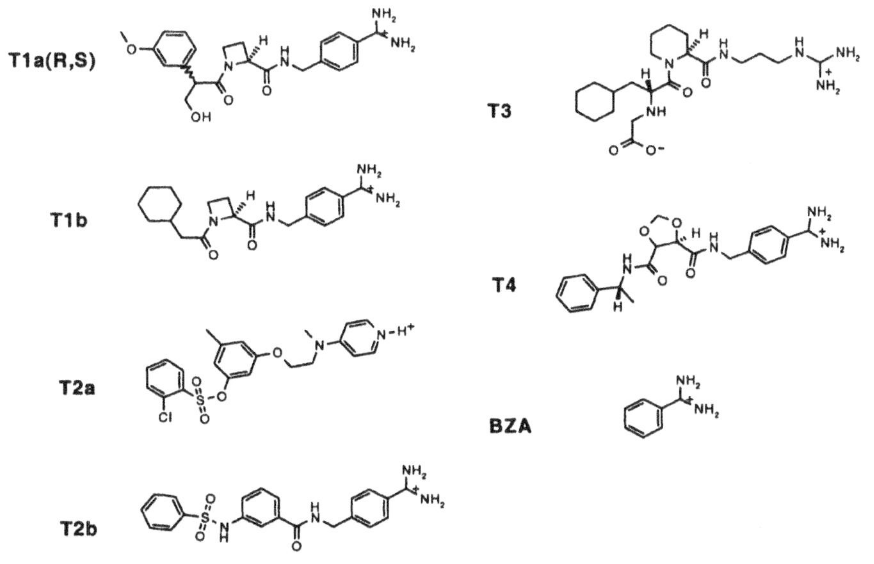

Figure 3. Chemical structures of the thrombin inhibitors for which calculations were carried out in Ref. 46 (BZA is benzamidine).

but rather confirms its predictivity. Reoptimization of only α, with β=½ and γ=0, as suggested in Ref. 43 yields a worse model with α=0.28 and a mean unsigned error of 0.95 kcal/mol.

It is noteworthy that the MD simulations, even without reparametrization of the LIE equation, gives a good ranking of the rather diverse set of thrombin inhibitors. The calculations thus reproduce, e.g., the large affinity difference between the *S*-form of T1a, which is a potent inhibitor in the nM

Table 1. Average ligand-surrounding interaction energies, electrostatic corrections and observed binding free energies (kcal/mol) for the eight thrombin inhibitors (from Ref. 46).[a, b]

Ligand	$\langle V^{vdW}\rangle_{bound}$	$\langle V^{vdW}\rangle_{free}$	$\langle V^{el}\rangle_{bound}$	$\langle V^{el}\rangle_{free}$	$\Delta G_{el.\ corr}$	ΔG_{obs}
T1a (*R*)	-63.67	-32.86	-153.96	-151.27	0.28	-9.9[c]
T1a (*S*)	-62.88	-35.00	-162.24	-154.34	0.28	-11.3
T1b	-62.31	-34.87	-133.65	-133.70	0.33	-9.3
T2a	-70.58	-38.72	-111.02	-106.18	0.38	-10.7
T2b	-68.71	-33.54	-142.68	-143.10	0.35	-7.4
T3	-61.96	-32.92	-229.46	-222.29	0.02	-10.8
T4	-65.55	-34.84	-142.37	-146.20	0.22	-7.0
Bza	-21.05	-8.30	-122.63	-116.93	0.04	-8.1

[a]The typical convergence errors for the columns are ±0.3, ±0.5, ±0.8 and ±1.0 kcal/mol, respectively, yielding errors in ΔG_{calc} of about ±1.0 kcal/mol (the errors for compound T3 are about twice as large). [b]$\Delta G_{el.corr}$ is the correction for the ligand's interactions with distant neglected charged residues using Coulomb's law with ε=80. [c]Estimate based on comparison with T1a(*S*) in a thrombin clotting time assay.

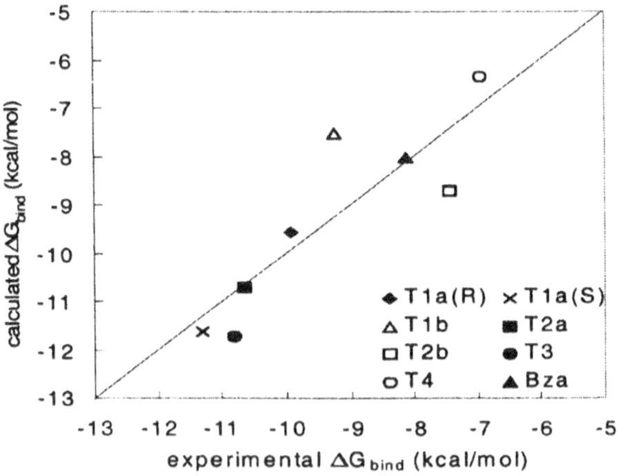

Figure 4. Calculated vs. observed free energies of binding for the eight thrombin complexes of Ref. 46.

region, and the least potent compound T4 that binds with only μM affinity. The recalibrated LIE equation gives binding free energies of -11.7 and -6.4 kcal/mol for these ligands, respectively, while the corresponding experimental results are -11.3 and -7.0 kcal/mol. The main reason for this large affinity difference can be traced to the different P2 and P3 fragments of the two ligands, while the benzamidine moiety (P1) unsurprisingly binds in an essentially identical pose. Figure 5 shows that the azetidine (T1a) and 1,3-dioxolane (T4) moieties reside in the same hydrophobic cavity (S2) in thrombin lined by His57, Tyr60A, Trp60D and Leu99 and it is likely that the larger desolvation effect associated with the more polar dioxolane group is partly responsible for its lower binding strength.

Furthermore, the CH_3 group in the phenylmethoxy fragment of T1a (*S*) is perfectly positioned (both in the MD and crystal structures) in another hydrophobic cavity (S3) created by Leu99, Ile174 and Trp215, while T4 lacks the methoxy substituent. However, the methyl group of T4 appears to play a similar role, but does not quite reach the bottom of the S3 cavity. The relatively flexible hydroxymethyl group of T1a(*S*) apparently also contributes to the affinity by several possible H-bond interactions, both to the protein and to water molecules

It is also encouraging that the calculations reproduce the stereospecificity of the two enantiomers of T1a. Here the predicted binding free energy difference is 2.1 kcal/mol in favor of the *S*-enantiomer while the experimental difference is roughly 1.4 kcal/mol. The structures of these

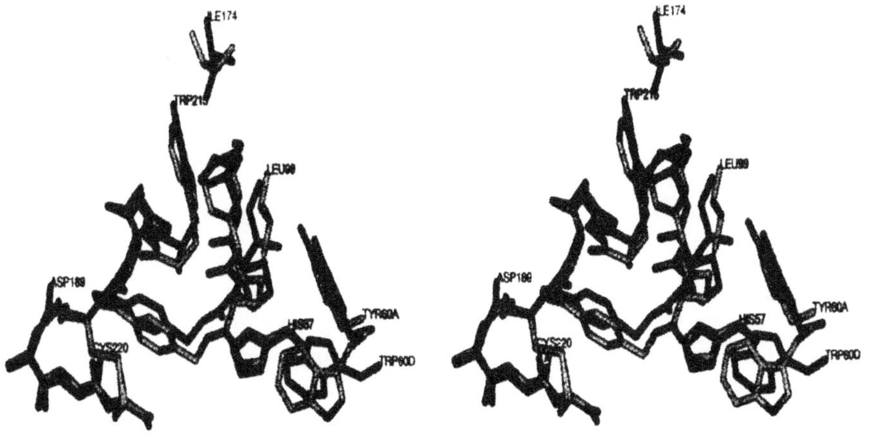

Figure 5. Stereo view of two MD snapshots showing the thrombin complexes with T1a(S) (light) and T4 (dark). The azetidine and dioxolane groups reside in the cavity to the left of Tyr60A.[46]

complexes show that the two chain directions extending from the chiral center superimpose well while the main difference is found in the position of the hydroxymethyl moiety. The difference in affinity of the enantiomers does not appear due to different interactions with the enzyme which are essentially the same, but rather to increased intramolecular strain in the *R*-enantiomer. That is, the interactions of the rest of the inhibitor seem to dictate a conformation where the hydroxymethyl group is nearly in the same plane as the adjacent phenyl ring with accompanying strain as a result.

The results for thrombin show that our previous parametrization of the LIE coefficients holds rather well in this case, provided that a constant term of -2.9 kcal/mol is added. At present it is not clear to us why thrombin would require such a constant term while, e.g., trypsin does not, but this issue is currently under investigation (see also Ref. 47 for a discussion of thrombin versus trypsin). Furthermore, one should note that with our computational procedures and the Gromos87 force field the results for thrombin inhibitors differ from those of Ref. 35 as well as Ref. 43. That is to say, three independent studies involving thrombin inhibitors have arrived at significantly different parametrizations of the LIE equation, that in all cases reproduce the experimental data well. It therefore seems clear that the differences in the computational procedures have a definite effect on the parameters of the binding energy approximation.

5. SOME TECHNICAL ASPECTS

One of the most critical technical issues regarding LIE type of calculations seems to be the treatment of electrostatic interactions, at least for charged ligands. These problems were discussed in Ref. 19 using the trypsin-benzamidine complex as an illustration. The starting point was then that one should expect, or require, that the electrostatic free energies calculated with the LIE method have some physical meaning, even in absolute terms. This is basically equivalent to saying that $\beta\left\langle V^{el}_{l-l}\right\rangle$ for the bound and free states should be a good approximation to the corresponding (intermolecular) "charging free energies" that would be obtained if sufficiently accurate FEP calculations could be carried out. Of course, one can treat the whole LIE approach as a QSAR-like model and not ascribe any deeper meaning to the binding contribution from the electrostatic term, and this might work fine, but it seems worth to try to elicit the actual energetic relationships involved.

There are several separate problems involved in trying to obtain accurate electrostatic free energies for charged groups from simulations of the type discussed herein. The most important of these can tentatively be summarized as follows. (*i*) In order for $\Delta\langle V_{l-s}^{el}\rangle$ for a charged compound to be a meaningful quantity, the net charge of the medium surrounding the ligand, within its interaction range (e.g., a possible cutoff), in the bound and free states must be equal.[10, 19] This is because the contribution to the electrostatic solvation energy from the medium outside the ligand "interaction sphere" (Born terms) otherwise will be unequal in the two simulations. So, if the receptor has a non-zero net charge within this interaction sphere then this charge must either be neutralized or ions must be added in the free (solvent) simulation.[10, 19] (*ii*) The use of cutoffs for, not only the ligand interactions, but also those of the surroundings (protein-protein, protein-water and water-water) can have a significant effect on the energetics. Cutoffs lead to an artificial overpolarization of the surrounding of a charge[19, 48] and the resulting effects on binding energetics are usually predictable (anti-binding). (*iii*) If a truncated or finite size system is used, which is often the case when one wants to speed up the calculations, it is dangerous to have charged groups to close to the boundary of the system. This is simply because there will not be enough polar medium (water or protein dipoles) around such groups to screen the electric fields properly, resulting in exaggerated (vacuum-like) effects from such charges. (*iv*) For flexible ligands with several ionized groups the convergence problems associated with the electrostatic term can be severe, in particular in the free state where the flexibility really dictates the amount of conformational sampling needed. An example of this is provided by the DHFR simulations in Ref. 29 where convergence in the bound state is reasonable due to the restricted space available, but simulations of methotrexate in solution require very long trajectories.

It may be appropriate here to point out that the points above are not specific for LIE type of calculations but apply equally well to FEP simulations. It is our feeling that the differences between some of the various parametrizations of the LIE equation reported in the literature may in part have their origin in varying computational procedures, particularly with respect to points (*i-iii*) above.

While we have attempted to ascribe a physical meaning to the electrostatic β coefficient of the LIE method, rather than using it as a free parameter, some authors have chosen such an alternative.[35, 36, 49] From the QSAR or empirical scoring function perspective, any value of the weight coefficients in the scoring equation should be allowed and it may well be that such a model could turn out to be optimal in some cases. To take the argument a step further one may, of course, question whether just the electrostatic and van der Waals interaction energies are the 'best' quantities to extract from MD or MC simulations in order to estimate binding affinities. In the original formulation of the LIE method the terms in Equation 3 were arrived at by the (at least seemingly) logical reasoning outlined in the beginning of this chapter, but there might be other useful descriptors for binding energetics.

Wall et al. have recently examined the LIE method in the context of neuraminidase inhibitor binding.[49] Of particular interest in that study is that a statistical analysis of different possible descriptors for predicting the binding affinity was carried out. These included both inter- and intramolecular energies as well as changes in accessible surface areas of the ligand and receptor. It is noteworthy that the results showed that neither intramolecular energies nor surface areas were significant descriptors for the binding free energy. Instead the statistical analysis yielded precisely the intermolecular electrostatic and van der Waals terms as the most predictive components.

6. CONCLUSIONS

In this chapter we have tried to give an overview of ligand-receptor binding affinity calculations using the linear interaction energy approach. This method was developed as an alternative to more time-consuming free energy perturbation calculations, in particular for predicting affinities of sets of ligands that are too diverse for falling into the "small perturbation" category required by the FEP method. The LIE approach was originally based on the linear response assumption for electrostatics together with an empirical scaling of non-polar interaction energies intended to capture non-polar or hydrophobic binding contributions. Regardless of whether one ascribes significance to the non-empirical linear response considerations behind the electrostatic part of the calculated binding free energy, or whether one simply regards the parameters of the LIE equation (Equation 3) as freely optimizable, like in QSAR approaches, the method reveals some rather unexpected features. That is, it came as somewhat of a surprise that (*i*) binding free energies could be so reasonably predicted by just considering the *intermolecular* interactions of the ligand and (*ii*) that absolute affinities, and not only relative ones, could be reasonably well predicted from MD or MC simulations, something that is not really within the scope of FEP calculations.

The fact that only intermolecular energies are needed for the binding estimate has often been interpreted in such a way that intramolecular relaxation/strain, entropy, receptor desolvation etc. are neglected. We have tried to illustrate here that, at least formally, this is not the case. Using ion binding to crown ethers as an example one can easily see that these effects are in principle embedded in the linear response approximation. Whether the approximations involved in the LIE type of equation are accurate enough is another matter. However, from the various reports on the method published so far it appears that most systems, irrespective of force fields, simulation procedures etc., lend themselves to fairly accurate parametrizations by the LIE equation.

While binding in a variety of different receptor-ligand systems appears to be well modelled by our LIE parameters given in Ref. 26, there seem to be exceptions that are not due to different force fields or simulation setups. A notable case here is the more or less entirely hydrophobic binding of retinoids to RBP (cf. Figure 2), but also the reported calculations on P450cam[38] and avidin[41] seem to point in the same direction. To summarize, these results basically suggest that binding which is dominated by hydrophobic interactions requires a different and significant higher value of the non-polar scaling factor α, or the addition of a constant γ. It therefore seems worthwhile to try to understand whether such a scaling factor, or constant, behaves in a systematic way and could be predicted without using simulations on a training set. In this respect, the efforts of Kollman and coworkers[43] seem quite promising.

We have also argued here that in some cases the difference in simulation procedures used in LIE studies are too drastic to allow any conclusions regarding transferability of parameters to be drawn. This particularly pertains to treatments of electrostatic interactions for charged ligands, since they tend to be very large and sensitive to cutoff procedures and related issues. On the other hand, if one regards the LIE coefficients just as free parameters without significance for absolute solvation energies, the need to get all the electrostatics modeled correctly becomes less crucial. Whether this type of QSAR-like interpretation of the method or a more systematic scheme of parameters will turn out to be most useful remains to be seen. At any rate, it appears that this type methodology can be valuable in the ligand screening/design process, when dealing with a limited number of compounds, and give rather detailed information regarding the interactions involved in binding.

Acknowledgements

We wish to thank Ellen Kindlund, Kajsa Ljungberg and Johan Modig for carrying out some of the reported calculations. Support from the TFR, SSF and NFR is gratefully acknowledged.

7. REFERENCES

1. H. J. Böhm and M. Stahl, Rapid empirical scoring functions in virtual screening applications, Med. Chem. Res. **9**:445 (1999).
2. T. I. Oprea and G. R. Marshall, Receptor based prediction of binding affinities, Perspect. Drug Discov. Design **9**:35 (1998).

3. M. L. Lamb and W. L. Jorgensen, Computational approaches to molecular recognition, Curr. Opin. Struct. Biol. **1**:449 (1997).
4. H. J. Böhm, The development of a simple empirical scoring function to estimate the binding constant for a protein-ligand complex of known three-dimensional structure, J. Comput.-Aided Mol. Design **8**:243 (1994).
5. A .N. Jain, Scoring noncovalent protein-ligand interactions: A continuous differentiable function tuned to compute binding affinities, J. Comput.-Aided Mol. Design **10**:427 (1996).
6. M. D. Eldridge, C. W. Murray, T. R. Auton, G. V. Paolini, and R. P. Mee, Empirical scoring functions: I. The development of a fast empirical scoring function to estimate the binding affinity of ligands in receptor complexes, Comput.-Aided Mol. Design **11**:425 (1997).
7. I. Muegge and Y. C. Martin, A general and fast scoring function for protein-ligand interactions: A simplified potential approach, J. Med. Chem. **42**:791 (1999).
8. H. Gohlke, M. Hendlich, and G. Klebe, Knowledge-based scoring function to predict protein-ligand interactions, J. Mol. Biol. **295**:337 (2000).
9. P. A. Kollman, Free energy calculations: Applications to chemical and biochemical phenomena, Chem. Rev. **93**:2395 (1993).
10. T. Hansson and J. Åqvist, Estimation of binding free energies for HIV protease inhibitors by molecular dynamics stimulations, J. Prot. Eng. **8**:1137 (1995).
11. V. Helms and R. C. Wade, Hydration energy landscape of the active site cavity in cytochrome P450cam, J. Am. Chem. Soc. **120**: 2710 (1998).
12. J. Åqvist, C. Medina, and J. E. Samuelsson, A new method for predicting binding affinity in computer-aided drug design, Prot. Eng. **7**:385 (1994).
13. J. Åqvist and T. Hansson, On the validity of electrostatic linear response in polar solvents, J. Phys. Chem. **100**:9512 (1996).
14. W. Blokzijl and J. B. F. N. Engberts, Hydrophobic effects. Opinions and facts, Ang. Chem. Int. Ed. Engl. **32**: 1545 (1993).
15. W. F. van Gunsteren and H. J. C. Berendsen, Groningen Molecular Simulation (GROMOS) Library Manual, Biomos B.V., Groningen, The Netherlands (1987).
16. H. A. Carlson and W. L. Jorgensen, An extended linear response method for determining free energies of hydration, J. Phys. Chem. **99**: 10667 (1995).
17. N. A. McDonald, H. A. Carlson, and W. L. Jorgensen, Free energies of solvation in chloroform and water from a linear response approach, J. Phys. Org. Chem. **10**:563 (1997).
18. J. Åqvist and S. L. Mowbray, Sugar recognition by a glucose/galactose receptor. Evaluation of binding energetics from molecular dynamics simulations, J. Biol. Chem. **270**:9978 (1995).
19. J. Åqvist, Calculation of absolute binding free energies for charged ligands and effects of long-range electrostatic interactions, J. Comput. Chem. **17**:1587 (1996).
20. J. Hultén, et al., Cyclic HIV-1 protease inhibitors derived from mannitol: Synthesis, inhibitory potencies, and computational predictions of binding affinities, J. Med. Chem. **40**:885 (1997).
21. J. Marelius, T. Hansson, and J. Åqvist, Calculation of ligand binding free energies from the molecular dynamics stimulations, Int. J. Quant. Chem. **69**:77 (1998).
22. H. Schäfer, W. F. van Gunsteren, and A. E. Mark, Estimating relative free energies from a single ensemble: Hydration free energies, J. Comput. Chem. **20**:1604 (1999).
23. S. Vajda, Z. Weng, R. Rosenfeld, and C. DeLisi, Effect of conformational flexibility and solvation on receptor-ligand binding free energies, Biochemistry **33**:13977 (1994).
24. A. R. Ortiz, M. T. Pisabarro, F. Gago, and R. C. Wade, Prediction of drug binding affinities by comparative binding energy analysis, J. Med. Chem. **38**:2681 (1995).
25. V. N. Viswanadhan, M. R. Reddy, A. Wlodaver, M. D. Varney, and J. N. Weinstein, An approach to rapid estimation of relative binding affinities of enzyme inhibitors:

Application to peptidomimetic inhibitors of the human immunodeficiency virus type 1 protease, J. Med. Chem. **39**:705 (1996).
26. T. Hansson, J. Marelius, and J. Åqvist, Ligand binding affinity prediction by linear interaction energy methods, J. Comput.-Aided Mol. Design **12**:27 (1998).
27. A. Ben-Naim and Y. J. Marcus, Solvation thermodynamics of non-ionic solutes, Chem. Phys. **81**:2016 (1984).
28. A. Ben-Naim and R. M. Mazo, Size dependence of the solvation free energies of large solutes, J. Phys. Chem. **97**:10829 (1993).
29. J. Marelius, M. Graffner-Nordberg, T. Hansson, A. Hallberg, and J. Åqvist, Computation of affinity and selectivity: binding of 2,4-diaminopteridine and 2,4-diaminoquinazoline inhibitors to dihydrofolate reductases, J. Comput.-Aided Mol. Design **12**:119 (1998).
30. J. Marelius, K. Kolmodin, I. Feierberg, and J. Åqvist, Q: A molecular dynamics program for free energy calculations and empirical valence bond simulations in biomolecular systems, J. Mol. Graph. Model. **16**:213 (1998).
31. L. Banaszak, N. Winter, Z. Xu, D. A. Bernlohr, S. Cowan, and T. A. Jones, Lipid-binding proteins: A family of fatty acid and retinoid transport proteins, Adv. Prot. Chem. **45**:89 (1994).
32. F. S. Lee, Z. T. Chu, M. B. Bolger, and A. Warshel, Calculations of antibody-antigen interactions: microscopic and semi-microscopic evaluation of the free energies of binding of phosphorylcholine analogs to McPC603, Prot. Eng. **5**:215 (1992).
33. G. S. Del Buono, F. E. Figueirido, and R. M. Levy, Intrinsic pKas of ionizable residues in proteins: an explicit solvent calculation for lysozyme, Proteins **20**:85 (1994).
34. Y. Y. Sham, Z. T. Chu, and A. Warshel, Examining methods for calculations of binding free energies: LRA, LIE, PDLD-LRA, and PDLD/S-LRA calculations of ligands binding to an HIV protease, Proteins **39**:393 (2000).
35. D. K. Jones-Hertzog and W. L. Jorgensen, Binding affinities for sulfonamide inhibitors with human thrombin using Monte Carlo simulations with a linear response method, J. Med. Chem. **40**:1539 (1997).
36. M. L. Lamb, J. Tirado-Rives, and W. L. Jorgensen, Estimation of the binding affinities of FKBP12 inhibitors using a linear response method, Bioorg. Med. Chem. **7**:851 (1999).
37. W. L. Jorgensen, D. S. Maxwell, and J. Tirado-Rives, Development and testing of the OPLS all-atom force field on conformational energetics and properties of organic liquids, J. Am. Chem. Soc. **118**:11225 (1996).
38. M. D. Paulsen and R. L. Ornstein, Binding free energy calculations for P450cam-substrate complexes, Prot. Eng. **9**:567 (1996).
39. P. Dauber-Osguthorpe, V. A. Roberts, D. J. Osguthorpe, J. Wolff, M. Genest, and A. T. Hagler, Structure and energetics of ligand binding to proteins: Escherichia coli dihydrofolate reductase-trimethoprim, a drug-receptor system, Proteins **4**:31 (1988).
40. A. D. Gorse and J. E. Gready, Molecular dynamics simulations of the docking of substituted N5-deazapterins to dihydrofolate reductase, Prot. Eng. **10**:27 (1997).
41. J. Wang, R. Dixon, and P. A. Kollman, Ranking ligand binding affinities with avidin: A molecular dynamics-based interaction energy study, Proteins **34**:69 (1999).
42. J. Åqvist and J. Marelius, The linear interaction energy method for predicting ligand binding free energies, Comb. Chem. High T. Scr. in press (2001).
43. W. Wang, J. Wang, and P. A. Kollman, What determines the van der Waals coefficient β in the LIE methods to estimate binding free energies using molecular dynamics stimulations?, Proteins **34**:395 (1999).
44. J. Xu, Q. Deng, J. Chen, K. N. Houk, J. Bartek, D. Hilvert, and I. A. Wilson, Evolution of shape complementarity and catalytic efficiency from a primordial antibody template, Science **286**:2345 (1999).

45. H. Adalsteinsson and T. C. Bruice, Generation and evaluation of putative neuroregenerative drugs. Part 2: screening virtual libraries of novel polyketides which possess the binding domain of rapamycin, Bioorg. Med. Chem. **8**:617 (2000).
46. K. B. Ljungberg, J. Marelius, D. Musil, P. Svensson, B. Nordén, and J. Åqvist, Computational modelling of inhibitor binding to human thrombin, Eur. J. Pharm. Sci. **14**:441 (2001).
47. C. W. Murray, T. R. Auton, and M. D. Eldridge, Empirical scoring functions. II. The testing of an empirical scoring function for the prediction of ligand-receptor binding affinities and the use of Bayesian regression to improve the quality of the model J. Comput.-Aided Mol. Design **12**:503 (1998).
48. J. Åqvist, Comment on transferability of ion models, J. Phys. Chem. **98**:8253 (1994).
49. I. D. Wall, A. R. Leach, D. W. Salt, M. G. Ford, and J. W. Essex, Binding constants of neuraminidase inhibitors: An investigation of the linear interaction energy method, J. Med. Chem. **42**:5142 (1999).
50. F. A. Quicho and N. K. Vyas, Novel stereospecificity of the L-arabinose binding protein, Nature **310**:381 (1984).
51. B. H. Oh, J. Pandit, C. H. Kang, K. Nikaido, S. Gokcen, G. F. Ames, and S. H. Kim, Three-dimensional structures of the periplasmic lysine/arginine/ornithine-binding protein with and without a ligand, J. Biol. Chem. **268**:11348 (1993).
52. A. C. M. Young, G. Scapin, A. Kromminga, S. B. Patel, J. H. Veerkamp, and J. C. Sacchettini, Structure **2**:523 (1994).
53. D. S. Goodman and A. J. Raz, Extraction and recombination studies of the interaction of retinol-binding protein, Lipid Res. **13**:338 (1972).

Chapter 10

New Free Energy Based Methods for Ligand Binding from Detailed Structure-Function to Multiple-Ligand Screening

Shinichi Banba,[†] Zhuyan Guo,[‡] and Charles L. Brooks, III[*]
[†]*Mitsui Chemical, Inc. Sodegaura-City, Chiba 299-0265, Japan*
[‡]*Schering-Plough Research Institute, Kenilworth, NJ 07033*
[*]*Dept. of Molecular Biology, The Scripps Research Institute, La Jolla, CA 92037*

1. INTRODUCTION

As an integral component in the design of new therapeutics, computational approaches that rapidly evaluate the ranking, or relative binding free energy, of ligands play an important role in the search for lead compounds and the subsequent refinement of these leads. More specifically, the quantitative prediction of free energy changes of binding for an ensemble of slightly varying inhibitors is of considerable assistance in guiding synthetic strategies. For this purpose, methods associated with the prediction of free energy differences of multiple compounds have been developed in many laboratories.[1-15]

In this chapter we focus our review primarily on the free energy-based, multiple-ligand screening methods, which are known as λ-dynamics[1-5] or Chemical Monte Carlo/molecular dynamics (CMC/MD).[6, 7] In Section 2 we provide a brief background of the basic formulation of conventional free energy calculation methods with highlights on problems associated with this approach and potential solutions. We then provide an overview of the recently developed computational methods for ligand screening in Section 3. Practical details for the application of free energy-based, multiple-ligand screening methods follow in Section 4, including enhancements to these methods such as the use of umbrella sampling, iterative techniques employing the weighted histogram analysis method (WHAM), approaches to restrict the configurations of the unbound ligands, and the incorporation of continuum solvation models. Finally, applications of free energy-based, multiple-ligand screening methods are commented on.

2. CONVENTIONAL FREE ENERGY METHODS

2.1 Background

All thermodynamic properties can be obtained from knowledge of the free energy and its first derivative. Thus, one should focus on the free energy of the molecular system when aiming to quantitatively predict the interactions between putative inhibitors and a receptor. In this chapter we frame our discussion of computational approaches to calculating free energies in the canonical ensemble. In this ensemble the Helmholtz free energy, which we denote as A throughout this chapter, is the appropriate thermodynamic potential and is given by

$$A = -k_B T \ln \int \exp(-\beta V(X))dX \tag{1}$$

where $V(X)$ is the potential energy, k_B is the Boltzmann constant, T is the absolute temperature and X represents the conformational space of the ligand, receptor and "environment", e.g., solvent present in the calculation. The exponential dependence of the Boltzmann factor on energy makes the configurational integral notoriously slow to converge. To see this problem more clearly, we re-write Equation 1 as follows[16] equation

$$A = -k_B T \ln\langle \exp(-\beta V)\rangle \tag{2}$$

The angle bracket in this expression symbolizes the configurational integral of the canonical ensemble. In principle, Equation 2 provides a means of calculating (excess) free energy from a single conventional simulation. However, conventional simulations predominately sample the low-energy regions of conformational space, i.e., in accordance with the Boltzmann factor given in Equation 1, and never adequately sample higher energy states that contribute most significantly to the ensemble average of the free energy (as given by the "inverse-Boltzmann" factor in Equation 2). Therefore, the calculation of free energy using a conventional simulation leads to poorly converged, and consequently inaccurate, free energy estimates. Fortunately, it is the free energy difference that is generally of greatest interest, and this can be calculated using a coupling-parameter approach when the states are similar, i.e. the energy difference is small for all important configurations.

2.2 Basic Formulation of Conventional Free Energy Methods

Conventional free energy calculation methods,[17-19] such as the free energy perturbation (FEP) and thermodynamic integration (TI) approaches, can be utilized to evaluate the relative binding free energy between two ligands according to the following thermodynamic cycle

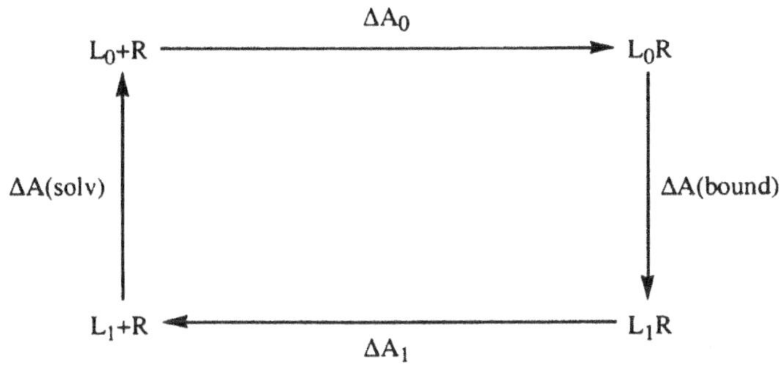

where L_0 and L_1 represent the free ligands in aqueous solution and L_0R and L_1R represent the corresponding ligands complexed with the protein receptor R. The relative binding free energy of the two ligands, $\Delta\Delta A_{(bind)}=\Delta A_1 - \Delta A_0=\Delta A_{(bound)} - \Delta A_{(solv)}$, is the difference between the relative free energy of the ligands in the complexed state and that of the free ligands. The free energy difference in each half of the thermodynamic cycle can be calculated using the potential energy of a hybrid system, written as a linear function of the two endpoint states and connected through a "chemical coordinate", the coupling parameter λ

$$\begin{aligned} V(x,\lambda) &= (1-\lambda)V_0(x)+\lambda V_1(x) \\ &= V_0(x)+\lambda\Delta V(x) \end{aligned} \tag{3}$$

The free energy difference between a state with the value of λ and the initial state ($\lambda=0$) is given by the FEP method[17]

$$\Delta A = A(\lambda)-A(0) = -k_BT\ln\langle\exp(-\beta\Delta V)\rangle_0. \tag{4}$$

The angle bracket denotes that the configurational integral is taken over the initial state. The conformational sampling indicated by Equation 4 is generated according to the Boltzmann probability associated with the initial state potential. As discussed in Section 2.1, convergence of conformational sampling has been a major issue in free energy calculations, and the relationship shown in Equation 4 doesn't always lead to converged free energy estimates. To ensure adequate sampling of the important conformations, FEP calculations are generally limited to free energy differences of less than 2 kcal/mol.[18-20] However, the free energy differences for many chemical and biological systems are larger than this. Therefore, a multi-step approach is generally adopted. By summing over the intermediate states along the λ coordinate, the total free energy change is determined by

$$\Delta A = \sum_{i=1}^{n} \Delta A(\lambda_{i-1} \rightarrow \lambda_i) \tag{5}$$

where the interval $\lambda=[0,1]$ has been divided into n small increments of $\Delta\lambda$.

An alternative approach to free energy calculations is the thermodynamic integration (TI) method,[18, 20] which considers the ensemble average of the first derivative of the hybrid potential with respect to λ at various values of λ

$$\begin{aligned}\Delta A &= \int_0^1 \left(\frac{\partial A}{\partial \lambda}\right) d\lambda = \int_0^1 \left\langle \frac{\partial V}{\partial \lambda} \right\rangle_\lambda d\lambda \\ &= \sum_{i=1}^{n} \left\langle \partial V / \partial \lambda \right\rangle_i \Delta\lambda_i .\end{aligned} \tag{6}$$

Although these methods have been successfully applied to assess the relative binding free energy in a number of protein-ligand systems,[18, 21-26] they are computationally expensive because of time-consuming sampling of nonphysical intermediate states. A typical calculation may take days to weeks to complete. This computational expense has hindered the broad application of conventional free energy based approaches to the drug design process.

2.3 Formulation of the Umbrella Sampling Technique

While FEP and TI are, in principle, umbrella sampling methods, the more conventional format in which we consider umbrella sampling is that used to obtain the free energy along a "reaction coordinate", ξ, typically a configurational coordinate. Here the "reactant" configuration and the "product" configuration are represented by $\xi = 0$ and $\xi = 1$, respectively. The Helmholtz free energy difference for a continuous coordinate, or the reversible work required to carry the system from the reactant configuration to the product configuration, is often referred to as the "potential of mean force", $W(\xi)$ and is derived from $\rho(\xi)$, the probability density of the system.[27]

$$W(\xi) = -k_B T \ln \rho(\xi). \tag{7}$$

As discussed earlier, inadequate sampling may occur if $W(\xi)$ differs by more than a few kcal/mol over the range of ξ. To overcome this problem, the umbrella sampling technique[28, 29] has frequently been used to enhance the sampling of conformational space along a reaction coordinate. In this approach, the original potential $V(x)$ is replaced by the modified potential $V(x)+U(\xi)$. The auxiliary potential, or umbrella potential, $U(\xi)$ is used to flatten out the energy barriers along the reaction coordinate ξ, or to restrict the sampling of the coordinate to a specific region of conformational space. In the former case, a more uniformly distributed density function $\rho^*(\xi)$ can be generated with a fixed amount of sampling because transitions between the reactant and product configurations are now more facile. In the latter case, the statistical sampling of important regions in the configurational space of the reaction coordinate can be better controlled. In both cases, the true probability density is recovered from the following equation

$$\rho(\xi) = \frac{\rho^*(\xi)\exp\{\beta U^*(\xi)\}}{\langle \exp\{\beta U^*(\xi)\}\rangle^*} \tag{8}$$

where the notation <...>* emphasizes that the ensemble average is being taken over conformations biased by the modified potential function.

In many applications, a single biasing potential is not sufficient to cover the whole range of ξ and simultaneously produce good sampling. Thus a set of restraining potentials, $U_i^*(\xi)$, are used to shift the local minima in the desired direction. In this "windowing" approach, the potential of mean force, $W_i(\xi)$, in each window takes the form

$$W_i(\xi) = -k_B T \ln \rho_i^*(\xi) - U_i^*(\xi) + C_i. \tag{9}$$

The difference constant C_i is

$$C_i = \beta^{-1} \ln\langle \exp\{\beta U_i^*(\xi)\}\rangle^*.$$

In order to achieve a uniformly good estimate of the potential of mean force, the difference constants from successive simulation windows has to be perfectly matched so as to make $W_i(\xi)$ agree in the overlapping regions.[18, 20, 30] Optimal data combining methods such as the weighted histogram analysis method (WHAM) can be used to optimize links between simulations and produce the best possible estimation of free energies.[28, 29, 31-35] It is clear that umbrella sampling is a powerful technique and should be of use in sampling "chemical space" as well as configurational space. We discuss an extension of umbrella sampling to "chemical coordinates" in Section 4.2.

3. OVERVIEW OF APPROXIMATE APPROACHES FOR MULTIPLE-LIGAND SCREENING

When relative binding free energies of multiple ligands are to be evaluated, FEP and TI require considerable amounts of time. This is because of the multi-step approach necessary to compute incremental free energy changes and the many pairwise comparisons that must be done. Alternatively, methods based on favorable interaction energies have been developed to rapidly approximate the free energy.[36-43] Although such approaches are relatively rapid, they are inherently incomplete since the entropy contribution to the free energy is (at least partially) ignored. Because of the potential importance of such entropic effects in chemical and biological systems, the development of new methodology for free energy calculations is an area of active research. Recently, statistical mechanical "ab initio" free energy-based computational methods have been developed to screen out the better binders from a group of candidate compounds.[1-7, 9-15] These methods will be briefly reviewed in Section 4. In the following we provide an overview of the interaction energy based approaches.

3.1 Linear Interaction Approximation

The linear interaction approximation (LIA) was introduced by Åqvist and co-workers[44-48] to calculate absolute binding free energies via MD simulations. A version of the LIA equation takes the following form

$$\Delta A_{(bind)} = \beta\langle \Delta E_{Coulomb}\rangle + \alpha\langle \Delta E_{L-J}\rangle + \gamma\langle \Delta SA\rangle. \tag{10}$$

In this expression, $\langle \Delta E \rangle$ is the energy difference between average contributions from ligand-solvent and ligand-protein interactions for the bound and unbound states. The scaling parameter for electrostatic interactions, β, is taken to be 0.5, from theories of ion solvation in which there is a linear response of the solvent to the electrostatic field of the ion. The scaling coefficient on the Lennard-Jones terms, α, varies from system to system. Therefore, a large number of ligands with known binding affinity have to be used to obtain the proper coefficients. Carlson and Jorgensen[49] also added a penalty for cavitation, which is linearly related to the change in solvent-accessible surface area (SA) upon binding. This term was added to obtain positive free energies of hydration for molecules such as hydrocarbons. LIA has been successfully applied to a number of protein-ligands systems.[44-47, 49-51] However, the values of β and α seem to depend on both the system and the force field.

3.2 Extrapolation from a Single Reference

An alternative approach, which is rooted in the FEP methodology described above, involves the extrapolation of free energy differences from a single reference. The method was introduced by Liu, Mark, and van Gunsteren for the estimation of free energies of related compounds.[14, 52] In implementing this approach, the incorporation of a soft-core potential leads to an expansion in the sampling of configurations for related ligands and consequently the range of correctly estimated free energy differences.[12] Mordasini and McCammon demonstrated the usefulness of the extrapolation method for similar sized molecules and the difficulties of obtaining reliable results for different sized molecules.[8]

Radmer and Kollman have introduced an approach they call PROFEC (Pictorial Representation of Free Energy Components),[15] as a tool for optimizing ligand affinity based on extrapolations from a single dynamics simulation. The PROFEC contour maps can be used to visualize how the free energy changes when additional particles are added to a residue of the protein or to the ligand. The contour map is generated by evaluating the insertion free energy of a test particle at various grid points near the residue of interest, using coordinates from a molecular dynamics simulation

$$\Delta A(g) = -k_B T \ln\langle \exp(-\beta \Delta V(g)) \rangle_0 \tag{11}$$

where g is the coordinate of a grid point, ΔA is the free energy cost to insert the test particle at that point, $\Delta V(g)$ is the interaction energy between the test particle and the system and the angle bracket indicates an average with respect to the reference state. PROFEC has been used to modify a ligand to improve its binding affinity[7, 15] and selectivity,[53] as well as to increase protein stability.[54] Recently, Pearlman developed a variant of this approach, the floating independent reference frame (FIRF), which may be applicable to flexible ligands.[11]

3.3 LIA with Continuum Solvent

The molecular-mechanics with Poisson-Boltzmann/surface area approach (MM/PBSA)[13] is a semi-empirical method to calculate free energy differences between protein-ligand complexes,[10] protein-protein complexes,[9] and different forms of DNA and RNA.[13, 55] The basic approach used in MM/PBSA follows the procedures used to analyze peptide and protein conformations as outlined by Yang and Honig and Osapay et al.[56, 57] Shen and co-workers demonstrated that the use of a single conformation together with the Poisson-Boltzmann electrostatic and surface-area-dependent terms in MM/PBSA can lead to successful estimation of the binding free energies for a number of ligands.[58, 59] In the MM/PBSA method, the binding free energies are estimated from

$$\Delta A_{(bind)} = \Delta A_{(bound)} - (\Delta A^{ligand}_{(solv)} + \Delta A^{receptor}_{(solv)}) \tag{12}$$

and

$$\Delta A_{(bind)} \approx \langle \Delta E_{gas} \rangle + \langle \Delta A_{PB} \rangle + \langle \Delta A_{SA} \rangle - T\langle \Delta S \rangle. \tag{13}$$

The solute configurations are sampled as "snap-shots" from a molecular dynamics simulation calculated using explicit solvent. For each solute configuration, the gas-phase energy, E_{gas}, is calculated without any solvent. Free energies of solvation are then re-introduced by using a Poisson-Boltzmann calculation for the electrostatic term (A_{PB}) and a surface-area-dependent term (A_{SA}) for non-electrostatic contributions. Solute entropy contributions are estimated from (quasi-) harmonic analysis. The differences (ΔE_{gas}, ΔA_{PB}, and ΔA_{SA}) are calculated between the bound state and unbound states as shown in Equation 12. Variants of the MM/PBSA approach such as "computational alanine scanning"[9] or "computational fluorine scanning"[10] were also introduced and shown to be useful techniques to explore sensitivity of a given receptor site (or amino acid site in a protein) to changes in composition.

4. FREE ENERGY BASED MULTIPLE-LIGAND SCREENING METHODS: λ-DYNAMICS AND CMC/MD

4.1 Basic Formulation

The free energy-based, multiple-ligand screening methods are an extension of the coupling-parameter approach used in thermodynamic cycles. They differ, however, in the following aspects: (1) In the coupling-parameter approach a single λ is used to transform one ligand into another, whereas in the free energy-based, multiple-ligand screening methods, multiple λs (each corresponding to a given ligand) are used. Because of this feature, the binding free energies of multiple ligands are evaluated simultaneously. (2) In FEP/TI, λ is fixed during the simulation. In the free energy-based, multiple-ligand screening methods, the λs evolve according to "equations of motion" via molecular dynamics or Monte Carlo methods. In this section, we will briefly elaborate on the main ideas and formulas associated with these approaches.

For a protein and a total of L ligands, a hybrid potential function is constructed as follows

$$V_0(X,\{x\},\{\lambda\}) = V_{env}(X) + \sum_{i=1}^{L} \lambda_i^2 (V_i(X,x_i) - F_i) \tag{14}$$

with

$$\sum_{i=1}^{L} \lambda_i^2 = 1$$

In Equation 14, i indicates the ith ligand, X and x_i denote the coordinates of environment atoms and ligand i respectively, V_{env} is the potential energy involving the environment atoms only (i.e., those atoms which are common to all protein-ligand pairs), V_i is the interaction energy involving ligand i in the protein-ligand complex state and λ_i is the coupling parameter associated with ligand i. The coupling parameter λ is replaced by λ^2 in order to avoid non-physical negative values in the λ-dynamics simulations.[3]

By properly coupling the system to a heat bath, the configurational partition function of the hybrid potential is canonical

$$Z_0(X,\{x\},\{\lambda\}) = \int \exp\left[-\beta\left\{V_{env}(X) + \sum_{i=1}^{L} \lambda_i^2 \left(V_i(X,x_i) - F_i\right)\right\}\right] dxdX. \tag{15}$$

F_i is a precalculated biasing potential, which may correspond to the relative free energy of ligand i in the unbound state (relative solvation free energy). In many cases, F_i may be rapidly evaluated using continuum solvation models such as the Poisson-Boltzmann or generalized Born methods.[59-65] Note that F_i can also serve as a biasing potential to achieve better sampling of the phase space of interest, and faster convergence of the calculations as shown later.

The difference in binding free energy ($\Delta\Delta A_{j\to i}$) between arbitrarily chosen ligands i and j can be obtained from[3, 66] $P_0(\lambda_i^2=1,\{\lambda_{m\neq i}^2=0\})$, which corresponds to the amount of time ligand i occupies $\lambda_i^2=1$ during the simulation, is an indicator of the binding free energy of that ligand to the protein receptor. Because the reference free energy appears in the hybrid potential of the protein-ligand complexed state, the resulting free energy from Equation 16 directly corresponds to the binding free energy difference. Therefore, these methods tend to provide better sampling for ligands that have more favorable binding free energies. Furthermore, such calculations often result in smaller statistical errors for the most favored compounds. There exists an analogy between this formalism and competitive binding experiments carried out in the laboratory. In fact, a competitive binding experiment usually consists of different ligands and a single receptor in solution, and the best ligands are determined by the probability that a ligand is bound to the receptor.

$$\begin{aligned}
\Delta\Delta A_{j\to i} &= \Delta A_{(bound)} - \Delta A_{(solv)} \\
&= -k_B T \ln \frac{\int \exp\{-\beta(V_{env}(X)+V_i(X,x_i))\}}{\int \exp\{-\beta(V_{env}(X)+V_j(X,x_j))\}} - (F_i - F_j) \\
&= -k_B T \ln \frac{\int \exp\{-\beta(V_{env}(X)+V_i(X,x_i)-F_i)\}}{\int \exp\{-\beta(V_{env}(X)+V_j(X,x_j)-F_j)\}} \\
&= -k_B T \frac{\ln Z_0(\lambda_i^2=1,\{\lambda_{m\neq i}^2=0\})}{\ln Z_0(\lambda_j^2=1,\{\lambda_{m\neq j}^2=0\})} \\
&= -k_B T \frac{\ln P_0(\lambda_i^2=1,\{\lambda_{m\neq i}^2=0\})}{\ln P_0(\lambda_j^2=1,\{\lambda_{m\neq j}^2=0\})}
\end{aligned} \tag{16}$$

In the λ–dynamics method,[1] both λ variables (coupling parameters) and atomic coordinates are propagated using molecular dynamics (MD). The dynamics of the system is generated from an extended Hamiltonian[67, 68]

$$H_0(X,\{x\},\{\lambda\}) = T_x + T_\lambda + V_0(X,\{x\},\{\lambda\}) \text{ and}$$

$$T_\lambda = \sum_{k=1}^{L} \frac{m_\lambda}{2} \dot{\lambda}_k^2 \tag{17}$$

T_x and T_λ are the kinetic energies of the atomic coordinates and λ variables, respectively. The λs are treated as volumeless particles with mass m_λ. Since the λ variables are associated with the "chemical reaction coordinates", the λ-dynamics method can utilize the power of specific biasing potentials in the umbrella sampling method to overcome sampling problems that require conventional FEP calculations to be performed in multiple steps.

Instead of using MD, the λ variables may also be sampled stochastically. In the hybrid CMC/MD approach, Metropolis Monte Carlo[69] is used to evolve the λ variables and molecular dynamics is used to evolve the atomic coordinates. The Metropolis Monte Carlo criteria leads to the generation of a canonical ensemble of the ligands when the following transition probability is used

$$T_{i\to j} - \min\left(1, \exp(-\beta \Delta V_{i\to j})\right) \tag{18}$$

where $\Delta V_{i\text{->}j}$ is defined by $(V_j - V_i)$ and $T_{i\text{->}j}$ is the transition probability of a move from ligand *i* to *j*. Both the λ-dynamics method and the hybrid MC/MD method give the same configurational partition functions (Equation 15). Therefore Equation 16 can be applied to CMC/MD as well. The MC/MD method was originally presented by Bennett[70] and Tidor.[66, 71] The straightforward extension of this approach to multiple ligands, which is called Chemical MC/MD, was carried out by Pitera & Kollman.[6]

4.2 Enhancements to the Free Energy Based Multiple-Ligand Screening Methods

4.2.1 Efficient Sampling of the Chemical Coordinates

In the MC/MD method, the stochastic sampling by MC steps permit one to restrict the sampling of chemical space, i.e., the space of {λ}. For example, Kollman and co-workers[6, 7] limit their chemical sampling only to transitions between the end points in their CMC/MD simulations

$$\sum_{i=1}^{L} \lambda_i^2 = 1 \text{ and } \lambda_i^2 = \{0,1\} \tag{19}$$

This condition allows sampling of the end states of interest exclusively. However, inefficient sampling of the chemical states, such as trapping in one end state, may occur. This is prevalent when there is a large free energy gap between the ligands. Trapping may, however, be partially avoided by the addition of a few chosen intermediate states to bridge the end points. In the λ-dynamics method, the λ-variables are treated as continuous variables, so smooth transitions between the end points are expected, and generally observed.

As shown in Equation 5, the conventional free energy calculation methods such as FEP and TI require the introduction of intermediate states to obtain converged free energy differences between the end points. In the free energy-based, multiple-ligand screening methods, some ligands can serve as the intermediate states connecting otherwise dissimilar end points. Nevertheless, there are potential problems. For example, if the relative free energy of the intermediate states is lower than that of the end points, most of the computational time will be spent exploring unphysical intermediate states. Therefore the relative free energy of the end points will be less well determined. Conversely, higher energy intermediate states require rare transitions in the chemical coordinates and therefore result in slowed convergence.

The umbrella sampling technique[28, 29] can be utilized to overcome these difficulties. An umbrella potential along the λ coordinate can be expressed as

$$V_{um} = V_0(X,\{x\},\{\lambda\}) + \sum_{i=1}^{L} B_i(\lambda_i) \tag{20}$$

where the λ-dependent potential term, $B_i(\lambda_i)$, will serve as an umbrella (or a biasing) potential to limit the range of {λ} and to increase the rate of transitions among potential wells separated by high-energy barriers. A harmonic potential is commonly used to flatten the energy surface and enhance sampling along the chemical coordinate λ

$$B_i(\lambda_i) = k_i\left(\lambda_i^2 - B_i^0\right)^2 \tag{21}$$

where $0 < B_i^0 < 1$. The condition ($k_i > 0$) can be used to increase the transition between the end points, while $k_i < 0$ tends to increase sampling of the end points. The unbiased probability of the bound states can be calculated by using the umbrella sampling formalism (Equation 8)

$$
\begin{aligned}
P_0(\lambda_i^2 = 1, \{\lambda_{m\neq i}^2 = 0\}) &= \frac{\int \exp(-\beta V_0)\theta(\lambda_i^2 - 1)}{\int \exp(-\beta V_0)} \\
&= \frac{\int \frac{\exp\{-\beta(V_0 + \sum B_k)\}}{\int \exp\{-\beta(V_0 + \sum B_k)\}} \exp(\beta \sum B_k)\theta(\lambda_i^2 - 1)}{\int \frac{\exp\{-\beta(V_0 + \sum B_k)\}}{\int \exp\{-\beta(V_0 + \sum B_k)\}} \exp(\beta \sum B_k)} \\
&= \frac{\left\langle \exp(\beta \sum_{k=1}^{L} B_k)\theta(\lambda_i^2 - 1) \right\rangle_{um}}{\left\langle \exp(\beta \sum_{k=1}^{L} B_k) \right\rangle_{um}}
\end{aligned}
\quad (22)
$$

where the angle bracket denotes the ensemble average over the biased distribution and $\theta(x)$ is a step function, which is unity when its argument is greater than zero but is otherwise zero. By using the probability of the bound states, $\Delta\Delta A$ is obtained from

$$
\begin{aligned}
\Delta\Delta A_{j\rightarrow i} &= -k_B T \ln \frac{P_0(\lambda_i^2 = 1, \{\lambda_{m\neq i}^2 = 0\})}{P_0(\lambda_j^2 = 1, \{\lambda_{m\neq j}^2 = 0\})} \\
&= -k_B T \ln \frac{\left\langle \exp(\beta \sum_{k=1}^{L} B_k)\theta(\lambda_i^2 - 1) \right\rangle_{um}}{\left\langle \exp(\beta \sum_{k=1}^{L} B_k)\theta(\lambda_j^2 - 1) \right\rangle_{um}} \\
&= -k_B T \ln \frac{\left\langle \theta(\lambda_i^2 - 1) \right\rangle_{um} \exp\left\{\beta \sum_{k\neq i}^{L} B_k(\lambda_k = 0)\right\} \exp\{\beta B_i(\lambda_i^2 = 1)\}}{\left\langle \theta(\lambda_j^2 - 1) \right\rangle_{um} \exp\left\{\beta \sum_{m\neq j}^{L} B_m(\lambda_m = 0)\right\} \exp\{\beta B_j(\lambda_j^2 = 1)\}} \\
&= -k_B T \ln \frac{P_{um}(\lambda_i^2 = 1, \{\lambda_{m\neq i}^2 = 0\})}{P_{um}(\lambda_j^2 = 1, \{\lambda_{m\neq j}^2 = 0\})} - \left(\Delta B_i - \Delta B_j\right)
\end{aligned}
\quad (23)
$$

$\Delta B_i = B_i(\lambda_i^2 = 1) - B_i(\lambda_i = 0)$ and P_{um} is the probability function of the hybrid potential with the umbrella potential. If ($\Delta B_I = \Delta B_j$), the effect of the umbrella potential will be canceled completely. An iterative procedure is sometimes required to produce complete sampling of important configurations along the chemical coordinates. In such cases, WHAM, discussed in the next section, can be used to process the sampling data in an efficient and general way.

Another approach for efficient sampling along the chemical coordinates was suggested by Tidor.[66] In his work, simulated annealing was used to sample the chemical variables on the free energy surface of the system (e.g., high temperature for the chemical variables and low temperature for Cartesian variables). In a demonstration calculation, the method was applied to simple molecules to select the one with the most favorable solvation energy. Agreement between the observed and calculated trends was obtained

4.2.2 Iterative Techniques Using WHAM

An iterative procedure using WHAM[31-35] was developed to improve sampling of the chemical space, and therefore to make free energy calculations converge more rapidly. The use of this method in conjunction with λ-dynamics is described below.[3] It's extension to MC/MD is straightforward.

Since {λ} is treated as a dynamic variable, just as the atomic coordinates, we use X_{Tot} to denote the phase space that encompasses X, $\{\lambda\}$, and $\{x\}$. Thus the hybrid potential in Equation 14 can be rewritten as

$$V_0(X_{Tot}) = V_{ref}(X_{Tot}) - \sum_{i=1}^{L} F_i \lambda_i^2 \tag{24}$$

where

$$V_{ref}(X_{Tot}) = V_{env}(X) + \sum_{i=1}^{L} \lambda_i^2 V_i(X, x_i) \tag{25}$$

When this potential is utilized in a series of λ-dynamics, or CMC/MD simulations, the WHAM equations for multiple reaction coordinates and at constant temperature can be readily applied to obtain the best estimate of free energy using all of the data from n previous simulations

$$P^n_{\{F^n\}}(\{\lambda^2\}) = \frac{\sum_{k=1}^{n} N_k(\{\lambda^2\}) \exp\left(-\beta \sum_{i=1}^{L} (-F_i^n)\lambda_i^2\right)}{\sum_{m=1}^{n} n_m \exp\left(f_m - \beta \sum_{i=1}^{L} (-F_i^m)\lambda_i^2\right)} \tag{26}$$

$$\exp(-f_m) = \sum_{\{\lambda^2\}} P^m_{\{F^m\}}(\{\lambda^2\}) \tag{27}$$

After the nth iteration, the estimated free energy relative to the reference free energy $\{F^n\}$ is

$$A_{i\{F\}} = -k_B T \ln P^{n}_{\{F^n\}}(\lambda_i^2 = 1, \{\lambda_{m\neq i}^2\} = 0) \tag{28}$$

and a new biasing potential for the next iteration is estimated as

$$F_i^{n+1} = A_{i,\{F=0\}}. \tag{29}$$

The above procedure can be used to extract the free energy, A_i, of each ligand.

Since this approach biases the sampling of different ligands in the receptor by successively better estimates of their relative binding free energy, the bound conformations of all the ligands are expected to be sampled equally well after some number of cycles of simulation. Sometimes an additional term Δ_i may also be added to Equation 29 to either enhance or reduce sampling of a state dominated by ligand *i*. As in all iterative procedures, an initial trial value of F_i must be given. If a poor initial free energy is used, then the states with $F_I < A_i$ will be sampled less frequently than they would with $F_I = A_i$. Similarly, states with $F_I > A_i$ will be sampled more frequently. The approach was also applied with the MC/MD method by Kollman and co-workers, which they renamed adapted Chemical MC/MD method.[7]

In the iterative approach using a constant bias $\{F\}$, the free energy barrier is reduced each successive iteration and therefore produces complete sampling of important configurations along the coupling parameter (i.e., reaction coordinates). Furthermore, more complicated umbrella potentials (e.g. see Equation 21) may also be applied with the iterative procedure.

4.2.3 Sampling of the Unselected Ligands in the Multiple Topology Model

When a single-topology representation of the ligand is used, i.e., one in which changing atoms are all connected to a common framework, the configuration of the unbound ligands is determined automatically. No ambiguity exists regarding the choice of the configuration of the unbound ligands with λ equal to zero, making the choice of proper MC steps straightforward. To see this, consider the detailed balance condition of MC for moves between ligand *i* and *j* using the single topology

$$\begin{gathered} P_0(\lambda_i^2 = 1, \{\lambda_{m\neq i}^2 = 0\})\alpha_j T_{i\to j} = P_0(\lambda_j^2 = 1, \{\lambda_{m\neq j}^2 = 0\})\alpha_i T_{j\to i} \\ \exp(-\beta V_i)\frac{1}{L-1}T_{i\to j} = \exp(-\beta V_j)\frac{1}{L-1}T_{j\to i} \end{gathered} \tag{30}$$

$$\frac{T_{i\to j}}{T_{j\to i}} = \exp\{-\beta(V_i - V_j)\} \tag{31}$$

α_i is the probability of selecting the ligand i and $T_{i\text{->}j}$ is the transition probability of a move from ligand i to j. It is straightforward to demonstrate that the basic Metropolis scheme shown in Equation 18 obeys this condition.[69] However, the assignment of a single topology model for multiple ligands is a complicated problem, especially when the ligands have the different shapes. Furthermore, inadequate assignment of the common features in a single topology leads to small overlap of the important configurations between the selected ligand and unselected ones. In general, multiple independent topologies have been used for the multiple-ligand screening methods to avoid these problems.[4-7]

In the multiple-topology model, the unbound ligands tend to move significantly from their preferred binding orientations and explore high-energy regions of conformation space.[4] This results in inefficient sampling of chemical space. In CMC/MD simulations,[7, 53] the problem was addressed by the addition of a harmonic potential between the centers of mass of all ligands and the imposition of "ghost" forces on the unbound ligands. The ligand "ghost" forces are those exerted on the unbound ligands by the environment atoms. However, in the approach of Kollman and co-workers, the unbound ligands remain invisible to the environment atoms. The effect of the harmonic potential is cancelled out in the calculation of ΔΔA, and thus moves with only this added potential satisfy the detailed balance condition. On the other hand, it is difficult to correct for the effect of the "ghost" forces, since they do not have a physical origin. To understand the effect of "ghost" forces on MC/MD steps and statistical averages arising from such calculations, we consider the special condition in which all environment atoms are fixed. With this idealization, the "ghost" forces can be recognized as those coming from a restraining potential, (R_i), arising from the fixed environment. For this situation, the probability of selecting the ligand i with the coordinates x_i ($\alpha_i(x_i)$) becomes

$$\alpha_i(x_i) = \left(\frac{1}{L-1}\right)\cdot\frac{\exp(-\beta R(x_i))}{\int \exp(-\beta R(x_i))} \tag{32}$$

The condition of detailed balance for an MC step between ligand i with coordinate of x_i and ligand j with coordinate of x_j can be written as follows

$$
\begin{aligned}
& P_0(\lambda_i^2=1,\{\lambda_{m\neq i}^2=0\},x_i)\alpha_j(x_j)T_{i\to j}(x_i,x_j) \\
&\quad = P_0(\lambda_j^2=1,\{\lambda_{m\neq j}^2=0\},x_j)\alpha_i(x_i)T_{j\to i}(x_j,x_i) \\
&\quad \cdot\exp(-\beta V_i)\left(\frac{1}{L-1}\right)\cdot\frac{\exp(-\beta R_j(x_j))}{\int\exp(-\beta R_j(x_j))}T_{i\to j} \\
&\quad = \exp(-\beta V_j)\left(\frac{1}{L-1}\right)\cdot\frac{\exp(-\beta R_i(x_i))}{\int\exp(-\beta R_i(x_i))}T_{j\to i}
\end{aligned}
\tag{33}
$$

$$
\frac{T_{i\to j}}{T_{j\to i}}=\left\{\exp\left(\beta\Delta R_{i\to j}\right)\frac{\int\exp(-\beta R_j)}{\int\exp(-\beta R_i)}\right\}\exp\left(-\beta\Delta V_{i\to j}\right) \tag{34}
$$

where $\Delta R_{i\text{->}j} = R_j(x_j) - R_i(x_i)$, $\Delta V_{i\text{->}j} = V_j(x_j) - V_i(x_i)$ and $T_{i\text{->}j}$ is the transition probability of a move from ligand i with x_i to ligand j with x_j. One transition rule that obeys the detailed balance condition and yields the canonical ensemble in this case is

$$
T_{i\to j}=\min\left(1,\exp(\beta\Delta R_{i\to j})\frac{\int\exp(-\beta R_j)}{\int\exp(-\beta R_i)}\exp\left(-\beta\Delta V_{i\to j}\right)\right) \tag{35}
$$

If, in fact, the environment atoms are allowed to move, the rigorous estimation of $A_{i\text{->}j}$ is unclear. Kollman and co-workers have assumed that the effect of the "ghost" forces cancel for comparisons of similar ligands, and used a transition rule following Equation 18 instead of Equation 35 in their CMC/MD simulations.[7, 53] This approximation was demonstrated for some systems to be at least qualitatively reasonable.[6, 7]

In order to overcome the problems that can occur in sampling configurations of the unbound ligands when using a multiple topology model, we consider two types of restraining potentials. For simplicity, both restraining potentials disappear at the bound states

$$
R_i'(\lambda_i^2=1)=R''(\lambda_i^2=1)=0 \tag{36}
$$

The first type of the restraining potential for ligand i (R_i') is defined as function of X, x_i, and, λ_i.

$$
V'(X,\{x\},\{\lambda\})=V_0(X,\{x\},\{\lambda\})+\sum_{i=1}^{L}R_i'(X,x_i,\lambda_i) \tag{37}
$$

With a straightforward application of the umbrella sampling formalism, $\Delta\Delta A$ is obtained from

$$\Delta\Delta A_{j\to i} = -k_B T \ln \frac{\left\langle \exp\left(\beta \sum_{k=1}^{L} R_k'(X, x_k, \lambda_k)\right)\theta(\lambda_i^2 - 1)\right\rangle_{V'}}{\left\langle \exp\left(\beta \sum_{l=1}^{L} R_l'(X, x_l, \lambda_l)\right)\theta(\lambda_j^2 - 1)\right\rangle_{V'}}$$

$$= -k_B T \ln \frac{\sum_{(\lambda_i^2=1)} \exp\left(\beta \sum_{k\neq i}^{L} R_k'(X, x_k, \lambda_k = 0)\right)}{\sum_{(\lambda_j^2=1)} \exp\left(\beta \sum_{l\neq j}^{L} R_l'(X, x_l, \lambda_l = 0)\right)} \tag{38}$$

The summation is taken over the bound state (λ^2=1) of the λ-dynamics trajectory, including the restraining potential. Unfortunately, with this biasing potential the effect of the restraint (R_i') becomes too large to yield reasonable convergence as the number of the unbound ligands increases.

Another type of restraining potential for ligand *i* (R_i") is defined as a function of x_i and λ_i. In this case, the restraining potential does not depend directly on the environment atom coordinates

$$V''(X, \{x\}, \{\lambda\}) = V_0(X, \{x\}, \{\lambda\}) + \sum_{i=1}^{L} R_i''(x_i, \lambda_i) \tag{39}$$

Because $R_i'' = 0$ when $\lambda_i^2 = 1$, and none of the restraining potentials depend on the environment atoms, the partition function for the system when $\lambda_i^2 = 1$ can be expressed as follows

$$Z''_{\lambda_i^2=1} = \int \exp\{-\beta(V_{env} + V_i - F_i)\} dX dx_i \prod_{k\neq i}^{L} \int \exp\left(-\beta R_k''(x_k, \lambda_k = 0)\right) dx_k \tag{40}$$

Using this relationship, $\Delta\Delta A$ can be written as two terms. The first term involves the probability that a ligand is in the dominant state (λ^2=*1*) during the λ-dynamics simulation and in the presence of the restraining potential. The second term corresponds to the partition function of the restraining potential (the umbrella correction)

$$\begin{aligned}\Delta\Delta A_{j\to i} &= -k_B T \ln \frac{Z'_{\lambda_i^2=1} \prod_{k\neq j}^{L} \int \exp\left(-\beta R_k''(x_k,\lambda_k=0)\right) dx_k}{Z'_{\lambda_j^2=1} \prod_{l\neq i}^{L} \int \exp\left(-\beta R_l''(x_l,\lambda_l=0)\right) dx_l} \\ &= -k_B T \ln \frac{P'_{\lambda_i^2=1}}{P'_{\lambda_j^2=1}} - k_B T \ln \frac{\int \exp\left(-\beta R_i(x_i,\lambda_i=0)\right) dx_i}{\int \exp\left(-\beta R_j(x_j,\lambda_j=0)\right) dx_j}\end{aligned} \tag{41}$$

The second term in Equation 41 is constant and may be estimated using free energy simulation or semi-empirical methods. When the ligands are similar and the entropy terms associated with the restraining potential are expected to cancel, the second term can be approximated by an internal energy difference

$$\begin{aligned}&-k_B T \ln \frac{\int \exp\left(-\beta R_i''(x_i,\lambda_i=0)\right) dx_i}{\int \exp\left(-\beta R_j''(x_j,\lambda_j=0)\right) dx_j} = (U_i'' - U_j'') - (S_i'' - S_j'')T \\ &\approx U_i'' - U_j''\end{aligned} \tag{42}$$

This internal energy can be estimated by using the trajectory of the free energy-based, multiple-ligand screening simulations

$$\begin{aligned}U_i'' &= \int R_i''(x_i,\lambda_i=0) \exp\left(-\beta R_i''(x_i,\lambda_i=0)\right) dx_i \\ &\approx \frac{1}{n} \sum_{(\lambda_i=0)}^{m=1,n} R_m''\end{aligned} \tag{43}$$

The restraining potential should be chosen carefully since the important configurations for $\{R_i''\}$ should have large overlap with those for $\{V_i\}$. The interaction potential $\{V_i\}$ for the average structure of the environment atoms is a reasonable choice for $\{R_i''\}$. In fact, it is an optimal choice to bias the ligands that do not have large λ-values because it restrains these ligands to the vicinity of the receptor. To represent fluctuations in this mean-receptor potential field, soft-core representations of the van der Waals or electrostatic interactions can be used, or the overall potential field can be scaled. Such an approach has been implemented and used by Banba and Brooks in λ-dynamics calculations of binding free energies in a protein-ligand system.[4]

4.2.4 Incorporation of Continuum Solvent Models Using a Generalized Born Approach

The use of continuum solvent models will decrease the number of degrees of freedom in the system, and consequently accelerate the convergence of thermodynamic properties by eliminating the ensemble average of the solvent molecules. Moreover, the absence of collisions between the unbound ligands and mobile solvent enhances the overlap of the important configurations. However, conventional numerical solutions to the Poisson-Boltzmann equations are too slow for practical applications and one must use approximate analytical representations such as the generalized Born model originally proposed by Still and co-workers.[61, 72, 73] Since the electrostatic solvation energy and its derivative can be calculated analytically in the generalized Born model, it may be applied to configurational sampling using molecular dynamics.[60, 74-80]

In the generalized Born model, the solvent polarization energy, G_{pol}, is approximated by the following equation

$$G_{pol} = -166\left(1-\frac{1}{\varepsilon}\right)\sum_i^N\sum_j^N \frac{q_i q_j}{\sqrt{\left(r_{ij}^2 + \alpha_i\alpha_j e^{-D_{ij}}\right)}} \tag{44}$$

with $D_{ij} = \dfrac{r_{ij}^2}{4\alpha_i\alpha_j}$ atoms i and j, respectively, r_{ij} represents the distance

between atoms i and j, and α_i and α_j are "generalized Born" radii of atoms i and j in a specific molecular environment. α_i is computed from the relationship

$$\alpha_i = -166 / G_{pol,i} \tag{45}$$

with $G_{pol,i}$ taken from a linearized form of Still's original empirical formula[74]

$$G_{pol,i} = \frac{1}{\lambda_a}\left(\frac{-166}{R_{vdW\,i}}\right) + P_1\left(\frac{166}{R_{vdW\,i}^2}\right) + P_2\sum_j^{bond}\frac{V_j}{r_{ij}^4} + P_3\sum_j^{angle}\frac{V_j}{r_{ij}^4} + P_4\sum_j^{nonbond}\frac{V_j}{r_{ij}^4}CCF. \tag{46}$$

The values of λ_a and P_1-P_5 can be determined by fits to Poisson-Boltzmann solvation energies for a database of compounds and

$$CCF = \begin{cases} 1; & \left(\dfrac{r_{ij}}{R_{vdW,i}+R_{vdW,j}}\right)^2 > \dfrac{1}{P_5} \\ \dfrac{1}{2}\left[1-\cos\left\{P_5\pi\left(\dfrac{r_{ij}}{R_{vdW,i}+R_{vdW,j}}\right)^2\right\}\right]^2; & \left(\dfrac{r_{ij}}{R_{vdW,i}+R_{vdW,j}}\right)^2 \le \dfrac{1}{P_5} \end{cases} \tag{47}$$

The incorporation of the generalized Born model into free energy calculation methods using FEP/TI and λ-dynamics was carried out by Banba and Brooks.[81] They define the electrostatic solvation energy for the hybrid system as follows

$$G_{pol} = -166\left(1-\frac{1}{\varepsilon}\right)\left\{\sum_i^{i\in env}\sum_j^{j\in env}\frac{q_i q_j}{\sqrt{\left(r_{ij}^2+\alpha_i\alpha_j e^{-D_{ij}}\right)}} + \sum_{k=1}^{L}\lambda_k^2\left(2\sum_i^{i\in env}\sum_j^{j\in k}\frac{q_i q_j}{\sqrt{\left(r_{ij}^2+\alpha_i\alpha_j e^{-D_{ij}}\right)}} + \sum_i^{i\in k}\sum_j^{j\in k}\frac{q_i q_j}{\sqrt{\left(r_{ij}^2+\alpha_i\alpha_j e^{-D_{ij}}\right)}}\right)\right\} \tag{48}$$

where *env* represents the environment atoms and *k* denotes the ligand number. The effective Born radius for the environment atoms can be calculated from Equations 45 and 49,and the effective Born radii for the atoms belonging to ligand *k* are calculated from Equations 45 and 50

$$\begin{aligned} G_{pol,i}^{i\in env} = {} & \frac{1}{\lambda_a}\left(\frac{-166}{R_{vdW,i}}\right) + P_1\left(\frac{166}{R_{vdW,i}^2}\right) \\ & + P_2\sum_{j\in env}^{bond}\frac{V_j}{r_{ij}^4} + P_3\sum_{j\in env}^{angle}\frac{V_j}{r_{ij}^4} + P_4\sum_{j\in env}^{nonbond}\frac{V_j}{r_{ij}^4}CCF \\ & + \sum_{k=1}^{L}\lambda_k^2\left(P_2\sum_{j\in k}^{bond}\frac{V_j}{r_{ij}^4} + P_3\sum_{j\in k}^{angle}\frac{V_j}{r_{ij}^4} + P_4\sum_{j\in k}^{nonbond}\frac{V_j}{r_{ij}^4}CCF\right) \end{aligned} \tag{49}$$

$$\begin{aligned} G_{pol,i}^{i\in k} = {} & \frac{1}{\lambda_a}\left(\frac{-166}{R_{vdW,i}}\right) + P_1\left(\frac{166}{R_{vdW,i}^2}\right) \\ & + P_2\sum_{j\in\left\{\substack{env\\k}\right.}^{bond}\frac{V_j}{r_{ij}^4} + P_3\sum_{j\in\left\{\substack{env\\k}\right.}^{angle}\frac{V_j}{r_{ij}^4} + P_4\sum_{j\in\left\{\substack{env\\k}\right.}^{nonbond}\frac{V_j}{r_{ij}^4}CCF \end{aligned} \tag{50}$$

These definitions are interpreted as follows: environment atoms recognize the average weighted states of the ligands; each ligand recognizes only the environment atoms and itself with the full scale for estimation of its effective Born radius. Alternate definitions of the electrostatic solvation energy and the effective Born radius at the intermediate states are also possible.[81]

Non-electrostatic terms, comprising the solvent-solvent cavity term and solute-solvent van der Waals term, may be linearly related to solvent-accessible surface area (SA)

$$V_{SA,i} = V_{cav} + V_{vdW} = \sum_{i \in X, x_i} \sigma_i SA_i \tag{51}$$

where SA_i is the total solvent-accessible surface area of atom i and σ_i is an empirical atomic solvation parameter for atom i. Although $V_{SA,i}$ and the electrostatic solvation energy can be calculated at intermediate states, the calculation of the SA and its first derivative at every MD step is significantly more time consuming.

4.3 Application of Multiple-Ligand Free Energy Methods

The basic formalism of the λ-dynamics method has taken various forms in its application to problems of interest. In an early prototype calculation to assess umbrella sampling in chemical coordinates, the λ-dynamics method was used to evaluate the relative free energy of hydration for a set of small molecules which included both nonpolar (C_2H_6,) and polar (CH_3OH, CH_3SH, and CH_3CN) solutes.[1] By assigning a separate λ variable to the Lennard-Jones and Coulomb interactions, a linear partition of the potential part of the hybrid Hamiltonian was constructed

$$\begin{aligned} V(x,\lambda_1,\lambda_2) = {} & (1-\lambda_1)V_{\mathrm{R}}^{\mathrm{Coul}}(x) + \lambda_1 V_{\mathrm{P}}^{\mathrm{Coul}}(x) \\ & + (1-\lambda_2)V_{\mathrm{R}}^{\mathrm{LJ}}(x) + \lambda_2 V_{\mathrm{P}}^{\mathrm{LJ}}(x) + V_{\mathrm{Env}}(x) \end{aligned} \tag{52}$$

In the situation where the transformation involved barrier crossing, e.g., associated with a nonpolar to polar transformation, the computational time was substantially reduced using the λ-dynamics formalism, compared with a standard FEP method. This is because λ-dynamics searches for alternative lower free energy pathways; the coupling parameters (λ_1 and λ_2) evolve in the canonical ensemble independently and find a smoother path then when constrained to move as $\lambda_1 = \lambda_2$. Furthermore, a biasing potential in the form of $K(\lambda_1-\lambda_2)^2$ was used to reduce the barrier between the two end states, and therefore enhances sampling and accelerates convergence.

The λ-dynamics method has also been successfully applied to evaluate the relative binding free energies for a number of biologically and chemically relevant systems. The calculation of the relative binding free energy of benzamidine and its derivatives to trypsin represent the first application of λ-dynamics to a biological system. The particular inhibitors studied were benzamidine, p-aminobenzamidine, p-methylbenzamidine, and p-chlorobenzamidine.[2, 3] Rapid screening and detailed structure-function paradigms for utilization of λ-dynamics were explored in both qualitative and quantitative calculations. The method yielded the correct ranking of binding affinities and required less than 100ps of simulation, even though the binding affinity between some of the inhibitors differed by only ~0.5 kcal/mol. The results were validated by comparison to conventional FEP calculations. The calculated binding free energy differed from FEP by less than 0.5 kcal/mol. The iterative technique using the multiple reaction coordinate WHAM discussed in 4.2.2. was applied to obtain an optimal estimate of the biasing potentials, which was directly related to the binding free energy differences. The results were compared with those from standard FEP calculations. For a similar level of precision, the λ-dynamics method was two times more efficient. The λ-dynamics method with the generalized Born model (see 4.2.4.) was also applied to the same system.[81] The implicit solvent model showed qualitative agreement with explicit water calculations. In this study, the continuum solvent model using the GB approach also accelerated the convergence of the free energies calculations since the average over the solvent degrees of freedom was implicitly incorporated. Non-electrostatic terms (Equation 51) were estimated from trajectories using the umbrella sampling.

In a recent application, the relative binding free energy and binding orientation of the ligands of cytochrome c peroxidase were evaluated.[4, 5] Using the revised hybrid potential energy function (Equation 40), relatively short simulations yielded reasonable estimates of the binding affinity of the ligands compared with both experiment and FEP calculations. Long-time λ-dynamics simulations revealed that better ligands tend to have smaller statistical errors in the estimates of their binding free energy, which is appropriate for screening out plausible ligands from all candidates. Furthermore, a λ-dynamics simulation starting from random initial orientations, in which some ligands take significantly different orientations as compared with those from the X-ray structure, successfully sampled the X-ray crystallographic orientations in all ligands. Ligands sampled by conventional MD remained trapped in the local minima from which they started. Analysis of the λ-dynamics trajectory of the ligands revealed that for some ligands alternative-binding orientations were also observed.

A variant of the λ-dynamics method, CMC/MD (Chemical Monte Carlo/Molecular Dynamics) has been applied to predict the binding of small molecules to Rebek's "tennis ball" host.[6] A total of nine guests binding to the host were evaluated. CH_2F_2 was predicted to bind the best, which is supported by standard free energy calculations. The method was also applied to rank 13 HIV-RT TIBO derivatives.[7] In this study, the iterative procedure using WHAM played an essential role for efficient sampling of the chemical space. The calculated ranking agreed well with experiment, with average errors in the binding free energies of 1 kcal/mol.

5. SUMMARY AND OUTLOOK

In this chapter, we have reviewed newly developed free energy-based, multiple-ligand screening methods. These methods may be used either to rapidly identify ligands with the most favorable binding free energy or to estimate specific changes in free energy within a congeneric series. Since λ-dynamics and the related family of methods work based on the binding free energy of the ligands instead of the interaction energy, they provide a more accurate assessment of binding affinity. Species whose binding free energies differ by more than a few kcal/mol from the most favorable binder can be rapidly screened out within a few tens of picoseconds of simulation because they do not compete for interactions with the receptor. Furthermore, computations utilizing this approach with multiple ligands are not anticipated to be overly demanding. The total computation time is not expected to increase with the total number of ligands because only the few favorable binders are able to compete for the $\lambda^2 = 1$ state. This situation is in contrast to that of conventional free energy calculation methods, where a typical calculation of relative binding free energy between two ligands takes hundreds of picoseconds of simulation time and increases in proportion to the number of ligands. Although the intrinsic problems of the FEP method, such as requiring proper overlap of the important configurations, still exists in the λ-dynamics-based methods, they can be minimized by using umbrella sampling and/or iterative procedures with WHAM. Moreover, iterative procedures such as WHAM may also be applied to yield quantitative free energy differences for all ligands.

Free energy-based, multiple-ligand screening methods should fill the gap between empirical methods, and theoretically rigorous but computationally intensive methods such as FEP and TI. For example, they can be applied to design a combinatorial library or funnel down the large number of hits detected by the empirical methods. The incorporation of continuum solvent representations such as the generalized Born model into free energy-based, multiple-ligand screening methods accelerates the computational screening process and has a promising future for drug lead optimization. Given this renewal of effort in "computational alchemy" and the encouraging findings from early studies, we can anticipate that rational free energy-based computational approaches to drug discovery and optimization will re-emerge from the tool chest and move to the desktop of the computational medicinal chemist.

Acknowledgments

The authors acknowledge valuable discussions with Dr K.V. Damodaran and comments on the draft manuscript by Dr D. Price. Financial support from the NIH (GM37554) and Novartis Pharmaceuticals is appreciated.

6. REFERENCES

1. X. Kong and C. L. Brooks, III, λ-dynamics: A new approach to free energy calculations, J. Chem. Phys. **105:**2414 (1996).
2. Z. Guo and C. L. Brooks, III, Rapid screening of binding affinities: Application of the λ-dynamics method to a trypsin-inhibitor system, J. Am. Chem. Soc. **120:**1920 (1998).
3. Z. Guo, C. L. Brooks, III, and X. Kong, Efficient and flexible algorithm for free energy calculations using the λ-dynamics approach, J. Phys. Chem. B **102**:2032 (1998).
4. S. Banba and C. L. Brooks, III, Free energy screening of small ligands binding to an artificial protein cavity, J. Chem. Phys. **113** (2000).
5. S. Banba, Z. Guo, and C. L. Brooks, III, Efficient sampling of ligand orientations and conformations in free energy calculations using the lambda-dynamics method, J. Phys. Chem. B **104:**6903 (2000).
6. J. Pitera and P. Kollman, Designing an optimum guest for a host using multimolecule free energy calculations: Predicting the best ligand for Rebek's "Tennis Ball", J. Am. Chem. Soc. **120**:7557 (1998).
7. M. A. L. Eriksson, J. Pitera, and P. A. Kollman, Prediction of the binding free energies of new TIBO-like HIV-1 reverse transcriptase inhibitors using a combination of PROFEX, PB/SA, CMC/MD, and free energy calculations, J. Med. Chem. **42**:868 (1999).
8. T. Z. Mordasini and J. A. McCammon, Calculations of relative hydration free energies: A comparative study using thermodynamic integration and an extrapolation method based on a single reference state, J. Phys. Chem. B **104**:360 (2000).
9. I. Massova and P. A. Kollman, Computational alanine scanning to probe protein-protein interactions: A novel approach to evaluate binding free eneries, J. Am. Chem. Soc. **121**:8133 (1999).
10. B. Kuhn and P. A. Kollman, A ligand that is predicted to bind better to avidin than biotin: Insights from computational fluorine scanning, J. Am. Chem. Soc. **122**:3909 (2000).
11. D. A. Pearlman, Free energy grids: A practical qualitative application of free energy perturbation to ligand design to OWFEG method, J. Med. Chem. **42**:4313 (1999).
12. H. Schaefer, W. F. van Gunsteren, and A. E. Mark, Estimating relative free energies from a single ensemble: Hydration free energies, J. Comput. Chem. **20**:1604 (1999).
13. J. Srinivasan, I. Cheatham, T. E. P. Cieplak, P. A. Kollman, and D. A. Case, Continuum solvent studies of the stability of DNA, RNA, and phosphoramidate-DNA helices, J. Am. Chem. Soc. **120**:9401 (1998).

14. R. J. Radmer and P. A. Kollman, Free energy calculation methods: A theoretical and empirical comparison of numerical errors and a new method for qualitative estimates of free energy changes, J. Comput. Chem. **18**:902 (1997).
15. R. J. Radmer and P. A. Kollman, The application of three approximate free energy calculations methods to structure based ligand design: Trypsin and its complex with inhibitors, J. Comput. Aided Mol. Design **12**:215 (1998).
16. J. C. Owicki and H. A. Scheraga, Monte Carlo calculations in the isothermal-isobaric ensemble. 1. liquid water, J. Am. Chem. Soc. **99**:7403 (1977).
17. R. W. Zwanzig, High-temperature equation of state by a perturbation method. I. Nonpolar gases, J. Chem. Phys. **22**:1420 (1954).
18. D. L. Beveridge and F. M. DiCapua, Free energy via molecular simulation: Application to chemical and biomolecular systems, Annu. Rev. Biophys. Chem. **92**:18 (1989).
19. M. R. Reddy, M. D. Erion, and A. Agarwal, Free energy calculations: Use and limitations in predicting ligand binding affinities, in: Reviews in Computational Chemistry, K. B. Lipkowitz and D. B. Boyd eds, VCH Publishers, New York, (2000), vol. 16, pp. 217-304.
20. C. L. Brooks, Thermodynamic calculations on biological molecules, Int. J. Quantum Chem. Quantum Biol. Symp. **15**:221 (1988).
21. P. A. Kollman, Free energy calculations: Applications to chemical and biochemical phenomena, Chem. Rev. **93**:2395 (1993).
22. U. C. Singh, F. K. Brown, P. A. Bash, and P. A. Kollman, An approach to the application of free energy perturbation methods using molecular dynamics: Applications to the transformations of $CH_3OH \rightarrow CH_3CH_3$, $H_3O^+ \rightarrow NH_4^+$, glycine $\rightarrow$ alanine, and alanine $\rightarrow$ phenylalanine in aqueous solution and to $H_3O{+}(H_2O)^3 \rightarrow NH_4{+}(H_2O)^3$ in the gas phase, J. Am. Chem. Soc. **109**:1607 (1987).
23. B. L. Tembe and J. A. McCammon, Ligand-receptor interactions, Comput. Chem. **8**:281 (1984).
24. C. L. Brooks, III and S. H. Fleischman, A theoretical approach to drug design. 1. Relative solvation thermodynamics for the antibacterial compound trimethoprim and ethyl derivatives substituted at the 3', 4', and 5' positions, J. Am. Chem. Soc. **112**:3307 (1990).
25. J. J. McDonald and C. L. Brooks, III, A theoretical approach to drug design. 2. Relative thermodynamics of inhibitor binding by chicken dihydrofolate reductase to ethyl derivatives of trimethoprim substituted at 3'-, 4'-, and 5'- positions, J. Am. Chem. Soc. **113**:2295 (1991).
26. J. J. McDonald and C. L. Brooks, III, A theoretical approach to drug design. 3. Relative thermodynamics of inhibitor binding by E. coli dihydrofolate reductase to ethyl derivatives of trimethoprim substituted at 3', 4', and 5' positions, J. Am. Chem. Soc. **114**:2062 (1992).
27. L. D. Landau and E. M. Lifschitz, Statistical Physics, 2nd ed., Pergamon, New York (1960).
28. J. P. Valleau and G. M. Torrie, Statistical Mechanics, Part A, B. J. Berne, ed., Plenum Press, New York (1977).
29. G. M. Torrie and J. P. Valleau, Nonphysical sampling distributions in Monte Carlo free-energy estimation: Umbrella sampling, J. Comput. Chem. **23**:187 (1977).
30. D. J. Tobias, S. F. Sneddon, and C. L. Brooks, III, Reverse turns in blocked dipeptides are intrinsically unstable in water, J. Mol. Biol. **216**:783 (1990).
31. S. Kumar, D. Bouzida, R. H. Swendsen, P. A. Kollman, and J. M. Rosenberg, The weighted histogram analysis method for free-energy calculations on biomolecules. I. The method, J. Comput. Chem. **13**:1011 (1992).
32. S. Kumar, P. W. Payne, and M. Vasquez, Method for free-energy calculations using iterative techniques, J. Comput. Chem. **17**:1267 (1996).
33. E. M. Boczko and C. L. Brooks, III, Constant-temperature free energy surface for physical and chemical processes, J. Phys. Chem. **97**:4509 (1993).

34. A. M. Ferrenberg and R. H. Swendsen, New Monte Carlo technique for studying phase transitions, Phys. Rev. Lett. **61**:2635 (1988).
35. A. M. Ferrenberg and R. H. Swendsen, Optimized Monte Carlo data analysis, Phys. Rev. Lett. **63**:1195 (1989).
36. I. Muegge and Y. C. Martin, A general and fast scoring function for protein-ligand interactions: A simplified potential approach, J. Med. Chem. **42**:791 (1999).
37. R. S. DeWitte and E. I. Shakhnovich, Smog de Novo design method based on simple, fast, and accurate free energy estimates. 1. Methodology and supporting evidence, J. Am. Chem. Soc. **118**:11733 (1996).
38. H. J. Bohm, The development of a simple empirical scoring function to estimate the binding constant for a protein-ligand complex of known three-dimensional structure, J. Comput.-Aided Mol. Design **8**:243 (1994).
39. H. J. Bohm, On the use of LUDI to search the fine chemicals directory for ligands of proteins of known three-dimensional structure, J. Comput.-Aided Mol. Design **8**:623 (1994).
40. M. D. Eldridge, C. W. Murray, T. R. Auton, and G. V. Paolini, Empirical scoring functions: 1. The development of a fast empirical scoring function to estimate the binding affinity of ligands in receptor complexes, J. Comput.-Aided. Mol. Design **11**:425 (1997).
41. I. D. Kuntz, J. M. Blaney, S. Oatley, R. Langridge, and T. Ferrin, A geometric approach to macromolecule-ligand interactions, J. Mol. Biol. **161** (1982).
42. I. D. Kuntz, E. C. Meng, and B. K. Shoichet, Structure-based molecular design accounts. Chem. Res. **27**:117 (1994).
43. R. L. DesJarlais, R. P. Sheridan, J. S. Dixon, I. D. Kuntz, and R. Venkataraghavan, Docking flexible ligands to macromolecular receptors by molecular shape, J. Med. Chem. **29**:2149 (1986).
44. J. Åqvist, C. Medina, and J. E. Samuelsson, A new method for predicting binding affinity in computer-aided drug design, Protein Eng. **7**:385 (1994).
45. J. Åqvist, Calculation of absolute binding free energies for charge ligands and effects of long-range electrostatic interactions, J. Comput. Chem. **17**:1587 (1996).
46. T. Hansson and J. Åqvist, Estimation of binding free energies for HIV proteinase inhibitors by molecular dynamics simulations, Protein Eng. **8**:1137 (1995).
47. T. Hansson, J. Marelius, and J. Åqvist, Ligand binding affinity prediction by linear interaction energy methods, J. Comput.-Aided Design **12**:27 (1998).
48. J. Marelius, T. Hansson, and J. Åqvist, Calculation of ligand binding free energies from molecular dynamics simulations, J. Comput. Chem. **69**:77 (1998).
49. H. A. Carlson and W. L. Jorgensen, An extended linear response method for determining free energies of hydration, J. Phys. Chem. **99**:10667 (1995).
50. M. L. Lamb, J. Tirado-Rives, and W. L. Jorgensen, Estimation of the binding affinities of FKBP12 inhibitors using a linear response method, Bioorg. Med. Chem. Lett. **7**:851 (1999).
51. J. Wang, R. Dixon, and P. A. Kollman, Raning ligand binding affinities with avidine: A molecular dynamics-based interaction energy study, Proteins: Structure, Function, and Genetics **34**:69 (1999).
52. H. Liu, A. E. Mark, and W. F. van Gunsteren, Estimating the relative free energy of different molecular states with respect to a single reference state, J. Phys. Chem. **100**:9485 (1996).
53. J. W. Pitera, N. R. Nunagala, C. C. Wang, and P. A. Kollman, Understanding substrate specificity in human and parasite phosphoribosyltranferase through calculation and experiment, Biochemistry **38**:10298 (1999).
54. L. Wang, D. L. Veenstra, R. J. Radmer, and P. A. Kollman, Can one predict stability? An attempt to do so for residue 133 of T4 lysozyme using a combination of free energy

derivatives, PROFEC, and free energy perturbation methods, Proteins: Structure, Function, and Genetics **32**:438 (1998).
55. T. E. Cheatham, III, J. Srinivasan, D. A. Case, and P. A. Kollman, Molecular dynamics and continuum solvent studies of the stability of polyG-polyC and polyA-polyT DNA duplexes in solution, J. Bioml. Struct. Dyn. **16**:265 (1998).
56. A.-S. Yang and B. Honig, Free energy determinants of secondary structure formation: I. λ-Helixes, J. Mol. Biol. **252**:351 (1995).
57. K. Osapay, W. S. Young, D. Bashdord, C. L. Brooks, III, and D. A. Case, Dielectric continuum models for hydration effects on peptide conformational transitions, J. Phys. Chem. **100**:2698 (1996).
58. J. Shen and J. Wendoloski, Electrostatic binding energy calculation using the finite difference solution to the linearized Poisson-Boltzmann equation: Assessment of its accuracy, J. Comput. Chem. **17**:350 (1996).
59. J. Shen and F. A. Quiocho, Calculation of binding energy differences for receptor-ligand systems using the Poisson-Boltzmann method, J. Comput. Chem. **16**:445 (1995).
60. M. Schaefer and M. Karplus, A comprehensive analytical treatment of continuum electrostatics, J. Phys. Chem. **100**:1578 (1996).
61. D. Qiu, P. S. Shenkin, F. P. Hollinger, and W. C. Still, The GB/SA continuum model for solvation. A fast analytical method for the calculation of approximate Born radii, J. Phys. Chem. A **101**:30005 (1997).
62. I. Klapper, R. Hagstrom, R. Fine, K. Sharp, and B. Honig, Focusing of electric fields in the active site of Cu-Zn superoxide dismutase: Effects of ionic strength and amino-acid modification, Proteins: Structure, Function, and Genetics **1**:47 (1986).
63. G. D. Hawkins, C. J. Cramer, and D. G. Truhlar, Pairwise solute A screening of solute charges from a dielectric medium, Chem. Phys. Lett. **246**:122 (1995).
64. G. D. Hawkins, C. J. Cramer, and C. D. Truhlar, Parametrized models of aqueous free energies of solvation based on pairwise descreening of solute atomic charges from a dielectric medium, J. Phys. Chem. **100**:19824 (1996).
65. S. R. Edinger, C. Cortis, P. S. Shenkin, and R. A. Friesner, Solvation free energies of peptides: Comparison of approximate continuum solvation models with accurate solution of the Poisson-Boltzmann equation, J. Phys. Chem. B **101**:1190 (1997).
66. B. Tidor, Simulated annealing on free energy surfaces by a combined molecular dynamics and Monte Carlo approach, J. Phys. Chem. **97**:1069 (1993).
67. J. Ji, T. Cagin, and B. M. Pettitt, Dynamics simulations of water at constant chemical potential, J. Chem. Phys. **96**:1333 (1992).
68. S. Nose, A unified formulation of the constant temperature molecular dynamics methods, J. Chem. Phys. **81**:511 (1984).
69. N. R. Metropolis, A. W. Rosenbluth, M. N. Rosenbluth, A. N. Teller, and E. Teller, Equation of state calculations by fast computing machines, J. Chem. Phys. **21**:1087 (1953).
70. C. H. Bennett, Efficient estimation of free energy differences from Monte Carlo data, J. Comp. Phys. **22**:245 (1976).
71. C. Jarque and B. Tidor, Simulated annealing on coupled free energy surfaces: Relative solvation energies of small molecules, J. Phys. Chem. B **101**:9362 (1997).
72. W. C. Still, A. Tempczyk, R. C. Hawley, and T. F. Hendrickson, Semianalytical treatment of solvation for molecular mechanics and dynamics, J. Am. Chem. Soc. **112**:6127 (1990).
73. D. Bashford and D. A. Case, Generalized Born models of macromolecular solvation effect, Annu. Rev. Phys. Chem. **51**:129 (2000).
74. B. N. Dominy and C. L. Brooks, III, Development of a generalized Born model parameterization for proteins and nucleic acids, J. Phys. Chem. B **103**:3765 (1999).

75. B. Jayaram, Y. Liu, and D. L. Beveridge, A modification of the generalized Born theory for improved estimates of solvation energies and pK shifts, J. Chem. Phys. **109**:1465 (1998).
76. B. Jayaram, D. Sprous, and D. L. Beveridge, Solvation free energy of biomacromolecules: Parameters for a modified generalized Born model consistent with the AMBER force field, J. Phys. Chem. B **102**:9571 (1998).
77. R. Luo, M. S. Head, J. Moult, and M. K. Gilson, pKa shifts in small molecules and HIV protease: Electrostatics and conformation, J. Am. Chem. Soc. **120**:6138 (1998).
78. A. Onufriev, D. Bashford, and D. A. Case, Modification of the generalized Born model suitable for macromolecules, J. Phys. Chem. B **104**:3712 (2000).
79. X. Zou, Y. Sun, and I. D. Kuntz, Inclusion of solvation in ligand binding free energy calculations using the generalized-Born model, J. Am. Chem. Soc. **121**:8033 (1999).
80. C. S. Rapp and R. A. Friesner, Prediction of loop geometries using a generalized Born model of solvation effects, Proteins: Structure, Function, and Genetics **35**:173 (1999).
81. S. Banba, K. V. Damodaran, and C. L. Brooks, III, Free energy simulations using the generalized Born approach, in preparation (2001).

Chapter 11

Ligand Interaction Scanning Using Free Energy Calculations

Mark D. Erion and M. Rami Reddy
Metabasis Therapeutics, Inc., San Diego, CA 92121

1. INTRODUCTION

The free energy perturbation (FEP) approach in conjunction with molecular dynamics represents a theoretically precise method for obtaining relative binding free energy differences.[1, 2] The high accuracy of FEP-based calculations relative to less rigorous approaches has led to its use in numerous structure-function studies designed to identify the key molecular factors that influence ligand binding affinity and enzyme catalysis. For example, Bash et al. showed that solvation plays a large role in the binding affinity of phosphorus-containing inhibitors of the bacterial protease thermolysin.[3] Erion and Reddy showed that FEP calculations could accurately predict the extent of heteroaromatic base hydration, which proved to be a dominating factor in the binding of certain purine analogues to adenosine deaminase.[4] FEP calculations also confirmed that a hydroxyl group was essential for high binding affinity for certain HIV-1 protease inhibitors[5] and a series of adenosine deaminase inhibitors.[6] Last, FEP calculations have been used to confirm the importance of certain active-site residues. For example, Kollman studied the role of several active-site residues of the serine protease subtilisin using the FEP approach. The calculated free energy differences for both the Asn155Ala mutant[7] and the Thr220Ala mutant[8] relative to wild-type enzyme were consistent with previously reported site-directed mutagenesis studies, which showed that both Asn155 and Thr220 were important for substrate binding and enzyme catalysis.

Despite these successes, the FEP method has not gained widespread use within the pharmaceutical industry. Probably the greatest factor that limits its use is that high accuracy is achieved only when calculating binding free energy differences for structurally-related inhibitors. Accordingly, the technique has little value in de novo drug design or for comparative analysis of compounds that differ by more than a few atoms. Another limitation is the substantial computer power required for FEP calculations, which prevents analysis of large sets of compounds in a high throughput manner.

To rectify this problem, efforts are underway to develop methods that predict binding free energies, at least qualitatively, for a medium-sized compound library (100-200 compounds) over a relatively short time period. The most promising methods to date include Linear Interaction Energy (LIE) analysis[9], Molecular Mechanics Poisson Boltzmann Surface Area (MM-PBSA)[10], the lambda dynamics approach[11] and the chemical Monte Carlo/Molecular dynamics approach[12]. Success using these methods will depend on their ability to accurately discriminate between structurally diverse compound series and thereby help prioritize the compound series for the medicinal chemists.

Recently, we reported an alternative strategy for using the FEP method in drug design and lead optimization.[13] The strategy entails using a series of FEP calculations to scan binding site interactions and obtain a highly accurate rank-ordering of their strength and overall contribution to ligand binding affinity. The binding site map generated from these calculations is expected to help pinpoint specific interactions within the binding site that contribute the most to ligand binding affinity. This information can help guide the design of more potent and specific ligands, since it reveals sites on the ligand that form interactions important for binding affinity and therefore should be retained in some form in the new analogue as well as sites on the ligand that interact weakly with the binding site and therefore can be changed in order to optimize the binding site interactions and improve overall ligand binding affinity.

1.1 Experimental Methods for Characterizing Binding Site Interactions

A high resolution X-ray structure of a protein-ligand complex provides a detailed map of the binding site interactions, but fails to reveal information regarding their relative contributions to binding affinity. This is especially true for hydrogen bonds whose strength varies from 0-5 kcal/mol depending on the bond distance and bond angle, as well as the electronic nature of the donor and acceptor groups.[14] In addition, the local environment is often an important factor with the strongest hydrogen bonds formed in poorly solvated regions of the binding site cavity.

Since the contribution of individual hydrogen bonds to ligand binding affinity is considered valuable information for the design of new ligands, analogues of the lead inhibitor are frequently prepared wherein the individual heteroatoms that form hydrogen bonds with the protein are replaced with non-hydrogen bonding atoms or substituents. For similar reasons, site-directed mutagenesis is often used to replace binding site residues containing side-chains that interact with the ligand through a hydrogen bond with residues whose side-chains are incapable of forming hydrogen bonds. Data generated from the analogues and site-directed mutants can be used to determine the role of specific binding site interactions in enzyme catalysis and ligand binding affinity. While this information is very valuable for optimizing lead candidates, it also entails an enormous amount of time and manpower to prepare and characterize the analogues and site-directed mutants.

The power of this information and the resource commitment required to obtain it are illustrated by work conducted in the 1980s and early 1990s on purine nucleoside phosphorylase (PNP). The X-ray structure of the PNP-formycin B complex suggested that PNP binds its substrates (guanosine and phosphate) through a complex network of hydrogen bonds involving the side-chains of 13 active-site residues.[15] Analysis of the extensive guanosine SAR[16] helped to characterize the importance of these interactions based on differences in inhibitory activity between guanosine and analogues of guanosine wherein a hydroxyl, oxygen or nitrogen was replaced with a non-hydrogen bonding group. Similarly, site-directed mutagenesis was used to replace the 13 active-site side-chains with a non-hydrogen bonding side-chain (e.g. methyl or phenyl).[17] The catalytic efficiency of each mutant was subsequently evaluated relative to wild-type PNP. These studies showed that many of the hydrogen bonds identified in the X-ray structure had little importance to catalysis or binding affinity. On the other hand, a hydrogen bond between ^{7}N and the side-chain amido group of Asn243 proved to be an important contributor to overall affinity and catalysis[18] despite initial guesses to the contrary based on the electron density associated with the Asn243 side-chain in the X-ray structure of the initial PNP-inhibitor complex and the distance and angle of the hydrogen bond. The results of both the pseudosubstrate kinetic studies and the mutagenesis studies enabled generation of a ligand interaction map wherein the heteroatoms of guanosine and the PNP active-site residues were classified based on the strength of their overall contribution (Figure 1). Accordingly, hydrogen bonds formed to the guanine base by Asn243 and Glu201 were exploited in the design of a series of potent PNP inhibitors whereas the weak hydrogen bonds formed between PNP and the ribose moiety were largely ignored leading to the replacement of the ribose with a hydrophobic arylalkyl group.[19]

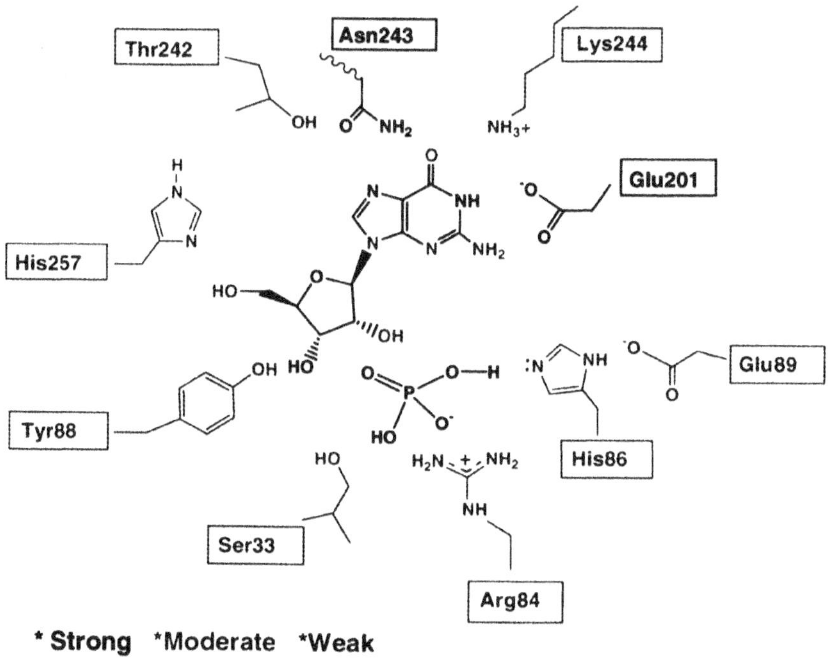

Figure 1. PNP Ligand Interaction Map. Ligand atoms or residues classified by their importance to ligand binding affinity.

2. INTERACTION SCANNING USING FREE ENERGY CALCULATIONS

Free energy calculations represent an alternative strategy for scanning ligand binding sites and identifying interactions important for drug design. The calculations are fast relative to either the synthesis of analogues or the preparation and characterization of active-site mutants. In addition, computational studies have the ability to evaluate certain interactions that are impossible to test experimentally. For example, interactions involving backbone amides are impossible to assess by mutagenesis. Moreover, replacement of ligand heteroatoms with the corresponding carbon atom can on occasion require an enormous effort due either to synthetic difficulties or compound instability.

Computational analysis of binding site interactions is achieved by calculating the relative binding free energy ($\Delta\Delta G_{bind}$) for a ligand L with an

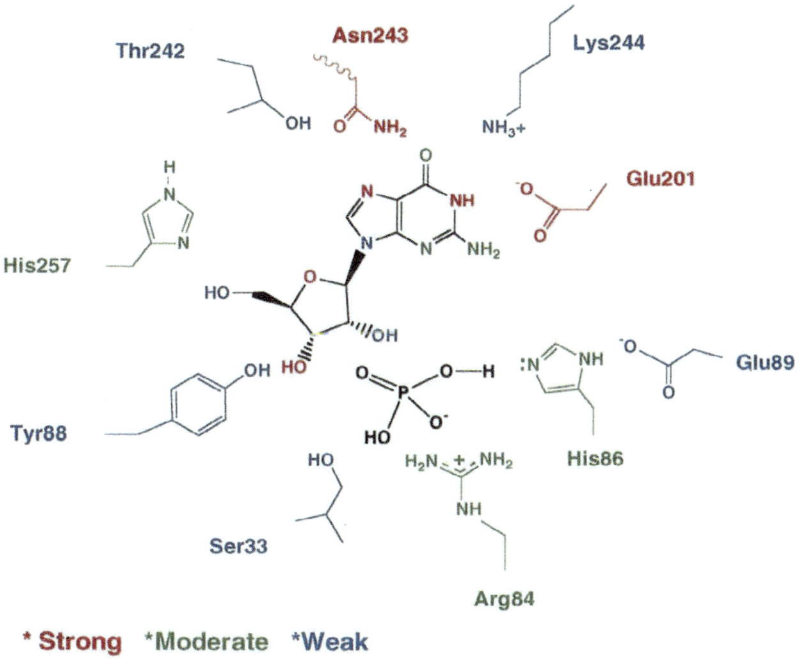

Figure 1. PNP Ligand Interaction Map. Ligand atoms or residues classified by their Importance to ligand binding affinity.

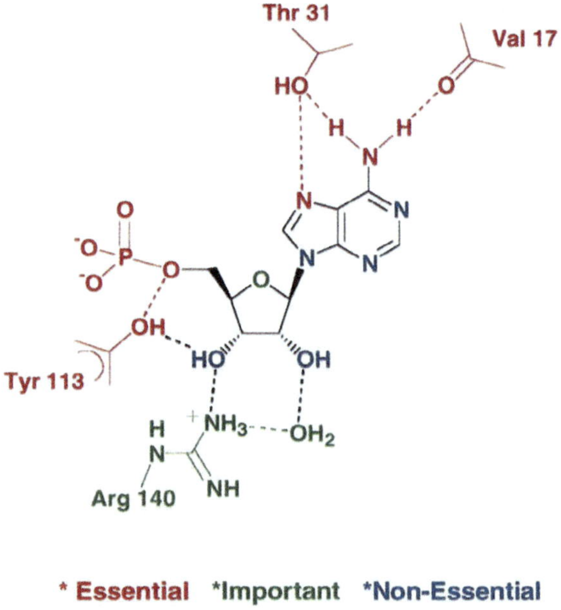

Figure 5. Strength of AMP interactions with FBPase (C4).

enzyme E relative to either a modified ligand L' with E or a mutated enzyme E' with L wherein both L' and E' differ from L and E, respectively, by a simple structural modification that primarily affects the interactions of the mutated group. Accurate results are obtained using the free energy perturbation (FEP) methodology,[1, 2] which computationally transforms L into L' or E into E' in the complex and in solvent in order to calculate ΔG_{com} and ΔG_{aq} the difference of which is $\Delta\Delta G_{bind}$. Accuracy, however, requires the calculations to reach satisfactory convergence, which for all practical purposes is achieved using molecular dynamics simulations with explicit solvent only when the transformation entails a small structural change or in some cases a slightly larger change coupled with long simulation times.[20] Since hydrogen bonds formed between E and L are usually eliminated by a simple structural modification in either the protein or ligand, free energy calculations are well suited for determining the relative intrinsic strength of each hydrogen bond and the effect of solvation on its overall contribution to ligand binding affinity.

3. SCANNING THE AMP BINDING SITE OF FBPASE

The AMP binding site of fructose 1,6-bisphosphatase (FBPase) represents a good target for evaluating the possible use of free energy calculations for scanning binding site interactions. First, FBPase is an important new target for type II diabetes drugs based on its central role within the gluconeogenesis pathway[21] and the association of this pathway with the excessive production of glucose by the livers of non-insulin dependent diabetes mellitus patients.[22] Second, our efforts to discover FBPase inhibitors[23] focused on an allosteric binding site in which AMP binds and induces a protein conformational change that results in potent inhibition of enzymatic activity. Unfortunately, the discovery of AMP mimetics that bind with high potency and specificity to AMP binding sites is considered to be extraordinary challenging with only a few success reported to date. A third reason to choose FBPase for our studies is revealed in the high resolution X-ray structure of the FBPase-AMP complex, which showed that AMP formed up to 13 hydrogen bonds with human FBPase (Figure 2).[24]

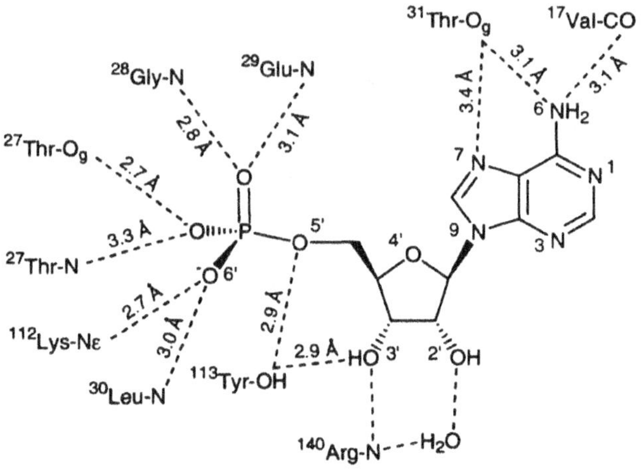

Figure 2. Binding site interactions of Human FBPase (C4) with AMP.

3.1 Computational Details

The FEP approach and its use for computing the relative binding free energy changes is well described.[1-4] The initial report describing our computational studies is also detailed elsewhere.[13] In brief, all molecular dynamics, mechanics and FEP calculations were carried out with the AMBER program using an all atom force field[25] and the SPC/E water potential[26] to describe water interactions. Electrostatic charges and parameters for the standard residues were taken from the AMBER database. For non-standard solute atoms, partial charges were obtained by using CHELPG to fit the ab initio 6-31G* basis set level wave functions calculated with Gaussian94.

Solvation free energies were computed by solvating the solute with SPC/E water and using the AMBER box option. All solvent molecules > 15.0 Å or < 2.5 Å from the closest solute atom were removed. Aqueous phase molecular dynamics simulations were carried out in a rectangular box using periodic boundary conditions in all directions. Binding free energies were computed using an FBPase model[27] generated from the coordinates of the FBPase-ZMP complex. The total charge on the FBPase tetramer complex was +4 e. No counterions or changes in the customary charge of protein residues were used. The entire system was immersed in a 25.0 Å radius sphere of solvent centered on the mutating group and subjected to a half-harmonic restraint near the boundary to prevent evaporation. During the simulation, all atoms of the protein were fixed beyond 25.0 Å. All non-bonded interactions involving the inhibitors and the charged residues of the protein were computed with infinite cutoff. A 15.0 Å non-bonded residue based cutoff was used for other residues of the system. The algorithm for the complex simulation was identical to the solvent simulation, except for the absence of periodic boundary conditions in the former.

Relative solvation ($\Delta\Delta G_{sol}$) and relative binding ($\Delta\Delta G_{bind}$) free energies between AMP and its analogues complexed to FBPase as well as AMP with FBPase mutants were calculated using the thermodynamic cycle perturbation approach in conjunction with molecular dynamics (time step = 2 fs) simulations. In all free energy simulations, the system was initially equilibrated for 20 ps followed by 2.5 ps of equilibration and 5 ps of data collection for each window. A total of 51 windows were used for each mutation. Free energies reported for each mutation represent the average of four calculations, i.e., forward and reverse mutations starting from AMP (L1) and the corresponding AMP analogue (L2). Error bars are estimated for each window by dividing the window statistics into four groups and computing the standard deviation. The root-mean-square of these window errors is reported as a measure of the statistical uncertainty in the results for each complete mutation.

3.2 Scanning Results vs. Experimental Relative Binding Affinities

The calculated relative binding free energies for AMP relative to AMP analogues were compared with relative binding free energies derived from the IC_{50} (AMP)/IC_{50} (AMP analogue) ratio. AMP and AMP analogues inhibit FBPase by non-competitive kinetics, which is best represented by a two site Hill model where $K_d(1) << K_d(2)$ and the ratio of the dissociation constants for site 1 and site 2 for AMP (K_d) and its analogue (K_d') are the same (Equation 1). Under these conditions, the ratio of the IC_{50}s for AMP and the AMP analogue is proportional to the ratio of the dissociation constants, which is proportional to the relative binding free energy (Equation 2).

$$K_d(1)/K_d'(1) \approx K_d(2)/K_d'(2) \tag{1}$$

$$\Delta\Delta G_{bind} \propto K_d(1)/K_d'(1) \propto IC_{50}/IC_{50}' \tag{2}$$

Similarly, differences for the binding affinity of AMP for the AMP binding site of wild-type FBPase vs. mutant FBPase were calculated and compared with experimental data derived from the ratio of the corresponding IC_{50}s. The calculated results generated from the scan were compared with the experimentally-determined relative binding affinities and reported in our earlier communication.[13]

4. ANALYSIS OF FBPASE-AMP INTERACTIONS

4.1 Purine Binding Site

The computer model of the FBPase-AMP complex showed 3 potential hydrogen bonds between the purine base and nearby residues (Figure 2). The hydrogen bonds between the $^6NH_2$ group and both the carbonyl oxygen of Val17 and the side-chain hydroxyl of Thr31 were well within the ranges reported for ideal hydrogen bonds and therefore expected to represent relatively strong interactions. In contrast, the hydrogen bond between the ^7N and the Thr31 hydroxyl appeared to be less ideal with a distance of 3.4 Å. No apparent hydrogen bond interactions were observed between FBPase and ^1N, ^3N, and ^9N. Each of these interactions was scanned and the calculated results compared with available experimental data (Figure 3).

4.1.1 Scan of Base Heteroatoms

Consistent with the experimental data, mutation of the purine base nitrogens, i.e. ^1N, ^3N, ^7N and ^9N, showed that replacement of ^1N, ^3N and ^9N with CH had little effect on binding affinity whereas a similar replacement of ^7N led to a loss of 2.8 kcal/mol. The 0.6 kcal/mol gain in affinity for the 1-deaza and 3-deaza AMP analogues was consistent with the hydrophobic nature of this portion of the binding site cavity and the absence of hydrogen bond donors in the vicinity of either heteroatom.

Less apparent from the X-ray structure were the calculated results for the mutations of the $^6NH_2$ group, ^7N and ^9N. Mutation of the $^6NH_2$ group to hydrogen led to a loss of only 2.3 kcal/mol, which was in agreement with the experimental result of 2.8 kcal/mol. The results indicate that despite

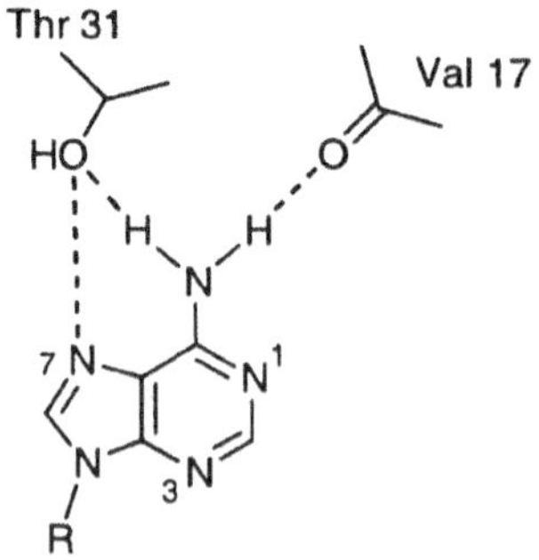

Mutation			ΔΔG_{bind} (kcal/mol) Calc	Exp
^1N	⟶	CH	-0.6±0.5	ND
^3N	⟶	CH	-0.5±0.5	ND
$^6NH_2$	⟶	H	2.3±0.8	2.7
^7N	⟶	CH	2.8±0.6	3.3
^9N	⟶	C	0.6±0.5	0.3
^{31}Thr	⟶	Ala	3.1±0.8	2.9

Figure 3. Scanning results for purine base interactions.

achieving near ideal hydrogen bond distances and angles in the complex, the $^{6}NH_2$ group contributes much less than the upper end of the range for a single hydrogen bond, i.e. 5 kcal/mol. One likely reason for the less than expected contribution of the $^{6}NH_2$ group is suggested by the FEP calculations, which predicted the relative solvation free energy for the mutation of the $^{6}NH_2$ group to hydrogen to be -4.0 kcal/mol and therefore that desolvation costs associated with the $^{6}NH_2$ group would compromise its overall contribution to the binding free energy.

In contrast to the $^{6}NH_2$ group, the calculated relative binding free energy for the mutation of ^{7}N to CH was much larger than expected based on the X-ray structure, which showed the interaction between ^{7}N and the hydroxyl of Thr31 (O_g) to span a distance (3.4 Å), i.e. a distance slightly outside the normal range for an optimal hydrogen bond. The calculated result of 2.8 kcal/mol agreed with the experimental result of 3.3 kcal/mol.[28] One possible explanation for the importance of ^{7}N despite the apparent suboptimal hydrogen bond distance is that there is an intervening water molecule. This possibility is supported by a recent X-ray structure by Novo Nordisk of the FBPase-ZMP complex.[29] While possible, our calculations using a model that was devoid of the intervening water molecule produced a relative binding affinity that agreed with the experimental result.[13] Similarly, as discussed in the next section, mutation of Thr31 to Ala showed a significant loss in binding affinity, which is best explained by the absence of interactions with both the $^{6}NH_2$ group and ^{7}N. It should be noted that the loss in relative binding free energy for the ^{7}N to CH mutation may not only be due to the loss of the ^{7}N-Thr31 interaction but also to the effects of the mutation on the partial atomic charges of other base heteroatoms, especially the $^{6}NH_2$ group. For example, the charge and desolvation costs associated with the $^{6}NH_2$ group are likely to change slightly with the mutation and thereby contribute to the effect of the ^{7}N to CH mutation on AMP binding.

Replacement of ^{9}N with a carbon atom converts ^{7}N from a hydrogen bond acceptor to a hydrogen bond donor. Thus, while no interactions are observed in the X-ray structure between FBPase and ^{9}N, the reversal of the hydrogen bonding role for ^{7}N might be expected to result in a decrease in relative binding affinity similar to or greater than that found for the ^{7}N to CH mutation. The calculated result, however, predicted only a slight loss in binding affinity (0.6 kcal/mol), which agreed with the experimental result (0.3 kcal/mol). This result suggests that the binding site can tolerate the dramatic change in hydrogen bonding pattern presumably by restructuring itself in a manner that allows the hydroxyl of Thr31 to accept a hydrogen bond from both ^{7}N and the $^{6}NH_2$ group. Such restructuring is precedented based on comparative studies with PNP and guanine-based inhibitors, which clearly indicated that the 9-deazaguanine inhibitor series was accommodated by the PNP binding site through movement of the amido group of the

Asn243 side-chain in a manner that converted it from a donor of a hydrogen bond to the ^{7}N of guanine to an acceptor of a hydrogen bond from the inhibitor ^{7}NH.[19] In our studies, 9-deaza AMP (formycin A monophosphate) showed no significant difference relative to AMP in binding affinity or protein interactions, which suggested that Thr31 can act as both a hydrogen bond acceptor and a donor with ^{7}N and that the strength of the interactions are approximately the same.

4.1.2 Scan of Residues in the Purine Base Binding Region

The Thr31 hydroxyl accepts a hydrogen bond from the $^6NH_2$ group and donates a hydrogen bond to ^{7}N. Mutation of Thr31 to Ala eliminates both hydrogen bonds and results in a large decrease in AMP binding affinity.[28] The calculated binding free energy for the mutation was 3.1 kcal/mol and therefore consistent with the experimental result of 2.9 kcal/mol. The proportion of the lost binding affinity attributed to the Thr31-$^6NH_2$ hydrogen bond relative to the Thr31-^{7}N hydrogen bond was not readily discernable from the calculations. The larger decrease in AMP binding affinity associated with the wild-type to Thr31Ala mutation compared to the loss incurred by the mutation of the $^6NH_2$ group to H, however, suggests that the hydrogen bond with ^{7}N is a significant contributor to AMP binding affinity.

4.2 Ribose 5-Phosphate Binding Site

The computer model of the FBPase-AMP complex showed a large hydrogen bonding network between the ribose 5-phosphate of AMP and FBPase (Figure 2). The phosphate moiety formed a full complement of hydrogen bonds within a very hydrophilic region of the binding pocket. In contrast, fewer hydrogen bonds were formed with the ribosyl moiety. The phenolic hydroxyl of Tyr113 appeared to donate a hydrogen bond to the 5'-oxygen and accept a hydrogen bond from the 3'-hydroxyl. No hydrogen bond was observed to the ribosyl oxygen. Hydrogen bonds to the 2' and 3' hydroxyls were dependent on the position of the guanidino group of Arg140, which varied across the subunits. In one subunit (C4), the guanidino group was within hydrogen bond distance to the 3'-hydroxyl. Moreover, an intervening water molecule between the guanidino group and the 2' hydroxyl was observed. In other subunits, however, the side-chain of Arg140 pointed out of the binding site and therefore was not in contact with AMP. Analysis of the interaction energy of AMP with each subunit showed the subunit designated C4 to exhibit the lowest energy possibly reflecting the Arg140 interaction. Accordingly, each of the interactions with AMP in the subunit C4 were scanned and the calculated results compared with available experimental data (Figure 4).

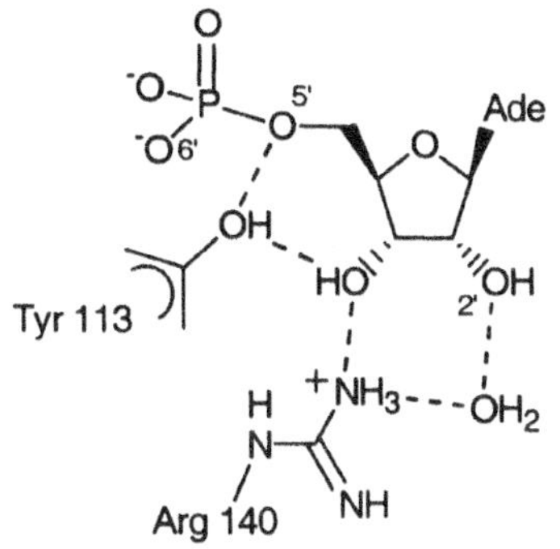

Mutation			$\Delta\Delta G_{bind}$ (kcal/mol) Calc	Exp
$^{2'}$OH	→	H	5.1±0.9	0
$^{3'}$OH	→	H	5.9±0.9	ND
$^{4'}$O	→	CH_2	1.1±0.7	≈ 0.6
$^{5'}$O	→	CH_2	4.6±0.8	5.4
$^{6'}$O	→	H	ND	> 5.4
^{113}Tyr	→	Phe	3.9	> 4.2

Figure 4. Scanning results for ribose 5-phosphate interactions.

4.2.1 Scan of Ribose 5-Phosphate Heteroatoms

As with the scan of the purine base heteroatoms, relative free energy calculations also provided insight into the contributions of the ribosyl 5-phosphate oxygen atoms to AMP binding affinity. Scanning the ribosyl oxygen and 5'-oxygen as well as a negatively charged phosphate oxygen yielded results that were expected from the X-ray structure and were consistent with experimental data. The modest loss in binding affinity for the mutation of the ribosyl oxygen (4') to methylene was not surprising given that no hydrogen bonds with the oxygen were formed in the complex. The modest loss in affinity (1.1. kcal/mol (calculated) and 0.6 kcal/mol (experimental)) may reflect either electronic effects caused by this change on the very important 5'-oxygen interactions or possibly it reflects detrimental effects associated with the placement of a methylene in a relatively hydrophilic region of the binding site.

In contrast to the mutation of the ribosyl oxygen, the $^{5'}$O → CH_2 mutation resulted in a large loss in binding affinity (4.6 kcal/mol). While this result properly reflected the importance of the $^{5'}$O interaction, it was less than the experimental value (> 5.4 kcal/mol) possibly because of an under estimation of the electrostatic differences between a phosphate and the less acidic phosphonic acid. Similarly, a large decrease in binding free energy was calculated for the $^{6'}$O → H mutation. It should be noted, however, that the calculated results were not readily quantifiable, since the reduction in atomic charge led to large changes in the solvation and complex free energies and poor overall convergence. Nevertheless, the results were clearly consistent with the experimental finding. Presumably, the H-phosphonate exhibits a substantially lower binding affinity relative to AMP due to the loss of interactions with the negatively charged phosphate oxygen as well as the reduced overall strength of the interactions with the remaining three phosphate oxygens, which as part of an H-phosphonate exhibit less negative charge.

Scanning the 2'-hydroxyl as well as the 3'-hydroxyl showed that the OH to H mutation in both cases is associated with a > 5 kcal/mol loss in binding affinity. The large loss in affinity was not surprising given the apparent strong hydrogen bonds formed between the Arg140 guanidino group and the ribosyl hydroxyls (via an intervening water molecule with the 2' hydroxyl). The calculated results, however, were inconsistent with experimental data. First, the IC_{50} for 2'-deoxyAMP was identical to the IC_{50} for AMP, which translates to an experimentally-determined relative binding free energy difference of ≈ 0 kcal/mol. Second, the Arg140Ala mutant showed only a modest decrease in AMP binding affinity.[28] One possible explanation for these results is that desolvation costs for the positively charged guanidino group completely offset any gains made in the complex through hydrogen bond donation to the ribosyl oxygens.

A simpler explanation for the modest effect of Arg140 to Ala mutation is that the minimized structure of the AMP binding site in the C4 subunit is incorrect and therefore that the Arg140 contacts with AMP are misrepresented. Indeed, analysis of the AMP interactions in the three remaining subunits showed that the amino acid side-chain orientations were nearly superimposable across the four subunits except for the Arg140 side-chain, which was pointing away from the ribosyl hydroxyls in all cases except for the subunit used (C4) for the initial free energy calculations. The finding of an alternative position for the Arg140 side-chain is plausible given that that the X-ray data showed the Arg140 side-chain to be very flexible and to be located at the protein-water interface.

To evaluate the effect of the Arg140 side-chain location on binding free energy, a second computer model was generated using the X-ray structure of the second subunit. A scan of this site (C2) gave nearly identical relative binding free energies for ^{7}N and $^{6}NH_2$, but resulted in much lower values for the ribosyl hydroxyls (Table 1).

Table 1. Scanning results for site 2.

Mutation			$\Delta\Delta G_{bind}$ (kcal/mol) Site 1	Site 2
$^{6}NH_2$	→	H	2.3±0.8	2.4±0.8
^{7}N	→	CH	2.8±0.6	2.5±0.7
$^{2'}OH$	→	H	5.1±0.9	-0.6±0.7
$^{3'}OH$	→	H	5.9±0.9	0.8±0.7

4.2.2 Scan of Residues in the Ribose 5-Phoshate Binding Region

Computational mutation of Tyr113 to Phe resulted in a 700-fold decrease in the binding affinity of AMP relative to wild-type FBPase. The calculated results were consistent with earlier experimental data.[28] Since Tyr113 hydrogen bonds to both the 5' oxygen and the 3' hydroxyl, it is unclear how much each hydrogen bond contributes to AMP binding affinity.

4.3 Classification of the AMP Interactions

The results of the ligand scan show that hydrogen bonds to ^{7}N and the $^{6}NH_2$ of the purine, the 5'-oxygen of the ribosyl moiety and the phosphate group are essential for high binding affinity (Figure 5). Interactions elsewhere on AMP appear to be less important and therefore may represent sites that could undergo structural modification in order to enhance inhibitor binding and or selectivity.

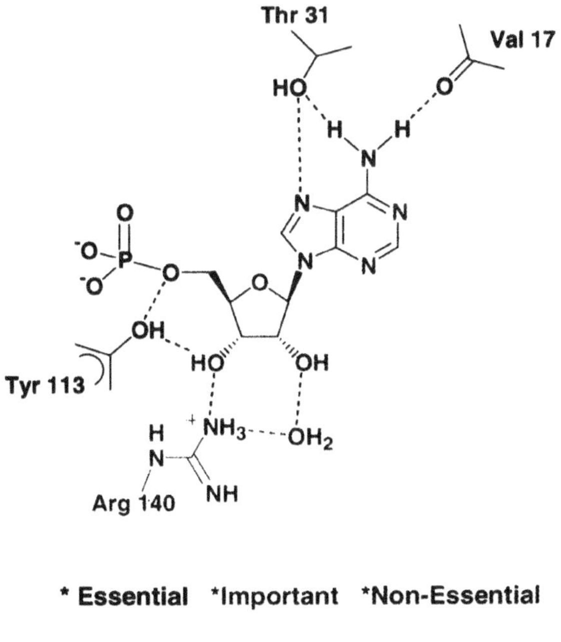

Figure 5. Strength of AMP interactions with FBPase (C4).

5. USE IN LIGAND DESIGN

Drug discovery usually entails two steps, namely an initial step that results in the identification of a lead inhibitor followed by an optimization step wherein analogues of the lead inhibitor are synthesized and evaluated in order to discover a development candidate with suitable potency and pharmacokinetics. The entire process can take from two to greater than seven years. Advances in X-ray crystallography and computational methods in the 1980s were expected to significantly shorten both phases by focusing medicinal chemistry efforts on targets that were "designed" to be potent and selective inhibitors based on analysis of the structure and binding energetics. Unfortunately, the promise of computational chemistry has largely gone unfulfilled with few examples clearly attributing these methods to the discovery of a novel and unanticipated inhibitor series or to a significantly shorter lead optimization period. While many factors are likely to account for this disappointment, the major reason for the declining emphasis of computational methods in drug discovery programs relates to the accuracy and reliability of the results.

The FEP method has shown good accuracy when used to determine the relative binding affinity for two structurally-related ligands. Unfortunately, the newly designed analogues requested by medicinal chemists frequently represent large structural changes from the lead compound and therefore are not well suited for the FEP method.

In the studies described earlier[13] and discussed above, we used the FEP method for determining the contribution of specific binding interactions to the overall ligand binding affinity. The studies entailed relatively small structural changes and therefore were associated with high accuracy and speed. The power of the strategy for drug design is that it provides a semi-quantitative map of the ligand interactions with the binding site. While the X-ray structure of the protein-ligand complex is essential for the identification of these interactions, this method enables analysis of their actual contribution to overall ligand binding affinity. Knowledge of the hydrogen bond strength is valuable in drug design, since it helps focus the design process on molecules that retain hydrogen bond interactions that contribute most to ligand binding affinity. Moreover, it reveals other areas of the molecule that contribute considerably less to binding affinity and therefore can be modified in order to produce more optimal interactions and a more potent and selective inhibitor.

6. CONCLUSIONS

The results from these studies showed that ligand interaction scanning accurately characterized the interactions of AMP with the AMP site of FBPase.[13] AMP, like other nucleotides, is bound to the AMP site predominantly through a large network of hydrogen bonds. Ligand interaction scanning correctly classified these interactions into strong, medium and weak. The information gained from these studies is currently being exploited in our design of AMP mimetics as FBPase inhibitors.[27] Moreover, these results are expected to enhance our ongoing and highly challenging quest to discover AMP mimetics that exhibit high binding affinity, high enzyme specificity and good cell penetration.[30]

7. REFERENCES

1. B. L. Tembe and J. A. McCammon, Ligand-receptor interactions, Comput. Chem. **8**:281 (1984).
2. D. A. Pearlman, Free energy calculations: Methods for estimating ligand binding affinities, in: Free Energy Calculations in Rational Drug Design, M. R. Reddy and M.D. Erion, eds., Kluwer/Plenum Press, New York (2001), pp. 9-35.
3. P. A. Bash, U. C. Singh, F. K. Brown, R. Langridge, P. A. Kollman, Calculation of the relative change in binding free energy of a protein-inhibitor complex, Science **235**:574 (1987).
4. M. D. Erion and M. R. Reddy, Calculation of relative hydration free energy differences for heteroaromatic compounds: Use in the design of adenosine deaminase and cytidine deaminase inhibitors, J. Am. Chem. Soc. **120**:3295 (1998).
5. M. R. Reddy and K. Appelt, HIV-1 protease: Structure-based drug design of inhibitors using the free energy perturbation approach, in: Free Energy Calculations in Rational Drug Design, M. R. Reddy and M. D. Erion, eds., Kluwer/Plenum Press, New York (2001), pp. 317-334.
6. L. M. Hansen and P.A. Kollman, Free energy perturbation calculations on models of active sites: Applications to adenosine deaminase inhibitors, J. Comp. Chem. **11**:994 (1990).
7. S. N. Rao, U. C. Singh, P. A. Bash, and P.A. Kollman, Free energy perturbation calculations on binding and catalysis after mutating Asn155 in subtilisin, Nature **328**:551 (1987).
8. N. Mizushima, D. Spellmeyer, S. Hirono, D. Pearlman, and P. Kollman, Free energy perturbation calculations on binding and catalysis after mutating threonine 220 in subtilisin, J. Biol. Chem. **266**:11801 (1991).
9. J. Aqvist and J. Marelius, The linear interaction energy method for computation of ligand binding affinities, in: Free Energy Calculations in Rational Drug Design, M. R. Reddy and M. D. Erion, eds., Kluwer/Plenum Press, New York (2001), pp. 171-194.
10. B. Kuhn, O. Donini, S. Huo, J. Wang, and P. A. Kollman, MM-PBSA applied to computer-assisted ligand design, in: Free Energy Calculations in Rational Drug Design, M. R. Reddy and M. D. Erion, eds., Kluwer/Plenum Press, New York (2001), pp. 243-251.
11. S. Banba, Z. Guo, and C. L. Brooks III, New free energy based methods for ligand binding from detailed structure-function to multiple-ligand screening, in: Free Energy

Calculations in Rational Drug Design, M. R. Reddy and M. D. Erion, eds., Kluwer/Plenum Press, New York (2001), pp. 195-223.

12. A. L. Eriksson, J. Pitera, and P. A. Kollman, Prediction of the binding free energies of new TIBO-like HIV-1 reverse transcirptase inhibitors using a combination of PROFEC, PB/SA, CMC/MD, and free energy calculations, J. Med. Chem. **42**:868 (1999).
13. M. D. Erion, P. D. van Poelje, and M. R. Reddy, Computer-assisted scanning of ligand interactions: Analysis of the fructose 1,6-bisphosphatase-amp complex using free energy calculations, J. Am. Chem. Soc. **122**:6114 (2000).
14. A. R. Fersht, J. P. Shi, J. Knill-Jones, D. M. Lowe, A. J. Wilkinson, D. M. Blow, P. Brick, P. Carter, M. M. Waye, and G. Winter, Hydrogen bonding and biological specificity analysed by protein engineering, Nature **314**:235 (1985).
15. S. E. Ealick, S. A. Rule, D. C. Carter, T. J. Greenhough, Y. S. Babu, W. J. Cook, J. Habash, J. R. Helliwell, J. D. Stoeckler, R. E. Parks, Jr., S. F. Chen, and C. E. Bugg, Three-dimensional structure of human erythrocytic purine nucleoside phosphorylase at 3.2 A resolution, J. Biol. Chem. **265**:1812 (1990).
16. J. D. Stoeckler, in: Developments in Cancer Chemotherapy R. I. Glazer, Ed., CRC Press Boca Raton, FL. (1984), pp. 35-60.
17. M. D. Erion, K. Takabayashi, H. B. Smith, J. Kessi, S. Wagner, S. Honger, S. L. Shames, and S. E. Ealick, Purine nucleoside phosphorylase. 3. Reversal of purine base specificity by site-directed mutagenesis, Biochemistry **36**:11725 (1997).
18. M. D. Erion, J. D. Stoeckler, W. C. Guida, R. L. Walter, and S. E. Ealick, Purine nucleoside phosphorylase. 2. Catalytic mechanism, Biochemistry **36**:11735 (1997).
19. S. E. Ealick, Y. S. Babu, C. E. Bugg, M. D. Erion, W. C. Guida, J. Montgomery, and J. A. Secrist, III, Application of crystallographic and modelling methods in the design of purine nucleoside phosphorylase inhibitors, Proc. Natl. Acad. Sci. USA **88**:11540 (1991).
20. M. R. Reddy, and M. D. Erion, Calculation of relative solvation free energy differences by thermodynamic perturbation method: Dependence on free energy results on simulation length, J. Comp. Chem. **20**:1018 (1999).
21. D. Granner, and S. J. Pilkis, The genes of hepatic glucose metabolism, J. Biol. Chem. **265**:10173 (1990).
22. I. Magnusson, D. L. Rothman, L. D. Katz, R. G. Shulman, and G. I. Shulman, Increased rate of gluconeogenesis in type II diabetes mellitus. A 13C nuclear magnetic resonance study, J. Clin. Invest. **90**:1323 (1992).
23. M. R. Reddy and M. D. Erion, Calculation of relative free energy differences for fructose 1,6-bisphosphatase inhibitors using the thermodynamic cycle perturbation approach, J. Am. Chem. Soc. in press (2001).
24. H. M. Ke, Y. P. Zhang, and W. N. Lipscomb, Crystal structure of the neutral form of fructose 1,6-bisphosphatase complexed with the product fructose 6-phosphate at 2.1-A resolution, Proc. Natl. Acad. Sci. USA **87**:5243 (1990).
25. S. J. Weiner, P. A. Kollman, D. A. Case, U. C. Singh, C. Ghio, G. Alagoha, S. Profeta, Jr., and P. K.Weiner, J. Am. Chem. Soc. **106**:765 (1984).
26. M. R. Reddy and M. Berkowitz, The dielectric constant of SPC/E water, Chem. Phys. Lett. **155**:173 (1989).
27. M. R. Reddy and M. D. Erion, Fructose 1,6-bisphosphatase: Use of free energy calculations in the design and optimization of AMP mimetics, in: Free Energy Calculations in Rational Drug Design, M. R. Reddy and M. D. Erion, eds., Kluwer/Plenum Press, New York (2001), pp. 285-297.
28. M. Gidh-Jain, Y. Zhang, P. D. van Poelje, J. Y. Liang, S. Huang, J. Kim, J. T Elliott, M. D. Erion, S. J. Pilkis, M. R. El-Maghrabi, and W. N. Lipscomb, The allosteric site of human liver fructose 1,6-bisphosphatase. Analysis of six AMP site mutants based on the crystal structure, J. Biol. Chem. **269**:27732 (1994).

29. L. F. Iversen, M. Brzozowski, S. Hastrup, R. Hubbard, J. S. Kastrup, I. K. Larsen, L. Naerum, L. Norskov-Lauridsen, P. B. Rasmussen, L. Thim, F. C. Wiberg, and K. Lundgren, Characterization of the allosteric binding pocket of human liver fructose 1,6-bisphosphatase by protein crystallography and inhibitor activity studies, Protein Sci. **6**:971 (1997).
30. M. D. Erion, S. R. Kasibhatla, B. C. Bookser, P. D. van Poelje, M. R. Reddy, H. E. Gruber, and J. R. Appleman, Discovery of AMP mimetics that exhibit high inhibitory potency and specificity for AMP deaminase, J. Am. Chem. Soc. **121**:308 (1999).

Chapter 12

MM–PBSA Applied to Computer-Assisted Ligand Design

Bernd Kuhn[†], Oreola Donini[‡], Shuanghong Huo[†], Junmei Wang[†], and Peter A. Kollman[†]
[†]*Dept. of Pharmaceutical Chemistry, University of California, San Francisco, CA 94143*
[‡]*Kinetek Pharmaceuticals, Inc., Vancouver, BC, Canada V6P 6P2*

1. INTRODUCTION

Structure based ligand design, which involves the finding and optimization of small ligands that bind tightly and specifically to a macromolecular target structure, is a significant challenge. Computational methods that have been applied to this problem include rapid docking and scoring with heuristic potentials so that entire databases or combinatorial libraries (10^5–10^6 compounds) can be screened.[1] At the other end of the spectrum are free energy approaches using either molecular dynamics (MD) or Monte Carlo (MC) simulations, which can be used to compare the relative free energy of binding of two ligands closely related in structure.[2] These latter calculations are the most accurate that can be currently applied to this problem, but suffer from being very computer intensive and applicable only to the comparison of chemically related ligands.

It makes sense to consider a divide-and-conquer strategy, using the simpler methods early in the design process and then the more rigorous later, when the list of possible ligands has been pruned. However, there is a large gap between the number of compounds (10^6 vs. 2) commonly compared with the two classes of methods mentioned above. Thus there is a great need for methods that can usefully be applied to 10–100 molecules in a computationally efficient way. There are four such approaches that have recently appeared in the literature: Aqvist's Linear Interaction Energy (LIE) Analysis,[3] which has been further elaborated by Jorgensen,[4] the lambda-dynamics (LD) approach developed by Brooks and co-workers,[5] the

Chemical Monte Carlo/Molecular Dynamics (CMC/MD) method by Pitera et al,[6] and a methodology developed by Srinivasan et al,[7] and our lab,[8] which we call MM–PBSA (Molecular Mechanics–Poisson Boltzmann Surface Area). LD and CMC/MD are free energy calculation approaches which allow one to consider up to ~10–20 ligands in a single simulation, thus improving the efficiency of free energy approaches. LIE is a more empirical approach, which requires a molecular dynamics or Monte Carlo simulation on a free and bound ligand, which is then analyzed to separately calculate the average van der Waals and electrostatic interaction energies between ligand and its surroundings in both trajectories. MM–PBSA is closest in concept and implementation to LIE, albeit it can be employed by running a molecular dynamics or Monte Carlo simulation only on the ligand-protein complex. It currently uses a combination of explicit molecular mechanical energies, continuum solvation free energies and normal mode analyses to calculate the absolute binding free energies of different ligands without employing any empirical parameters. Nonetheless, this approach has successfully ranked relative binding free energies in a variety of protein-ligand complexes such as in avidin, a matrix metalloproteinase (MMP), cathepsin D, and HIV reverse transcriptase (HIVRT), which we describe below.

2. METHODS

The MM-PBSA approach assumes that the free energy $\langle G \rangle$ of a macromolecular system in solution can be adequately approximated by Equation 1

$$\langle G \rangle = \langle E_{MM} \rangle + \langle G_{PBSA} \rangle - T\langle S_{MM} \rangle \qquad (1)$$

where $\langle E_{MM} \rangle$ is the average molecular mechanical energy, $\langle G_{PBSA} \rangle$ is the continuum solvation free energy, T is the absolute temperature, and $\langle S_{MM} \rangle$ is the average entropy of the solute. The continuum solvation free energy is composed of an electrostatic term, usually calculated using a Poisson-Boltzmann (PB) model and a hydrophobic non-polar term, which is proportional to the solvent accessible surface area (SA). Whereas $\langle E_{MM} \rangle$ and $\langle G_{PBSA} \rangle$ can be evaluated using "snapshots" from a molecular dynamics or Monte Carlo trajectory directly, $\langle S_{MM} \rangle$ is either ignored if one is comparing similar ligands or estimated from normal mode analyses of a more limited number of structural snapshots. The trajectory is typically carried out with full inclusion of water and counterions in order to make the structural snapshots as representative of the real system as possible. To calculate the free energy of association of a ligand to the macromolecule $<\Delta G>$, we use

$$<\Delta G> = <G>_C - <G>_M - <G>_L \tag{2}$$

where the free energies of the complex (C), macromolecule (M), and ligand (L) are given by Equation 1.

In all the applications to date on protein-ligand complexes, we have employed Equation 2 using only the complex trajectory, calculating $<G>_M$ and $<G>_L$ from the snapshots of the complex and discarding the atoms of the ligand and macromolecule, respectively. This approach cannot be used if there are significant conformational changes upon binding such as in RNA-protein complexes[9] and in DNA intercalation.[10] In those instances, it is necessary to run separate trajectories of both uncomplexed reaction partners to evaluate their free energies. However, we have found the assumption of using only the complex trajectory reasonably in the avidin, MMP, cathepsin D, and HIVRT examples studied so far.

In the following, we give a brief description on how to evaluate the various terms in Equation 1. A more extensive summary of the computational details typically used in our studies as well as the CPU requirements of the MM/PBSA approach can be found elsewhere.[11] $<E_{MM}>$ simply evaluates the average potential energy of the system using the same force field (e.g. Equation 3 of Cornell et al.)[12] as used to propagate the molecular dynamics trajectory.

$$E_{total} = \sum_{bonds} K_r(r - r_{eq})^2 + \sum_{angles} K_\Theta(\theta - \theta_{eq})^2 + \sum_{dihedrals} \frac{V_n}{2}[1 + \cos(n\phi - \gamma] + \sum_{van\ der\ Waals}^{i<j} [\frac{A_{ij}}{R_{ij}^{12}} - \frac{B_{ij}}{R_{ij}^{6}}] + \sum_{electrostatic}^{i<k} \frac{q_i q_i}{\varepsilon R_{ij}} \tag{3}$$

When one employs Equation 2 using only the complex trajectory, the internal energy contributions cancel and only the non-bonded terms (van der Waals and electrostatic) between ligand and macromolecule contribute to $<\Delta G>$. However, if one runs separate trajectories of C, M, and L, all the terms in Equation 3 will contribute to $<\Delta G>$.

The second term in Equation 1, $<G_{PBSA}>$, involves carrying out a Poisson-Boltzmann calculation and evaluating the exposed surface area of all atoms for all the snapshots for C, M, and L. Currently, we use Hartree-Fock (HF)/6-31G* restrained electrostatic potential (RESP)[13] charges and PARSE[14] radii for the PB calculation within DELPHI[15] and the program

MSMS[16] to evaluate the surface area, with a surface area free energy for non-polar atoms taken from Sitkoff et al.[14]

It is less obvious how to efficiently evaluate the third term of Equation 3, $-T<S_{MM}>$. If one is comparing closely related ligands, one can ignore it, assuming that $<S_{MM}>$ is the same for them. For structurally different compounds, the change in solute entropy upon binding can vary considerably (up to 10 kcal/mol in our studies) and must be included. We have explored an approach, in which one takes a limited number (~5) of snapshots from the trajectories, discards the explicit solvent molecules used in the initial MD or MC simulation, minimizes the energy with a dielectric constant $\varepsilon = 4R_{ij}$ (R_{ij} = atom-atom distance) for C, M, and L, and, after reaching a suitably low gradient, carries out a normal mode and classical statistical analysis (rigid rotor-harmonic oscillator approximation) to determine the vibrational, rotational, and translational entropies.

3. RESULTS

We summarize the results from the various MM-PBSA studies comparing the free energy of binding of different ligands to protein targets. The full details and results can be found in the original papers and in a recently published review of the MM-PBSA approach.[17]

Kuhn and Kollman[11] have studied the interactions of avidin to seven biotin analogs, the aromatic dye 2-(4'-hydroxyazobenzene) benzoic acid (HABA), and a cyclic hexapeptide, for which the X-ray structures of their complexes with avidin or streptavidin are known (Figures 1A–B). The calculated absolute binding free energies, $<\Delta G>$, using Equations 1 and 2 on the seven biotin analogs were in reasonable agreement with experiment, but systematically slightly too positive (e.g., for biotin $<\Delta G>_{exp} = -20$ kcal/mol; $<\Delta G>_{calc} = -18$ kcal/mol). Nonetheless, the relative $<\Delta G>$'s correlated with experiment ($r^2 = 0.92$) significantly better than found with the same molecules using LIE ($r^2 = 0.55$).[18] Even more exciting was the fact that the $<\Delta G>$ calculated for both HABA and the peptide fell on the same regression line (r^2=0.92), even though these are structurally very different from biotin.

Donini and Kollman[19] have studied six MMP inhibitors, two charged and four neutral ones (Figure 1C). In this case, the presence of a Zn^{2+} ion as a common atom in contact with the ligands leads to absolute $<\Delta G>$ values considerably too positive. Nonetheless, the free energy of binding of the four neutral and two charged ligands were ranked correctly within each series, although the binding free energies of the charged ligands were too positive relative to the neutral ones. We have shown that running a single trajectory and then mutating the snapshots to each ligand in turn was an effective strategy to rank the $<\Delta G>$'s provided each snapshot was subjected to some minimization with local flexibility of the protein as well as the ligand. This strategy is analogous to that used by Kuhn and Kollman in another study of biotin analogs binding to avidin,[20] whose goal was to determine whether one could improve the binding affinity of biotin to avidin by structural modifications to the ligand. In this study, we considered fluoro substitutions on the $-(CH_2)_4-$ side-chain of biotin using limited minimization to find that

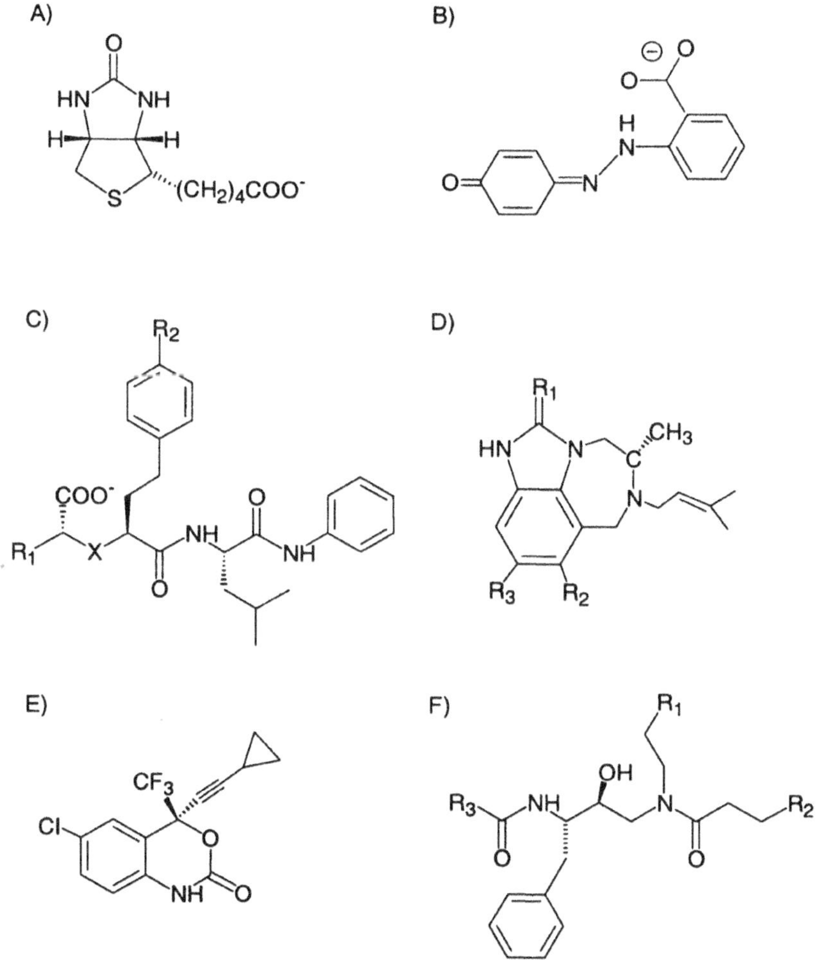

Figure 1. Chemical structures of representative ligands investigated: A) biotin, B) 2-(4'-hydroxyazobenzene) benzoic acid (HABA), C) charged ($X=CH_2$) and neutral ($X=NH_2^+$) carboxylate MMP inhibitors, D) TIBO scaffold, E) sustiva, and F) hydroxyethylamine scaffold. The biotin derivatives,[11] MMP inhibitors,[19] TIBO analogs,[21] and cathepsin D inhibitors[22] derived from structures A), C), D), and F), respectively, have been published elsewhere.

one particular monofluoro substitution should improve binding affinity, a result that was supported by "traditional" free energy calculations.

J. Wang and Kollman have studied the binding of 12 TIBO analogs (Figure 1D) binding to HIV reverse transcriptase using MM–PBSA.[21] All the analogs were relatively rigid and could be straightforwardly built into the binding site using the crystal structure of 8–Cl TIBO complexed to HIVRT. Encouragingly, we calculated not only the absolute binding free energies quite close to experiment, but also the relative free energies correlated quite well ($r^2 = 0.71$) with the experimental reference. Wang also used DOCK and MM–PBSA to predict the binding geometry of sustiva (Figure 1E) starting with the binding geometry of 8–Cl TIBO. The geometry and the absolute value of the binding free energy found with MM-PBSA were quite accurate.

Finally, Huo and Kollman[22] have found a good correlation ($r^2 \approx 0.8$) for the binding free energies of eight cathepsin D inhibitors (Figure 1F) to the protein. In this case, some docking was used to generate the placement of the inhibitor in the active site, except for the one ligand whose crystal structure in complex with the protein was known.

4. DISCUSSION AND CONCLUSIONS

Overall, the initial applications of MM-PBSA suggest that the method can effectively calculate the relative binding free energies in a number of protein-ligand complexes. However, there are a number of issues involving limitations or potential improvements in this methodology which we now discuss.

As noted above, our initial applications used HF/6-31G* RESP charges and PARSE atomic radii for the calculation of the continuum solvation free energy, $\langle G_{PBSA} \rangle$. The latter were derived by using another set of charges and adjusting the radii to reproduce the electrostatic free energy of solvation, based on Poisson-Boltzmann continuum calculations. Such a model had a limited number of atom types and, for example, had only a single C and N radius, which did not lead to an accurate free energy of solvation of cyano compounds when evaluated using HF/6-31G* RESP charges. Thus, J. Wang[21] derived new radii for sp-hybridized C and N atoms to reproduce the experimental solvation free energy of cyano compounds and this change of parameters improved the agreement with experiment for the binding of the TIBO analog with a C≡N group. Currently, J. Wang and S. Huo are deriving a complete new set of radii for the Poisson-Boltzmann calculations, which should improve its accuracy and applicability to a larger set of molecules.[23]

Currently, we evaluate the vibrational/rotational/translational entropy of the solute molecules using normal mode and classical statistical analyses. Although the use of a quasiharmonic analysis, as suggested by Schlitter[24] and applied to peptide folding by van Gunsteren[25] is a more accurate way to calculate entropies, it is slow to converge and, for the purposes of differences in entropy between C, M, and L, it is likely to be no more accurate and considerably less efficient than our approach. We should point out that at this point the $<S_{MM}>$ in Equation 1 only includes the vibrational entropy due to the local conformation and not the conformational entropy due to, e.g., the conformations in equilibrium in a floppy ligand free in solution. The assumption that this term is comparable for all the ligands in a given series seems quite reasonable for the applications described above.

One finds reasonable convergence in $<G>$ and $<\Delta G>$ for trajectories lasting a few hundred picoseconds in the cases discussed above. The rate-limiting step in the MM–PBSA approach is the time required to carry out the trajectory, although the time for determining the PB energies for the snapshots is not negligible. We use water, counterions, and in some cases periodic boundary conditions and Particle Mesh Ewald for the trajectories in order to make the snapshots as realistic as possible. If one had reliable and efficient Generalized Born or PB energies with gradients, one could use these to carry out the trajectories more efficiently.

One further application of MM–PBSA is to use it to ascertain ligand geometry in the complex structure. This is useful if docking gives a number of positions and orientations of the ligand that do not interconvert during molecular dynamics. MM–PBSA free energies can be used to determine which position is most favorable. Such MM-PBSA calculations have been applied to the HABA (Figure 1B) – avidin complex and have confirmed[11] that in HABA bound to avidin, the X-ray structure with its CO_2^- group "inside" the binding site,[26] is lower in free energy than the alternate orientation with the C=O inside, despite the fact that in biotin, the C=O is "inside" and the CO_2^- "outside".

Finally, we should put our MM-PBSA calculations in the context of others that have used electrostatic continuum models alone to calculate the $<\Delta G>$ of binding various ligands such as the study of Shen and Wendoloski.[27] Our approach for ligand binding has two new features: first, as shown by Srinivasan for DNA and RNA duplexes,[7] averaging over an MD trajectory using ≈100 snapshots, despite the very large electrostatic energies, allows a well converged average, particularly of $<E_{MM}^{elec}>$ + $<G_{PB}>$ (where $<E_{MM}^{elec}>$ is the electrostatic term in Equation 3 and $<G_{PB}>$ is the Poisson-Boltzmann solvation free energy). Secondly, and particularly important for hydrophobic/van der Waals dominated binding, which is an essential part of most drugs, is the inclusion of the explicit $<E_{MM}^{vdW}>$ (the van der Waals term in Equation 3). This enables the very large van der Waals term in most drug-receptor interactions to be robustly and accurately determined. In our model, the surface area dependent non-polar solvation energy term is generally small. Even if it were significantly larger, it plus the electrostatic term alone is not likely to be large enough to explain the structure-binding relationships of the various series we have considered here.

Acknowledgments

B.K. acknowledges support from the German Academic Exchange Service (DAAD) through a research scholarship and O.D. acknowledges an NSERC PDF for salary support. P.A.K., S.H., and J.W. are grateful for research support from the NIH (GM-29072). We thank the NSF supercomputer centers for providing computational resources.

5. REFERENCES

1. T. P. Lybrand, Ligand protein docking and rational drug design, Curr. Opin. Struct. Biol. **5**:224 (1995).
2. M. R. Reddy, M. D. Erion, and A. Agarwal, Free energy calculations: Use and limitations in predicting ligand binding affinities, in: Reviews in Computational Chemistry, K. B. Lipkowitz and D. B. Boyd eds, VCH Publishers, New York, (2000), vol. 16, pp. 217-304.
3. J. Marelius, T. Hansson, and J. Åqvist, Calculation of ligand binding free energies from molecular dynamics simulations, Int. J. Quant. Chem. **69**:77 (1998).
4. M. L. Lamb, J. Tirado-Rives, and W. L. Jorgensen, Estimation of the binding affinities of FKBP12 inhibitors using a linear response method, Bioorg. Med. Chem. Lett **7**:851 (1999).
5. Z. Guo, C. L. Brooks, and X. Kong, Efficient and flexible algorithm for free energy calculations using the lambda-dynamics approach, J. Phys. Chem. **102**:2032 (1998).
6. M. A. L. Ericksson, J. Pitera, and P. A. Kollman, Prediction of the binding free energies of new TIBO-like HIV-1 reverse transcriptase inhibitors using a combination of PROFEC, PB/SA, CMC/MD, and free energy calculations, J. Med. Chem. **42**:868 (1999).
7. J. Srinivasan, T. E. Cheatham, P. Cieplak, P. A. Kollman, and D. A. Case, Continuum solvent studies of the stability of DNA, RNA, and phosphoramidate - DNA helices, J. Am. Chem. Soc. **120**:9401 (1998).
8. I. Massova and P. A. Kollman, Computational alanine scanning to probe protein-protein interactions: A novel approach to evaluate binding free energies, J. Am. Chem. Soc. **121**:8133 (1999).
9. C. M. Reyes and P. A. Kollman, Structure and thermodynamics of RNA-protein binding: Using molecular dynamics and free energy analyses to calculate the free energies of binding and conformational change, J. Mol. Biol. **297**:1145 (2000).
10. P. Cieplak and P. A. Kollman, unpublished results.
11. B. Kuhn and P. A. Kollman, Binding of a diverse set of ligands to avidin and streptavidin: An accurate quantitative prediction of their relative affinities by a combination of molecular mechanics and continuum solvent models, J. Med. Chem. **43**:3786 (2000).
12. W. Cornell, P. Cieplak, C. I. Bayly, I. Gould, K. M. Merz, D. M. Ferguson, D. C. Spellmeyer, T. Fox, J. W. Caldwell, and P. A. Kollman, A second generation force field for the simulation of proteins, nucleic acids and organic molecules, J. Am. Chem. Soc. **117**:5179 (1995).
13. C. I. Bayly, P. Cieplak, W. D. Cornell, and P. A. Kollman, A well-behaved electrostatic potential based method using charge restraints for deriving atomic charges – the RESP model, J. Phys. Chem. **97**:10269 (1993).

14. D. Sitkoff, K. Sharp, and B. Honig, Accurate calculation of hydration free-energies using macroscopic solvent models, J. Phys. Chem. **98:**1978 (1994).
15. B. Honig and A. Nicholls, Classical electrostatics in biology and chemistry, Science **268:**1144 (1995).
16. M. F. Sanner, A. J. Olson, and J. C. Spehner, Reduced surface – An efficient way to compute moledular surfaces, Biopolymers **38**:305 (1996).
17. P. A. Kollman, I. Massova, C. Reyes, B. Kuhn, S. Huo, L. Chong, M. Lee, T. Lee, Y. Duan, W. Wang, O. Donini, P. Cieplak, J. Srinivasan, D. A. Case, and T. E. Cheatham, III, Calculating structures and free energies of complex molecules: Combining molecular mechanics and continuum models, Acc. Chem. Res. **33**:889 (2000).
18. J. Wang, R. Dixon, and P. A. Kollman, Ranking ligand binding affinities with avidin: A molecular dynamics-based interaction energy study, Proteins **34**:69 (1998).
19. O. Donini and P. A. Kollman, Calculation and prediction of binding free energies for the matrix metalloproteinases, J. Med. Chem. **43**:4180 (2000).
20. B. Kuhn and P. A. Kollman, A ligand that is predicted to bind better to avidin than biotin: Insights from computational fluorine scanning, J. Am. Chem. Soc. **122**:3909 (2000).
21. J. Wang, P. Morin, W. Wang, and P. A. Kollman, Use of MM-PBSA in Reproducing the binding free energies to HIV-RT of TIBO derivatives and predicting the binding mode to HIV-RT of efavirenz by docking and MM-PBSA, J. Am. Chem. Soc. in press (2001).
22. S. Huo and P. A. Kollman, unpublished results.
23. J. Wang and S. Huo, studies in progress.
24. J. Schlitter, Estimation of absolute and relative entropies of macromolecules using the covariance matrix, Chem. Phys. Lett. **215**:617 (1993).
25. H. Schaefer, A. E. Mark, and W. F. van Gunsteren, Absolute entropies from molecular dynamics simulation trajectories, J. Chem. Phys. **113**:7809 (2000).
26. O. Livnah, E. A. Bayer, M. Wilchek, and J. L. Sussman, The structure of the complex between avidin and the dye, 2-(4'-hydroxyazobenzene) benzoic acid (HABA), FEBS Lett. **328**:165 (1993).
27. J. Shen and J. Wendoloski, Electrostatic binding energy calculation using the finite difference solution to the linearized Poisson-Boltzmann equation – Assessment of its accuracy, J. Comp. Chem. **17**:350 (1996).

Chapter 13

Reaction Free Energy Profiles Using Free Energy Perturbation and Coordinate Coupling Methodologies: Analysis of the Dihydrofolate Reductase Catalytic Mechanism

U. Chandra Singh,[†] K. Ramnarayan,[‡] and Frederick H. Hausheer[*]
[†]*AM Technologies, Inc., San Antonio, TX 78218*
[‡]*ImmunoPharmaceutics, Inc. San Diego, CA 92127*
[*]*BioNumerik Pharmaceuticals, Inc. San Antonio, TX 78229*

1. INTRODUCTION

Dihydrofolate reductase (DHFR) catalyzes the NADPH-dependent reductions of folate and dihydrofolate to tetrahydrofolate (THF).[1] THF is a cofactor used by enzymes catalyzing 1-carbon donor reactions. The importance of these reactions and therefore DHFR catalysis in the synthesis of purine and pyrimidine bases has made it a primary drug target for a variety of chemotherapy agents, including anticancer drugs and antibiotics.[2-5] DHFR is inhibited by folate antagonists such as 2,4-diamino pteridine and its derivatives aminopterin (AMT) and methotrexate (MTX). The chemical, biological and medicinal aspects of MTX have been studied extensively for the past 35 years and MTX still remains as one of the most important agents for patients undergoing organ transplantation or treatment for cancer,[6-10] rheumatoid arthritis, psoriasis or asthma. Despite the significant utility of MTX as a therapeutic agent, its toxicity and limited activity against certain malignancies have provided an impetus for the development of less toxic drugs with broader oncolytic activity.

The importance of DHFR as a drug target and the availability of X-ray structures of DHFR from different animal species with and without the co-factor NADPH[11-15] have made DHFR an ideal target for studying various new structure-based drug design methodologies.[16, 17] Free energy calculations have been provided substantial insight into DHFR-ligand interactions. Earlier work from our group established the importance of the ionic interaction between Asp27 and the pteridine ring in a DHFR-MTX complex using the Free Energy Perturbation (FEP) technique.[18] The same technique was also used in a subsequent study examining the binding affinity of MTX to mutants of DHFR.[19] More recently, the FEP methodology was employed to study the binding of the drug Trimethoprim and its derivatives to the DHFR-NADPH complex.[20] Another FEP study[21] investigated the difference between the binding of reduced and oxidized NADP cofactor to *E.Coli* DHFR binary and ternary complex in water.

In addition to these studies, the DHFR catalytic mechanism has been studied using the local density function (LDF) approach.[22] A detailed understanding of the catalytic mechanism is often very useful in the design of high affinity enzyme inhibitors since it can shed light on the transition state structure and the key interactions used by the enzyme to stabilize it.

1.1 DHFR Catalytic Mechanism

Experimental evidence from a variety of sources suggests that folate binds to DHFR in the same site as MTX but that the pteridine ring is rotated by 180° with respect to the side-chain containing the C6-C9 bond.[11, 23, 24] A recent crystallographic study on the structure of 5-deaza folate-DHFR complex has unambiguously verified this rotation.[25] It is also generally accepted that the dihydropterin ring of the substrate must be protonated at N5 position to facilitate the hydride ion transfer from NADPH to C6. The proton donor, Asp27 is ≈ 6.4 Å away from N5 and forms key hydrogen bonds not with N5 but with N3 and the 2-amino group of the pteridine ring. It has also been proposed that the principal effect of the enzyme is to promote hydride ion donation by NADPH, rather than to facilitate hydride ion acceptance by H2-folate.[13] On the other hand, a recent kinetic study has shown that the rate determining step for the catalytic reaction must occur after the hydride ion transfer.[14] Thus, the precise mechanism by which DHFR assists in hydride ion transfer is unresolved at present.

The reduction of 7,8-dihydrofolate (H_2F) to 5,6,7,8-tetrahydrofolate (H_4F) has been analyzed extensively[14, 26-30] and a kinetic scheme for *E. Coli* DHFR was proposed in which the steady-state kinetic parameters as well as the full time course kinetics under a variety of substrate concentrations and pHs were determined. From these studies, the pK_a of Asp27 is 6.5 in the ternary complex between the enzyme, the cofactor NADPH and the substrate dihydrofolate. The second observation is that, contrary to earlier results,[27] the rate determining step involves dissociation of the product from the enzyme, rather than hydride ion transfer from the cofactor to the substrate.

To better understand the catalytic mechanism of DHFR and to use this information for the design of potent DHFR-specific inhibitors, we evaluated the proton and hydride transfers using an integrated ab initio Quantum Mechanics/Molecular Mechanics (QM/MM) approach in combination with FEP technology. The combinatorial application of these methods enabled us to propose a precise path along which the proton and hydride ion are transferred and to address the key structural and energetic changes associated with catalysis.

2. MAPPING THE PROTON AND HYDRIDE TRANSFER REACTIONS

The pterin moiety of 7,8-dihydrofolate exists in tautomeric equilibrium between the enolic form and the cyclic amide form as show in Figure 1. Experimental work has established that the equilibrium favors the cyclic amide form.[28] Site-directed mutagenesis experiments have established that Asp27 is the crucial amino acid for thc substrate protonation.[28] Thus, during the binding of the substrate with the enzyme, the catalytically important aspartic acid (Asp27 of *E. Coli* DHFR) is protonated, while the pterin ring is either in the cyclic amide form or the enolic form. In this state, the initial complex formation should involve hydrogen bond formation between O4 and N3 of the pterin ring and the carboxylic acid of Asp27 (Figure 2).

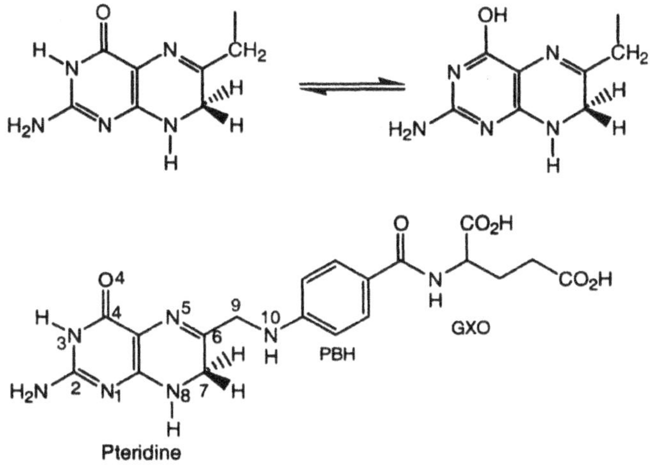

Figure 1. Tautomeric equilibrium of the pteridine ring and the structure of 7,8-dihydrofolate.

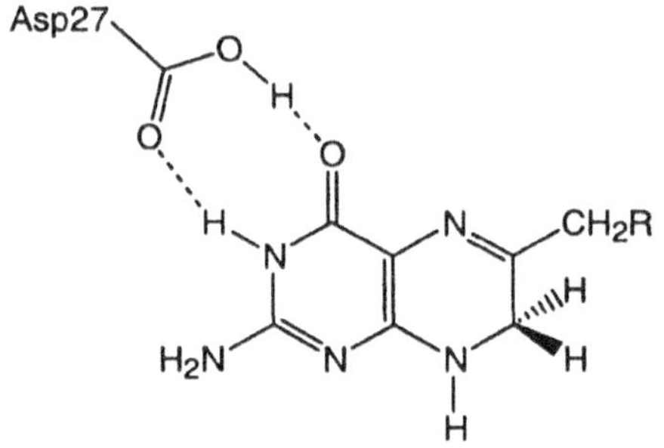

Figure 2. Asp27 interaction with the pteridine ring of dihydrofolate (R = PBH-GXO).

Alternatively, if Asp27 is negatively charged, then a hydrogen bond donor at N3 would be important thereby favoring the cyclic amide form. It is important to note that, if the substrate is protonated at O4 (enolic form) and the carboxyl group of Asp27 is unprotonated, then the substrate has three hydrogen bond donating sites to form two types of complexes as represented schematically in Figure 3a.

Assuming an initial complex to be of the form given in Figure 3a (protonated Asp27 and enolic form of the pterin ring), it is important to characterize the structural and energetic basis for how the proton travels from the Asp27 to the N5 position of pterin. Based on the X-ray structure, the proton travels a distance of nearly 5.0 Å. To address this question, the proton trajectory was mapped by a four stage transition as shown schematically in Figure 3. The four stages include: (i) proton transfer from Asp27 to the O4 atom of the pterin ring to form a pair of hydrogen bonds involving O4 and N3 (Figure 3a); (ii) hydrogen bond shift from O4 to N3 (Figure 3b); (iii) hydroxyl group rotation towards N5 (Figure 3c); and (iv) the transfer of the proton from O4 to N5 of the pterin ring (Figure 3d).

Using this proton transfer pathway as a starting point, we generated four snap shot geometries for characterizing the pathway. Each structure was then evaluated using a constraint applied to the HO4...N5 distance. Quantum mechanics was used for the acetic acid complex and 6-methyl 7,8 dihydropterin. Eighteen structures were generated along the proton transfer pathways from which 15 structures were selected for detailed analysis. The quantum mechanical optimizations were carried out using the Hartree-Fock 6-31G basis set, and the single point calculations were performed using 6-31 G* and 6-31G** to compute the quantum mechanical components of the free energies. The structures were subsequently mutated, ad seriatim, from one to another along the defined pathway using the coordinate coupled FEP approach to obtain the energetic contribution from the surroundings. The

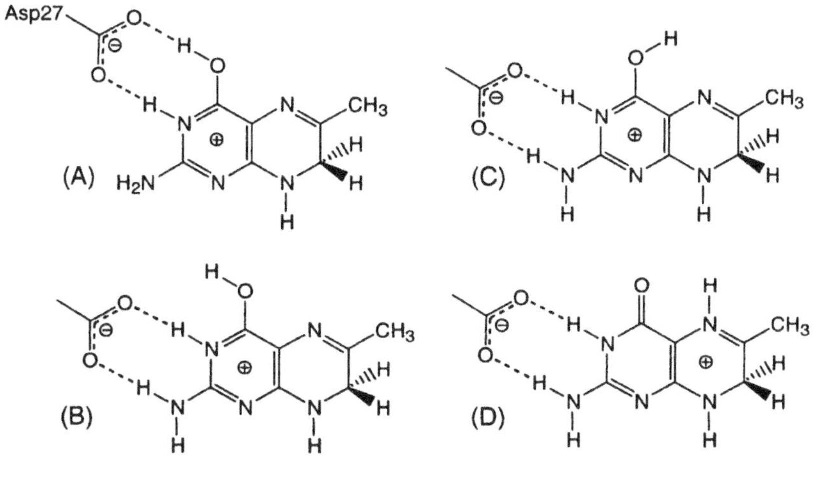

Figure 3. Proton transfer pathway.

equilibrium bond length, bond angle and dihedral parameters were evaluated based on ab initio quantum mechanical optimization and checked for proper duplication of the quantum mechanical structure in order to make sure that the parameters were accurate for each snap shot geometry.

Following the proton transfer, hydride from NADPH is added to C6 of the positively charged base. The trajectory of the hydride ion is shown schematically in Figure 4, and is based on the following intermediate stages: (i) C4 hydrogen of the nicotinamide base is in close proximity to C6 of the pterin ring in the ternary complex (Figure 4a); (ii) a hydride transfer entails a transition state in which the hydride ion is part way between C4 of nicotinamide and C6 of the pterin ring (Figure 4b); and (iii) products formed after hydride transfer (Figure 4c).

Using this hydride ion transfer pathway, we obtained snap shot geometries and subjected them to ab initio quantum mechanical optimization using a constraint for the C4 (H)...C6 distance. The ab initio quantum mechanical part of the system was composed of the nicotinamide ring, 6-methyl dihydropterin and acetic acid. A total of 9 quantum mechanically optimized structures were generated to characterize the hydride ion transfer using a 6-31G basis set. The free energy contribution due to the surroundings are calculated in the same manner as the proton transfer by successively mutating one snap shot geometry to the next within the active site, using the FEP approach.

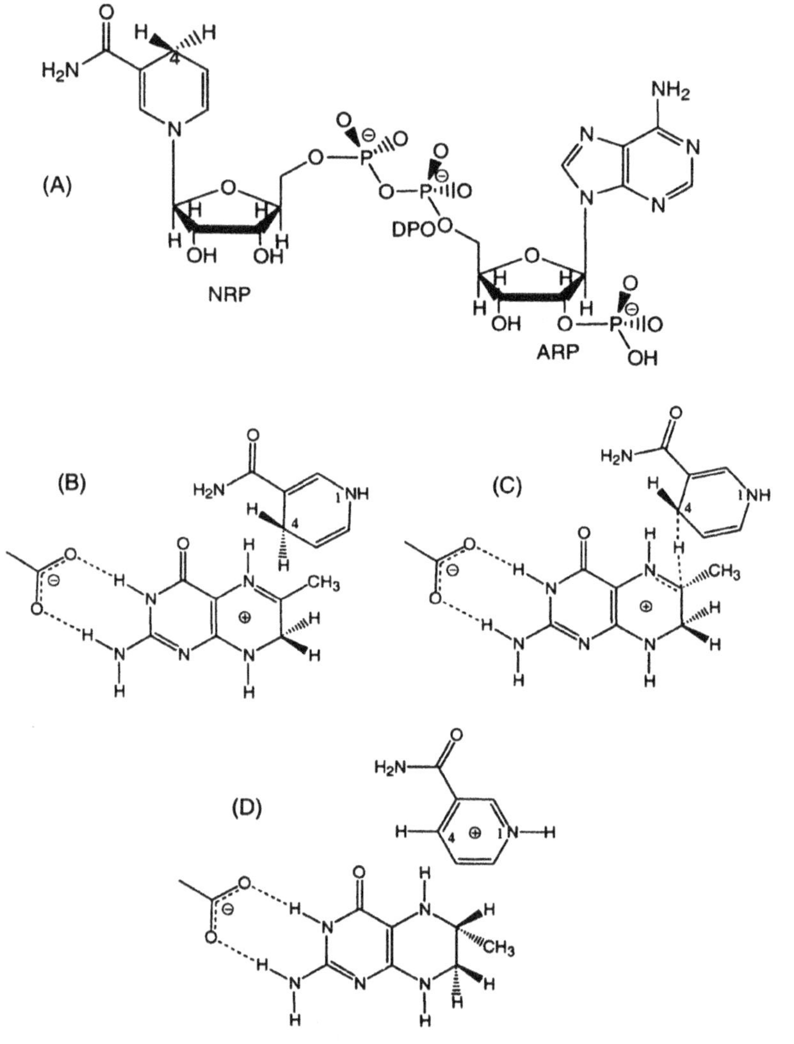

Figure 4. NADPH structure (A) and the hydride ion transfer pathway (B-D).

3. INTEGRATION OF QUANTUM AND MOLECULAR MECHANICAL METHODS

Ab initio quantum mechanical methods are needed to semi quantitatively study chemical reactions involving bond breaking and bond forming phenomena. Once a reaction path for a chemical process is defined, it is possible to obtain the key transition geometries and associated energies of the reacting species along this path. This combinatorial QM/MM approach has been successfully used to obtain transition, or snap shot, geometries of the reactants along the reaction coordinates. Using this approach, the molecular system is divided into two subsystems, namely the quantum subsystem and the classical subsystem. The quantum subsystem is established in a manner which is defined by all the bond reorganizations pertaining to the reaction pathway. The classical subsystem is constituted by the remainder of the surrounding system, which can include solvent and counter ions. This approach enables the optimal pairing of the strengths of QM and MM within a concurrent simulation of a molecular ensemble.

The total energy of the system is evaluated as,

$$E_{tot} = E_{QM} + E_{MM} + E_{QM/MM} \tag{1}$$

The functional form of E_{MM} is the same as in Galaxy and AMBER (version 3.3[31]). The form of E_{MM} is described as,

$$\begin{aligned} E_{MM} = &\sum_{bonds} \frac{K_r}{2}(r - r_{eq})^2 + \sum_{angles} \frac{K_\theta}{2}(\theta - \theta_{eq})^2 \\ &+ \sum_{dihedrals} \frac{V_n}{2}(1 + \cos(n\phi - \gamma)) \\ &+ \sum_{i<j}\left(\frac{A_{ij}}{R_{ij}^{12}} - \frac{B_{ij}}{R_{ij}^{6}} + \frac{q_i q_j}{\varepsilon R_{ij}}\right) + \sum_{H-bonds}\left(\frac{C_{ij}}{R_{ij}^{12}} - \frac{D_{ij}}{R_{ij}^{10}}\right) \end{aligned} \tag{2}$$

It should be noted that in the above expression the summation is only over the atoms from the Molecular Mechanics part of the system. The expression for $E_{\{QM/MM\}}$ is the same as above, but with a critical difference in summation. The summation over bond, angle and dihedral energies is for terms with at least one atom in both the QM and MM part of the system. To determine the non-bonded energy, the i summation is over the MM part of the system and the j summation is over the QM part of the system. The E_{QM} component of the energy is computed using the ab initio quantum mechanical SIVA[32] module of the program GALAXY using a 6-31G* basis[33] set.

The free energy profile of the enzymatic reaction is decomposed into two components as,

$$\Delta G_{tot} = \Delta G_{QM} + \Delta G_{protein} \tag{3}$$

where, ΔG_{QM} is the free energy contribution from the subsystem treated by quantum mechanical means, and $\Delta G_{protein}$ is the energetic contribution due to the protein surroundings which are not treated by quantum mechanics as the chemical reaction occurs inside the active site of the enzyme. This well defined non-quantum mechanical component provides the energetic contribution from the protein in the reaction, and, will be a measure of the catalytic efficiency of the enzyme in the particular reaction. In addition to this component, the enzyme surroundings can contribute indirectly to the component ΔG_{QM} by stabilizing or destabilizing the reactants relative to the gas or the solution phase structures.

3.1 Free Energy Perturbation Method

The statistical perturbation theory arising from the classical work of Zwanzig[34] and its detailed implementation in a molecular dynamics program for computation of free energies is described in detail elsewhere.[35, 36] We give a very brief description of the method for the sake of completeness. The total Hamiltonian of a system may be written as the sum of the Hamiltonian (H_0) of the unperturbed system and the perturbation (H_1):

$$H = H_0 + H_1 \tag{4}$$

The free energy contribution due to the perturbation is given by,

$$G_1 = -\frac{1}{\beta}\rangle \exp(-\beta H_1)\rangle_0 \tag{5}$$

where $\beta = 1/kT$ and the mean of exp $(-\beta H_1)$ is computed over the unperturbed ensemble of the system. In order to compute ΔG, the difference in free energy between the two solute states, the Hamiltonian for states A and B are linked by the coupling parameter λ such that,

$$H_\lambda = \lambda H_A + (1-\lambda)H_B \tag{6}$$

and λ is between 0 and 1.

In the above equation, H_A is the Hamiltonian for the system at state A, and H_B is that for state B. The above equation implies that, when $\lambda = 1$, $H_\lambda = H_A$ i.e., the system is purely in state A and when $\lambda = 0$, $H_\lambda = H_B$ at which point the system is purely in state B. During intermediate values of λ, the solute is a hypothetical mixture of A and B. This type of coupling ensures a very smooth conversion of two solutes A and B, allowing the system to readjust its configuration smoothly as a function of the chemical state. If we divide the range of λ into N windows, then at each window λ_i, the solute state is perturbed to λ_{i+1} and λ_{i-1} states by taking the reference state as $H_0(\lambda_i)$. The free energy difference between the two solute states A and B is a simple summation over all windows ($G_1(\lambda_i)$) as given by,

$$\Delta G = \sum_{i=1}^{N} G_1(\lambda_i) \tag{7}$$

The evaluation of G_i at λ_{i+1} and λ_{i-1} is a check for possible hysteresis in the calculation and is a measure of the statistical error for the free energy change.

3.2 Coordinate Coupling

In order to calculate the configurational free energy along a reaction coordinate, we used the coordinate coupling method, which has been used successfully in earlier calculations.[37-39] The structures along a reaction pathway are perturbed from one to another using stringent means to maintain the specified conformational properties, which is essential to get the free energy profile along a reaction coordinate. The coordinates of the system are coupled as,

$$X_\lambda = \lambda X_A + (1-\lambda)X_B \tag{8}$$

where X_λ represents the coordinates of the system at a given value of λ and X_A and X_B are the coordinates of the system in state A and B respectively. The coordinates of the system at λ_{i+1} have been extrapolated from those at λ_i using Equation 8.

3.3 Free Energy Decoupling

Even though the free energy difference is a path independent quantity, it is observed that certain sampling difficulties arise when a polar solute is transferred to a non polar solute accompanied by a large change in molecular volume. Under this circumstance, if one attempts mutation of both the partial charges and the non bonded parameters simultaneously, the solute-solvent energy increases enormously as a consequence of very close approach of some of the solvent molecules. This artificial limitation can be avoided by decoupling the total free energy as a sum of electrostatic and van der Waals contributions. Supposing we denote the energy due to bond, angle and dihedral as E_{bad}, the electrostatic energy as E_{ele} and the van der Waals interaction energy as E_{vdw}, we can write the total energy of the system as:

$$E_{tot} = E_{bad} + E_{ele} + E_{vdw} \tag{9}$$

The conversion of state A to B is achieved through an intermediate stage A' such that

$$A \rightarrow A' \rightarrow B$$

and the corresponding net free energy change is:

$$\Delta G_{AB} = \Delta G_{AA'} + \Delta G_{A'B} \tag{10}$$

Here the state A' has the charge distribution of B, but maintains the molecular geometry and van der Waals parameters of A. Hence $\Delta G_{AA'}$ corresponds to the electrostatic contribution to the free energy difference. The conversion of A to A' is achieved smoothly since the van der Waals parameters are the same for these two states. $\Delta G_{A'B}$ is the van der Waals contribution to the free energy which includes the bond, angle and dihedral contributions as well as the non bonded interactions.

3.4 Solvation

The complex of protein, crystallographic water and the counter ions are treated as a fully solvated system. Two key developments were made to treat the system in a more realistic manner during molecular dynamics: (i) water molecules were placed as a spherical shell with a radius 34 Å from the center of the protein. The outer boundary of the spherical shell was defined by means of an artificial wall with a potential of the type W defined as:

$$W = \left\{ \frac{A_{wall}}{\left(R_b - |\hat{r}_i|\right)^{12}} - \frac{C_{wall}}{\left(R_b - |\hat{r}_i|\right)^{6}} \right\} + A\left(|\hat{r}_i| - R_0\right) + B \tag{11}$$

where the constants A_{wall} and C_{wall} are computed as,

$$A_{wall} = \sqrt{\varepsilon_s * \varepsilon_w}\,(R_s + R_w)^{12} \quad \text{and} \quad C_{wall} = 2 * \sqrt{\varepsilon_s * \varepsilon_w}\,(R_s + R_w)^6$$

This potential was developed to ensure that the molecules inside the sphere never escape and maintain a fully solvated system during molecular dynamics. Here, ε_s , R_s , ε_w and R_w are the van der Waals constants for the solvent and the wall and r_i is the distance between the molecule i and the center of the water sphere, R_0 is the radius of the sphere. The quantities A, B and R_b are determined by imposing the condition that W and dW/dr_i vanish at $r_i = R_0$. The restraining potential W is set to zero for $r_i < R_0$. The van der Waals parameters ε_s, ε_w, R_s and R_w can also be specifically defined for different solvents. The constants A_{wall} and C_{wall} are computed using a well depth of $\varepsilon_s = \varepsilon_w = 0.1$ kcal and the radius of $R_s = R_w = 1.25$Å. For the other set of simulations, especially for the hydride ion transfer, we applied periodic boundary conditions by using a spherical boundary shell of 10.0 Å of TIP3P[40] water to cover the edges of the protein.

3.5 Computational Details

The calculations described in this chapter were performed using the modified version of AMBER (version 3.3) on a CRAY X-MP computer and part of the calculations were done using the software GALAXY on IBM RS-6000 machines. The high resolution crystallographic coordinates of the binary and ternary complexes of the E. Coli DHFR were kindly provided by Dr. J.E. Villafranca. The substrate was locked into the enzyme by flipping the pterin ring by 180° and the complex was energy minimized to relieve steric hindrances. The structure was neutralized by the addition of counter ions, based on the location of charged residues in the enzyme. Counter ions were not placed near potential sites for salt bridge formations. The total number of atoms in the system, including the protein, the substrate, the cofactor, the crystallographic as well as TIP3P water molecules and the counter ions was 24,886 atoms.

The FEP calculations were performed using the following general procedure. In the first stage, each structure was minimized by a conjugate gradient method for 2000 cycles by freezing the protein and substrate, i.e. allowing movement in only the water and the counter ions. In the second stage of the minimization, the whole system was minimized for 2000 cycles. In the last stage, the system was minimized again with the SHAKE[41] option for 100 cycles to constrain the bond lengths to their equilibrium values. The system was then equilibrated for 15 ps of MD using a time step of 0.001 ps, at a constant temperature of 300° K and at 1 atm pressure, in the case where the periodic boundary condition was applied. The mutation was achieved by using coordinate coupled free energy perturbation technique with the window method. 51 windows were used with 0.8 ps of equilibration and 0.8 ps of data collection. For the hydride ion transfer, the same protocol as above was used to obtain the free energy profile. Each structure was mutated to another over a period of 50 ps.

4. PROTON TRANSFER REACTION PATHWAY

4.1 Free Energy Profile

The ab initio quantum mechanical energies for the 15 snap shot geometries obtained during the proton transfer are reported in Table 1. Based on the profile of the relative ab initio energies, it is clear that three energy barriers are present along the reaction path and that the three energetic barriers are associated with HO4...N5 distances of 3.2 Å, 2.5 Å, and 1.4 Å, respectively. It is notable that the free energy contribution from the protein surroundings steadily decreases from 0.0 kcal/mol to -17.3 kcal/mol as the proton moves from Asp27 to N5. A plot of the quantum mechanical contribution to the free energy as a function of the HO4...N5 distance is given in Figure 5. As shown in Figure 5, a small energy barrier is observed for the proton transfer from Asp27 to N5 when the HO4...N5 distance is about 4.1 Å after which the energy rises to approximately 6.5 kcal/mol.

Table 1. Quantum mechanical energies for snap shot geometries

Structure Number	Distance in Å HO4...N5	Hartree Energy*	Relative Energy#	Free Energy#
1	4.50	-845.28	0.00	0.00
2	4.20	-845.27	6.14	-1.02
3	4.08	-845.27	9.13	-1.26
4	3.87	-845.27	9.80	-4.06
5	3.51	-845.26	18.05	-4.50
6	3.20	-845.24	25.74	-5.26
7	2.99	-845.25	24.41	-6.10
8	2.89	-845.25	23.11	-6.50
9	2.49	-845.25	19.72	-7.76
10	2.13	-845.26	16.32	-7.51
11	1.78	-845.25	22.86	-7.52
12	1.56	-845.23	32.50	-7.54
13	1.44	-845.22	39.60	-9.51
14	1.13	-845.23	36.95	-14.75
15	1.01	-845.24	30.14	-17.26

*Quantum mechanical energy at the 6-31G level
#Energy in kcal/mol

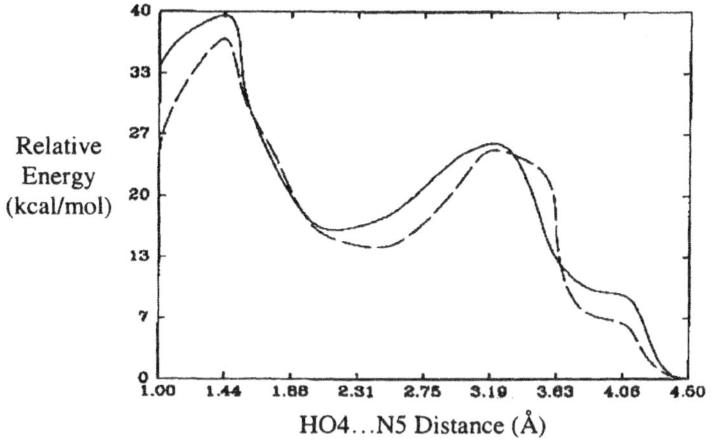

Figure 5. A plot of the quantum mechanical contribution to the free energy as a function of the HO4...... N5 distance. (– – –) is the energy from 6-31G basis set and ( ) is the energy from 6-31 G** basis set.

Using the larger basis set, the energy is somewhat higher and is around 9 kcal/mol. The hydrogen bond shifted structure (Figure 3b) is approximately 2 kcal/mol higher in energy than the structure containing the hydrogen bond involving O4 (Figure 3a). Rotation across the C4–O4 bond brings the proton within 2 Å of N5 (Figure 3c). This rotation has a calculated barrier of approximately 16 kcal/mol from both basis sets, as shown in Figure 5. This energetic barrier is due to the double bond character of the phenolic hydroxyl group and to the presence of the negatively charged Asp27 residue in the vicinity of the hydroxyl group. In the last stage, the proton jumps from O4 to N5. This corresponds to a decrease in the HO4...N5 distance from approximately 2.0 Å to 1.0 Å. For the transfer of the proton from O4 to N5, the quantum mechanical component of the barrier height is ≈ 23 kcal/mol from the larger basis and the smaller basis sets as shown in Figure 5. It is interesting to note that the quantum mechanical energy difference between the N5 protonated structure and the Asp27 protonated structure is approximately 33 kcal/mol using the 6-31G** basis set and about 25 kcal/mol using the lower basis set. This large energy difference is mainly due to the larger charge separation in the QM part of the system as can be seen from the variation in the dipole moment. The computed dipole moment for the ion pair varies significantly from 5.27D for the Asp27 protonated structure to a value of 19.56D in the N5 protonated substrate system. To generate this large charge separation, a significant amount of energy is required.

As the snap shot geometries are successively mutated within the protein, the free energy contribution from the protein surroundings measures the response of the protein to the reaction. Accordingly, we characterized the role of the protein in stabilizing or destabilizing the proton transfer. The contribution to the free energy from the protein surroundings as a function of the HO4...N5 distance is plotted in Figure 6.

The proton motion from Asp27 to O4 comprises a trajectory of approximately 0.6 Å (Figure 6). There is a decrease in the free energy due to the surroundings, which corresponds to nearly 4.5 kcal/mol of stabilization by the protein. Further stabilization occurs as the O4–HO4 bond rotates toward the N5 atom, which corresponds to a proton movement of about 1.4 Å (Figure 6b)

The proton jump (from O4 to N5 corresponding in Figure 6 (c)) in the initial phase, between 2.4 and 1.5 Å, is almost flat. This region corresponds to the transition state for the proton transfer from O4 to N5. Before the complex reaches its transition state, the net stabilization of the complex by the protein is approximately 7 kcal/mol. After the transition state is reached, there is a steep fall in the surrounding free energy as shown in Figure 6. The net stabilization after the transition state is about 9 kcal/mol. This simulation accurately characterizes the role of the enzyme in stabilizing the energetically large transition and charge separation by reconfiguring the surroundings.

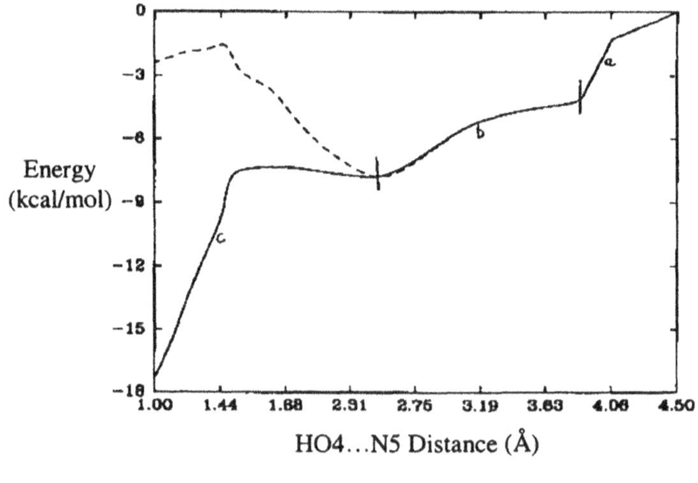

Figure 6. A plot of the free energy contribution arising from the protein surroundings as a function of HO4...N5 distance during the proton transfer mechanism.

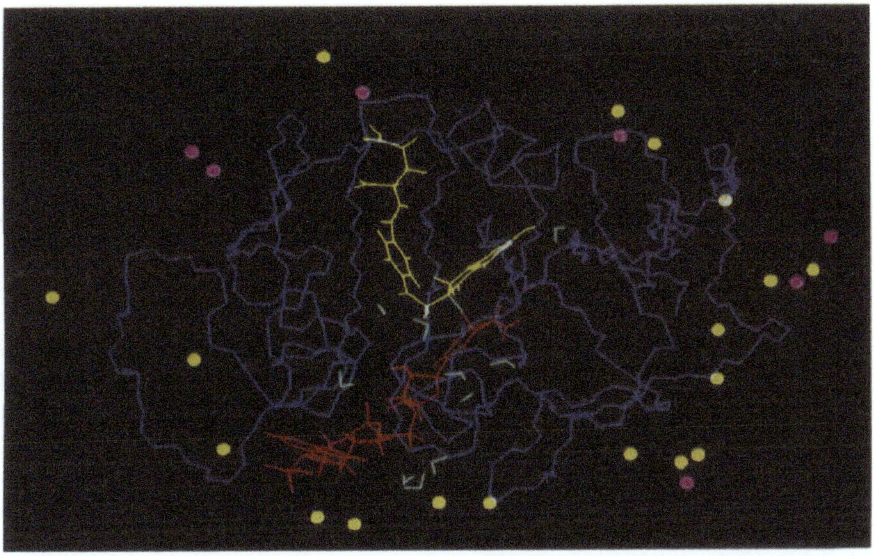

Figure 10. The ternary complex of the enzyme dihydrofolate reductase, the substrate and the cofactor during the transition state of the hydride ion transfer. The enzyme backbone atoms are shown alone for clarity and are colored blue. The substrate is shown in yellow and the cofactor is in red. The bond colored in light blue indicates the hydride ion being shared by both the cofactor and the substrate before the transfer to the substrate. Water molecules around the residue pteridine of the substrate and the nicotinamide ring of the cofactor alone are shown and colored in light blue. The yellow spheres represent the sodium ions and the pink spheres the chloride ions.

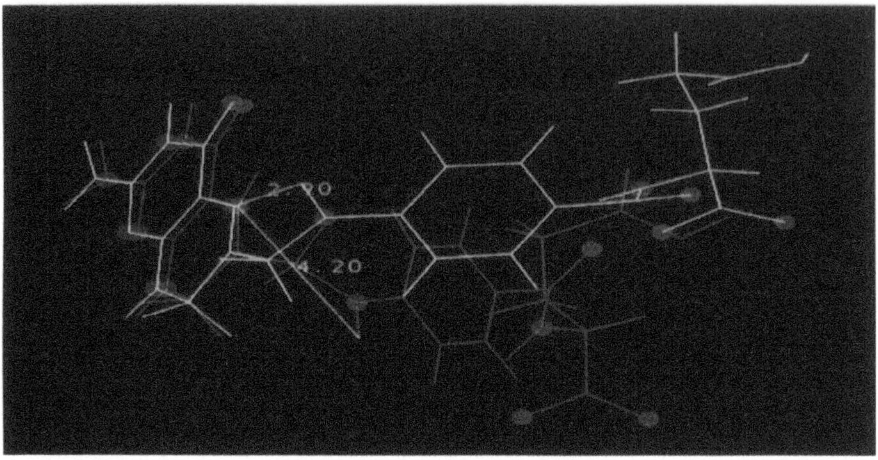

Figure 7. Superposition of the substrate alone before and after N5 protonation. The structure shown in yellow represents the substrate before N5 protonation and the one in red represents the substrate after the N5 protonation. The distances marked correspond to the contact N5 ... HIO before (2.9 Å) and after (4.2 Å) the proton transfer. The red spheres denote the oxygen atoms and the blue spheres the nitrogen atoms in the substrate.

As the proton is transferred form O4 to N5, there is a conformational transition in the substrate due to the rotation about the C9–N10 bond. Protonation of N5 of the substrate forces the N10 hydrogen to point away from it, resulting in the transition state. This was observed from the molecular dynamics data. To fully characterize the validity of this specific conformational transition, we performed another more complex simulation in which the substrate was not allowed to rotate about the C9–N10 bond. This simulation documented the presence of an unfavorable free energy contribution from the protein surroundings during proton transfer. This result is shown in Figure 6 as a dashed line. This series of simulations unambiguously established that a critical change in substrate conformation is a direct consequence of proton transfer (Figure 7). In this instance, the unprotonated substrate (yellow) is superimposed on the protonated substrate (red). The N5...H10 contact distance was 2.9 Å prior to N5 protonation and is 4.2 Å after substrate protonation due to the rotation about the C9–N10 bond. This conformational change is due to the steric overcrowding in the N5 protonated structure and also due to intermolecular electrostatic repulsion after the protonation. After the proton transfer, the PBH (para amino benzoyl) group is substantially displaced due to the conformational transition in the substrate.

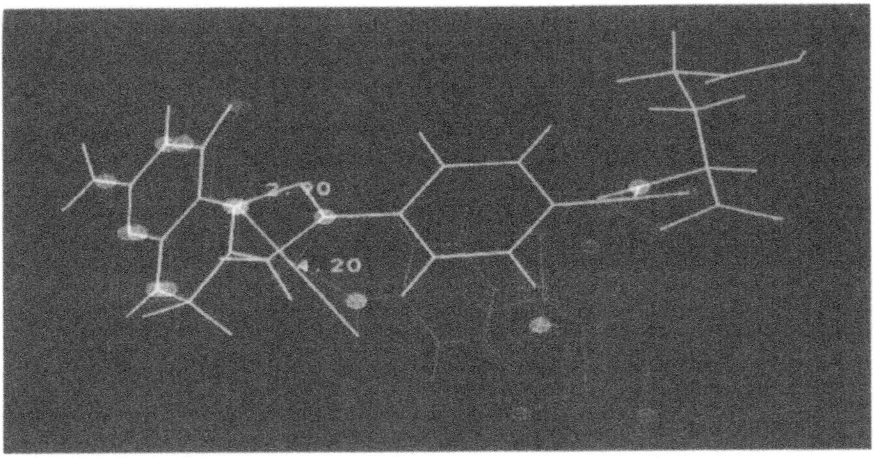

Figure 7. Superposition of the substrate alone before and after N5 protonation. The structure shown in yellow represents the substrate before N5 protonation and the one in red represents the substrate after the N5 protonation. The distances marked correspond to the contact N5 ... HIO before (2.9 Å) and after (4.2 Å) the proton transfer. The red spheres denote the oxygen atoms and the blue spheres the nitrogen atoms in the substrate.

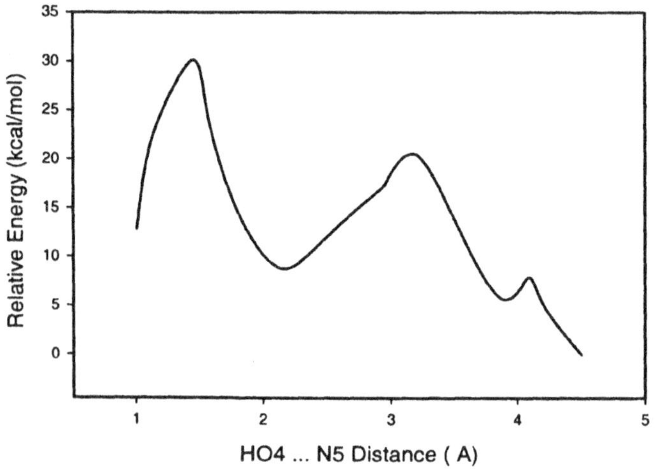

Figure 8. Total free energy profile for the transfer of the proton as a function of HO4...N5 distance.

The total free energy profile as a function of the HO4...N5 distance during the proton transfer from Asp27 to N5 is shown in Figure 8. The energy barrier of the proton transition from Asp27 to N5 is 8 kcal/mol. The rotation of the O4...HO4 group towards the N5 atom comprises the next stage of the mechanism corresponding to a 1.5 Å movement of the proton. This stage has a barrier of about 12 kcal/mol (Figure 8). The free energy change when the proton jumps from O4 to N5 has a barrier 19 kcal/mol. The total free energy change from the unprotonated to the protonated state is 12 kcal/mol. The N5 protonated species is expected to be present in very low concentrations as shown from these calculations.

4.2 Protein-Substrate Interactions During the Proton Transfer

Four of the snap shot geometries during the proton transfer mechanism were analyzed to further characterize the mechanism. Structure 1 corresponds to the protonated Asp27, which is depicted in Figure 2. Structure 3 represents the protonation of O4 in the pterin ring and the negatively charged Asp27, which forms hydrogen bonds with N3 and N2 as shown in Figure 3b. Structure 5 corresponds to one of the transition states for the transfer of the proton from O4 of the pterin ring. Structure 6 corresponds to the N5 protonated form of the pterin ring. The van der

Waals and hydrogen bond contacts associated with the substrate and the surroundings during the proton transfer were analyzed and described below.

4.2.1 Contacts with Pteridine Moiety

The contacts made by the pteridine moiety in the four structures (1, 3, 5, and 6) associated with the proton shuttle were identified. The pteridine residue forms a strong van der Waals interaction with Ala6, in structures 1 and 3, with distances 2.65 Å and 2.34 Å, respectively. In structures 5 and 6, pteridine has moved closer, resulting in hydrogen bond formation with the carbonyl group of Ala6. Pteridine also maintains van der Waals contact with the residue Ile94 in all four structures. The pteridine residue is hydrogen bonded to Tyr100 in structure 1 (distance = 2.15 Å). This hydrogen bond becomes a van der Waals contact in structure 3 and 6 (contact distances of 2.52 Å and 2.3 Å respectively), while in the structure 5, it becomes a shorter hydrogen bond (2.0 Å).

The pteridine residue is hydrogen bonded to the Tyr111 with a hydrogen bond length of 2.25 Å in structure 1. This bond contact is not observed in any of the other structures. The atom HN_2B of the residue pteridine adopts a bifurcated hydrogen bond configuration with atoms OG_1 and HOG of Thr113 in structure 1. This reduces to a single hydrogen bond between HN2B and HOG in structures 3 and 6 after developing a van der Waals contact (distance = 2.39 Å) in structure 5. It is notable that in structure 1 the proton is actually shared between both OD1 of Asp27 and O4 of the pteridine ring. This interaction consequently brings HN3 of the pteridine close to OD2 of Asp27 with a hydrogen bond distance of 1.79 Å. Furthermore, the Asp27 residue has two hydrogen bonds (structures, 3, 5 and 6) with minor differences in their values.

HN8 of the pteridine residue makes a hydrogen bond contact with the carbonyl oxygen of Ile5 (in structure 1). This hydrogen bond length reduces to 1.61 Å in structure 3, and increases to 1.98 Å in structure 5, and then reduces to 1.67 Å in structure 6. In structures 5 and 6, pteridine makes a van der Waals contact with Ala7. It is worthwhile to note that structure 5 shows two van der Waals, contacts which are not seen in any of the earlier three structures. These are the contacts with residues Phe31 and Thr46.

4.2.2 Contacts with *p*-Aminobenzoic Acid Moiety

Changes in the pteridine ring affect the contacts between DHFR and the *p*-aminobenzoic acid (PBH) moiety. PBH maintains van der Waals contact with Phe31 in all four structures (1, 3, 5 and 6) with contact distances ranging within 0.4 Å of one another. Contact is also observed with Leu54 in all four structures with distances ranging from 3.0 Å to 2.8 Å. Only in structure 5 does PBH develop an additional van der Waals contact with Ser49. The PBH residue develops a single hydrogen bond with residue Arg52 of the protein. In structure 1, the hydrogen bond is between oxygen and HN12 with a distance of 1.75 Å. The remainder of the structures (3, 5 and 6) each have two hydrogen bonds, namely O...HNE and O...HN21.

4.2.3 Contacts with Glutamate Moiety

The glutamate moiety of folate maintains van der Waals contact with the protein. Distances vary from 2.7 to 2.9 Å with the residue Leu28 in all four structures. Structure 1 exhibited an additional van der Waals contact with Phe31. The atom O2 of GXO is hydrogen bonded to HNZ1 of Lys32 in structure 1. This distance shortens significantly (to 1.65 Å) in structure 3, and lengthens in structure 5 (to 1.74 Å), and is broken in structure 6. Structure 1 also has a single hydrogen bond contact with atoms OE1 of glutamate and HN22 of Arg52. Both residues move away in the other three structures. O1 of glutamate forms a hydrogen bond with HN12 of Arg57 in all four structures with distances moving from 1.65 Å in structure 1 to 1.72 Å in structure 6. One other hydrogen bond contact persists in all the four structures, namely the hydrogen bond between O2 and HN22. Hydrogen bond distances are 1.74 Å, 1.65 Å, 1.80 Å and 1.89 Å in structures 1, 3, 5 and 6, respectively. Structure 3 develops a single hydrogen bond contact between OE2...HN22 of Arg52. Structure 6 develops two additional hydrogen bonds between O2 ...HNZ2 (Lys32) and O1 ... HN22 (Arg57).

5. HYDRIDE TRANSFER REACTION PATHWAY

5.1 Free Energy Profile

The hydride ion transfer catalyzed by DHFR was divided into three snap shot geometries selected from a total of nine pilot simulations. The three structures represent the initial state, the transition state and the final product of the hydride ion transfer, respectively. The structures obtained from gas phase simulations are denoted as GHT1, GHT2 and GHT3, and the structures obtained from the simulations in the presence of the protein are denoted as PHT1, PHT2 and PHT3. The quantum mechanical energy statistics are given in Table 2, along with the H41N...C6 distances. The variation of the quantum mechanical contribution to the free energy as well as the free energy from the surroundings to the hydride ion transfer are given in Figure 9.

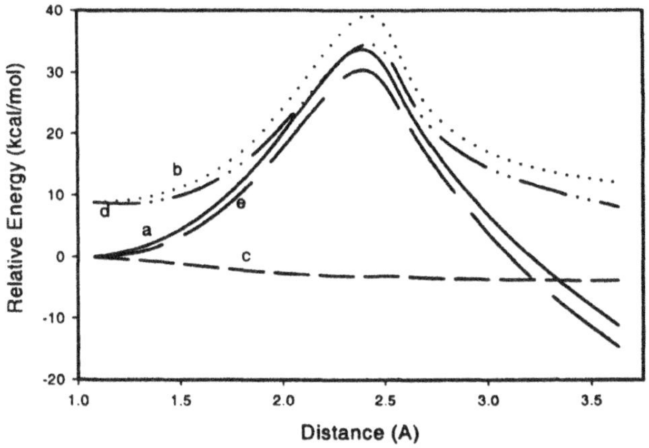

Figure 9. Plot of energy profiles during hydride ion transfer as a function of the H41N... C6 distance. (a) quantum mechanical energy contribution to the free energy obtained in the presence of the protein; (b) quantum mechanical energy contribution to the free energy obtained in gas phase; (c) free energy contribution from the protein surroundings; (d) resultant energy of (b)+(c); (e) resultant energy of (a)+(c).

The relative quantum mechanical energy is computed for GHT1 as the base value, thereby enabling straight forward characterization of the relative stabilization or destabilization by the protein during hydride ion transfer. During the initial state, the quantum mechanical energy contribution to the free energy in the gas phase is about 9 kcal/mol less than in the presence of the protein. This value characterizes the amount of destabilization by the protein in the initial state. At the transition state the destabilization by the protein has significantly decreased so that the energy difference compared to the gas-phase simulation is only 5 kcal/mol. In the final state where the

Table 2. Quantum mechanical energy obtained at the 6-31 G**

Structure	Distance [in Å] C4N ... H41N	Hartree Energy	Relative Energy #
Simulation in gas phase:			
GHT1	1.09	-994.43	0.0
GHT2	1.30	-994.38	32.7
GHT3	2.75	-994.45	-10.7
Simulation in protein:			
PHT1	1.08	-994.42	8.6
PHT2	1.30	-994.37	37.6
PHT3	2.73	-994.41	12.0

#Energy in kcal/mol

product is formed there is a large distortion of the complex by the protein surroundings. The quantum mechanical contribution to the free energy change at this stage in the gas phase is -11 kcal/mol, which is nearly 23 kcal/mol lower in energy compared to the simulation performed in the presence of the protein. These results clearly demonstrate that the protein destabilizes the initial state and the final product of the hydride ion transfer. It is important to note that there is greater energetic destabilization of the product compared to the initial state.

5.2 Protein-Substrate Interactions During the Hydride Transfer Reaction

The free energy associated with the hydride ion transfer from C4 of the nicotinamide ring to C6 of the pteridine ring was computed in a manner

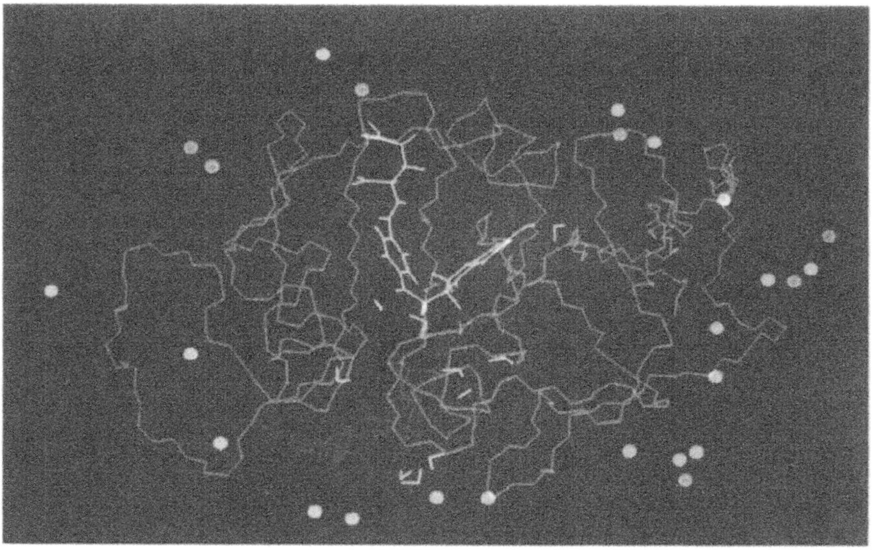

Figure 10. The ternary complex of the enzyme dihydrofolate reductase, the substrate and the cofactor during the transition state of the hydride ion transfer. The enzyme backbone atoms are shown alone for clarity and are colored blue. The substrate is shown in yellow and the cofactor is in red. The bond colored in light blue indicates the hydride ion being shared by both the cofactor and the substrate before the transfer to the substrate. Water molecules around the residue pteridine of the substrate and the nicotinamide ring of the cofactor alone are shown and colored in light blue. The yellow spheres represent the sodium ions and the pink spheres the chloride ions.

similar to the proton transfer, i.e. by mutating structural transitions generated from ab initio quantum mechanical optimized geometries. Three structures comprising the reaction pathway were used for the analysis of the molecular interactions. The first of these structures was designated as PHT1, representing the ternary complex between the protein, the protonated substrate and the reduced cofactor NADPH. The structure of the transition state for the transfer of hydride ion, denoted as PHT2 is shown in Figure 10. The third structure PHT3 represents the product of the hydride ion transfer, where the hydride ion is bound to the substrate. The van der Waals and hydrogen bond interactions are analyzed between the substrate, the cofactor and the remainder of the system during hydride ion transfer.

5.2.1 Contacts with the Pterin Ring Moiety

The van der Waals and hydrogen bond contacts between the pterin ring and both the nearby amino acids and water molecules in the three structures PHT1, PHT2 and PHT3 were evaluated in order to better understand the molecular factors controlling the hydride ion transfer. Ile5 makes a van der Waals contact with the pterin in PHT1, which becomes stonger in PHT2 and PHT3 where a hydrogen bond is established between HN8 and the O. Accordingly, the van der Waals distance for this pair of atoms in structure PHT1, 2.49 Å, becomes 1.66 Å in PHT2 and 2.09 Å in PHT3. This is an important interaction which results in hydrogen bond formation. The Ala7 residue develops a favorable nonbonded contact in PHT1 with pterin, which is not observed in PHT2, but reappears again in PHT3. Met20 gradually moves away from pterin, as shown by the successive increase in the van der Waals distances. Trp22 is in van der Waals contact with the pterin only in structure PHT1. The pterin residue is also in contact with Phe31 in all the three structures with little variation in the respective contact distance. Ile94 is in van der Waals contact with the pterin in PHT1 and comes in closer contact in PHT3, ultimately forming a hydrogen bond between H72 ...O. These residues move apart in PHT3 such that the hydrogen bond becomes a van der Waals contact. The pterin residue in PHT1 develops three additional van der Waals contacts with Tyr100, Thr113 and Wat787, which are not observed in PHT2 and PHT3. Special mention should be noted regarding the interaction between Asp27 and the pterin. In structure 1, Asp27 forms four hydrogen bonds with the pterin residue using both the main-chain and side-chain atoms. The side-chain atom OD2 of Asp27 shares two hydrogen bonds with HN2B and HN3 of the pterin. The other hydrogen bond involves the side-chain atom of Asp27 and the pterin residue (OD1 ... HN3). A fourth hydrogen bond involving Asp27 is the HN2B ...O hydrogen bond. In the PHT1 structure, only one hydrogen bond involving water is observed, i.e. the hydrogen bond between HN2A and O of water.

Asp27 had four hydrogen bonds with the pterin in PHT1 and two hydrogen bonds in structures PHT2 and PHT3. In PHT2, the two hydrogen bonds are between HN2B - OD2 and HN3 - OD1. In both instances, the hydrogens belong to pterin and the oxygens to Asp27. Both contacts are maintained in PHT3, with a small variation in length. It is important to note that there are several hydrogen bonds involving water in PHT2 and PHT3. The hydrogen bond between the pterin residue and Wat194 (HN2A...O) is observed in both PHT2 and PHT3. In PHT2, an additional hydrogen bond between the pterin and a water molecule (Wat682) involves atoms O4 and H2. In PHT3, the pterin residue forms a hydrogen bond with Wat755 (atoms O4 and H2). An important finding is that throughout the entire course of hydride ion transfer, the proton at the N5 position of pterin ring does not show direct contact either with the protein molecule or with any of the water molecules.

5.2.2 Contacts with *p*-Aminobenzoic Acid Moiety

The *p*-aminobenzoic acid moiety (PBH) of the substrate maintains van der Waals contact with Met20 in all the three structures (PHT1, PHT2 and PHT3). Ile50 and Leu54 also make van der Waals contact with PBH in all three structures. The contact distance between PBH and Leu54 is almost constant (at 2.7Å) whereas the contact distance with Ile50 varies slightly in the three structures. The van der Waals contact between PBH and Thr46 in structure PHT1 is not seen in PHT2, but is reestablished in PHT3 with an increased distance. In structure PHT1, the remainder of the three van der Waals contacts are between PBH and three water molecules (Wat198, Wat796 and Wat813). Wat479 is in van der Waals contact with PBH in PHT2. PHT3 has van der Waals contacts between PBH and Leu28, Arg52, and Ile94 which are not observed in PHT1 and PHT2. In PHT3, two additional van der Waals contacts between PBH and waters Wat537 and Wat823 are also noted. There are two hydrogen bonds to water in both PHT1 and PIM; none of these bonds involve the same water molecule. The hydrogen bond in PHT1 is between O and H2 of Wat294 and Wat813. Only PHT2 has a hydrogen bond with an amino acid Arg52. The other hydrogen bond present in PHT2 is between PBH and Wat1095. In PHT3, the hydrogen bond is between Wat818 and PBH between the atoms O and H1.

5.2.3 Contacts with Glutamate Moiety

The van der Waals contact between the glutamate (GXO) moiety of the substrate folate and Leu28 becomes stronger as the transition is made from PHT1 to PHT2 and PHT3. The van der Waals contact between GXO and Phe31 increases as the transition simulation proceeds from PHT1 to PHT2, and then decreases in PHT3. PHT1 has four additional van der Waals contacts, which are between GXO and Wat589, Wat727, Wat813 and Wat850. PHT2 exhibits only a van der Waals contact between GXO and Lys32 and no contacts between GXO and the water molecules. This observation is in contrast to PHT3, which exhibits two van der Waals contacts between GXO and Wat638 and Wat857. GXO forms hydrogen bonds with Arg52, Arg57 and water molecules in all three structures. In structure PHT3, an additional hydrogen bond develops between GXO and Lys32 (OE2...HNZ1). The hydrogen bond length between atoms O2 ... HN12 of GXO and Arg52 decreases as the transition proceeds from PHT1 to PHT2, and decreases further in PHT3. The hydrogen bond between atoms O2 and HN22 involving the same two residues increases in length from structure PHT1 to PHT2, and decreases slightly in PHT3. The hydrogen bond length involving atoms O1 and HN12 between GXO and Arg57 decreases relative to PHT1 in PHT2 and decreases further in PHT3. The hydrogen bond between O1 of GXO and HN22 of Arg57 remains essentially constant in PHT1, PHT2 and PHT3. PHT2 has an additional hydrogen bond (O1 ...HN12) between GXO and Arg52, which is not observed in the other two structures. PHT3 has two additional hydrogen bonds (OE2...HNZ1 and O2...HN22) between the residue GXO and Lys32 and Arg57. In addition, structures PHT1, PHT2 and PHT3 have three, four and one hydrogen bonds between GXO and surrounding water molecules.

5.2.4 Contacts with the Nicotinamide Moiety

The van der Waals and hydrogen bonded contacts associated with the nicotinamide moiety (NRP) and the surrounding amino acid residues of the protein and the water molecules were examined. In PHT1 and PHT3, Glu17 makes van der Waals contacts with NRP. NRP in PHT1 also exhibits van der Waals with Met20 and His45 whereas these contacts are not observed in PHT2 and PHT3. The contact between NRP and Trp22 persists throughout in PHT1, PHT2 and PHT3. The contact decreases from PHT1 to PHT2, and increases in PHT3. Ile94 also uniformly maintains in contact with NRP in all three structures. The contact between NRP and Tyr100 increases as the simulation proceeds from PHT1 to PHT2, and decrease in PHT3. PHT1 has only one van der Waals contact involving NRP and water (Wat200). PHT2 and PHT3 also exhibit unique contacts involving NRP with Gly15, Ala19, Thr46 and Thr123 in PHT2 and with Ala6, Thr46 and Ser49 in PHT3. PHT2 does not have any contacts between NRP and water molecules. This is in contrast to PHT3, which forms two contacts between NRP and Wat196 and Wat525. NRP exhibits hydrogen bonds with both Ala7 and Ile14 which are maintained in PHT1, PHT2 and PHT3. O7N of NRP forms a hydrogen bond with HN of Ala7 in PHT1, which decreases in PHT2, and then increases in PHT3. Also, H71N of NRP forms a hydrogen bond with O of Ile14 which is maintained in PHT1, PHT2 and PHT3. There is also a hydrogen bond between residues NRP and Ser49 (HO2'...OG) in PHT1 and PHT2. In PHT1, we find two additional hydrogen bonds involving residues NRP with Gly97 and Wat787. PHT3 also has two additional hydrogen bonds in which NRP is hydrogen bonded to Asn18 and Wat725.

6. DISCUSSION

In this study, we characterized the stepwise transformations constituting the mechanism by which DHFR catalyzes the reduction of dihydrofolate to tetrahydrofolate using a novel combinatorial application of advanced molecular simulation methods. By using this approach, we are able to make concise statements regarding the role of Asp27 in the catalytic mechanism of DHFR for the first time. Moreover, using the combined ab initio quantum mechanical and molecular mechanics approach, we are able to describe the mechanism in detail, including the electronic, energetic, and structural features associated with the proton transfer and the subsequent hydride ion transfer.

Based on experimental data, the initial state of the proton transfer involves a protonated Asp27 and the pterin ring in the enolic configuration. Our studies suggest that the proton moves ("shuttles") from Asp27 to O4 of the pterin ring, and is accompanied by hydrogen bond switching from O4 to N3 of the pterin. This step is followed by rotation of the O4-HO4 bond of the pterin moiety toward N5, which is a critical step facilitating the proton migration to N5. During this process, the proton shuttles from O4 to N5 of the pterin where there is a critical conformational transition in the substrate. This conformational transition subsequently drives the pteridine ring toward the nicotinamide ring resulting in facile hydride ion transfer. Based on our data, it is reasonable to believe that this conformational transition contributes approximately 2 to 3 kcal/mol for catalysis, however, an exact energetic contribution could not be fully characterized due to the complexity of the computational steps beyond that used in the present work.

During hydride ion transfer, careful analysis of the secondary structural features of the protein in the ternary complex reveals that the overall conformational features of the protein remain undisturbed. The conformational angles of the residues in the MET loop[25] were analyzed in detail due to its proximity to the substrate. The MET loop is composed of amino acid residues Ala9 through Leu24. Comparison of the initial, transition, and final states of the MET loop during hydride ion transfer reveals that: (a) between the initial state and the transition state, the Ψ angle of Gly15, Met16, Asn18 rotates by 30°, and (b) the Φangle of Glu17, Asn19 and Met21 also rotate by 30° These two factors distort the MET loop significantly so that the MET loop is compressed and adopts a more compact configuration in the transition state. However, when the final product of the hydride ion transfer forms, the MET loop "springs" back to nearly its original relaxed configuration. Between the initial and product states, the Φ Ψ angles of Asp11 and Val13 undergo approximately a 30° change. The single, largest conformational change occurs in the Ψ angle of Gly15 which flips 180°. The Φ angle of Met16 rotates by 80° and the Ψ angle 30° The conformational changes in the MET loop provide a significant and direct effect in bringing closer the substrate and the cofactor, in order to facilitate hydride ion transfer.

It is important to note that during these simulations, the solvent molecules do not access either the N5 position or the O4 position of the substrate. This fact supports the conclusion that the ternary complex in which the proton is bound to Asp27 is a kinetically productive complex.

It has been proposed[42, 43] that a protonated Asp27 can donate the proton to the substrate by means of intermediate solvent molecules. Another view[25] proposed that the proton source is an explicit solvent molecule and that the Asp27 acts as a proton relay. To resolve these contrasting views, we have conducted a series of simulations starting with the cyclic amide form of the pterin ring and a water molecule hydrogen bonding with HO4 and N5. Eight structures were generated along the path of the proton transfer from O4 to N5 using a 6-31G** geometry optimization of the transfer process. The hydrogen bond distances between O4-HO4 and the corresponding SCF energy for each of these structures are listed in Table 3. It is important to note that dhw0 corresponds to the state in which O4 is protonated, and dhw7 represents the final product of the proton transfer in which N5 is protonated.

Table 3. Ab initio energy profile during water mediated proton transfer.

Structure	**Ab Initio energy (Hartree Units)**	**Relative Energy (kcal/mol)**	**Distance N5-H2[a] (Å)**
dhw0	-693.58660	0.0000	2.165
dhw1	-693.58554	0.6672	1.906
dhw2	-693.58260	2.5088	1.609
dhw3	-693.57262	8.7773	1.361
dhw4	-693.58063	3.7462	1.302
dhw5	-693.58063	3.7495	1.231
dhw6	-693.58078	3.6551	1.291
dhw7	-693.59765	-6.9300	1.033

[a]The atom H2 belongs to the water molecule

Table 4. Ab initio energy profile during the proton transfer starting from the reduced tetrahydrofolate.

Structure	Ab Initio energy (Hartree Units)	Relative Energy (kcal/mol)	Distance HO4-O4 (Å)
h4fenl1	-617.61312	53.2390	0.965
h4fenl2	-617.60907	55.7788	1.005
h4fenl3	-617.60251	59.8975	1.130
h4fenl4	-617.64567	32.8124	2.444
h4fenl5	-617.69796	0.0	2.525

As the proton shuttles from the O4 to N5, there is a barrier of about 8.8 kcal/mol at which point the O4-HO4 distance is 1.4 Å. Once this barrier is crossed, the system becomes more stabilized. Following proton transfer the energy is reduced by -6.9 kcal/mol. This series of findings support the possibility that an energetically feasible mechanism can be proposed for proton transfer from O4 to N5 via an intermediate water molecule. But as has been discussed earlier, there was no solvent molecule in direct contact with either O4 or N5 of the substrate during these simulations.

We have also addressed the possibility to achieve hydride ion transfer without protonation of N5. The energy profile for proton transfer from N5 to O4 in the reduced tetrahydrofolate was obtained by means of five intermediate structures. The SCF energies for various distances of O4-HO4 are listed in Table 4. It is clear from this data that when the tautomer is in the keto form (hydrogen on N5), the energy difference is about 53 kcal/mol. On the other hand, the energy difference between the hydroxy and keto tautomer in the oxidized folate is 30 kcal/mol . There is no transition state for this reaction to occur. Therefore the energy difference is 50-60 kcal/mol , which is much greater than the free energy barrier height for hydride ion transfer when N1 is protonated. Thus it is critical for N5 to be protonated first, in order for hydride ion transfer to occur.

7. CONCLUSIONS

In this study, identification of the critical atomic and molecular determinants pertaining to the mechanism of dihydrofolate to tetrahydrofolate reduction was achieved by (i) ab initio quantum mechanics, (ii) molecular mechanics, and (iii) free energy perturbation techniques. For the first time, the complete free energy profile was calculated for the proton transfer from Asp27 of the enzyme *E. Coli* DHFR to the N5 position of the dihydropterin moiety of the substrate dihydrofolate. In addition, the free energy profile was calculated for the hydride ion transfer from NADPH to the pteridine ring. These calculations represent almost one nanosecond of molecular dynamics. The complete series of simulations required nearly 2000 cpu hours on the Cray-XMP supercomputer and 2000 cpu hours on the IBM RS-6000/570 machines. In addition to the free energy profiles, the chapter discussed the conformational changes associated with the catalysis. The function of the enzyme is to bring the reactants together and position them in the proper orientation so that the reaction can proceed through donation of a proton to the substrate which, in turn, facilitates hydride ion transfer. In principle the proton can come from the solvent, however, we have shown that proton donation from the enzyme is the most probable mechanism based on the energy profile. The rate of reaction is partially enhanced by the conformational changes in the substrate, and by the conformational changes in the protein. The catalytic rate of the forward and reverse reaction for hydride ion transfer is enhanced by the relative destabilization of the reactants and the products and the stabilization of the transition state.

A detailed understanding of the mechanism of DHFR at the molecular level enabled the identification of critical interactions between the substrate and the enzyme during the catalytic process. This information is ultimately expected to be useful in the design of potent and specific inhibitors of DHFR.

8. REFERENCES

1. M. H. Tattersall, R. C. Jackson, S. T. M. Jackson, and K. R. Harrap, Factors determining cell sensitivity to methotrexate: Studies of folate and deoxyribonucleoside triphosphate pools in five mammalian cell lines, Eur. J. Cancer **10**:819 (1974).
2. M. G. Nair and P. T. Campbell, Folate analogues altered in the C^9-N^{10} bridge region. 10-Oxafolic acid and 10-oxaaminopterin, J. Med. Chem. **19**:825 (1976).
3. C. A. Lewis and R. B. Dunlap, Topics in Molecular Pharmacology, A. S. V. Burgen and G. C. K. Roberts, eds., Elsevier, Holland (1981), pp. 170-219.
4. R. L Blakley, Molecular Actions and Targets for Cancer Therapeutic Agents; A. C. Sartorelli, J. S. Lazo, and J. R. Bertino, eds., Academic Press, New York (1981), pp. 302-332.
5. B. I. Schweitzer, A. P. Dicker, and J. R. Bertino, Dihydrofolate reductase as a therapeutic target, FASEB J. **4**:2441 (1990).
6. B. Roth and C. C. Cheng, Recent progress in the medicinal chemistry of 2,4-diaminopyrimidines, Prog. in Med. Chem. **19**:270 (1982).
7. B. A. Chabner, N. Clendeninn, G. A. Curt, and J. Jolivet, Methotrexate in Cancer Therapy, K. Kimura, and Y. M. Wang, eds., Raven Press, New York (1986), pp. 1-38.
8. J. N. Champness, L. F. Kuyper, and C. R. Beddell, Molecular Graphics and Drug Design; A. S. V. Burgen, G. C. K. Roberts, and M. S. Tute, eds.; Elsevier, Holland, (1986), pp. 336-363.

9. F. M. Huennekens, K. S. Vitols, J. M. Whiteley, and V. G. Neef, Methods in Cancer Research Academic Press, New York (1976), pp.199.
10. G. H. Hitchings and B. Roth, Enzyme Inhibitors as Drugs, M. Sandler, ed.; University Park Press, Baltimore, (1980), pp. 263-280.
11. D. A. Mathews, R. A. Alden, J. T. Bolin, S. T. Freer, R. Hamlin, N. Xuong, J. Kraut, M. Poe, K. Williams, and K. Hoogsteen, Dihydrofolate reductase: X-ray structure of the binary complex with methotrexate, Science **197**:452 (1977).
12. T. J. Bolin, D. J. Filman, D. A. Mathews, R. C. Hamlin, and J. Kraut, Crystal structures of escherichia coli and lactobacillus casei dihydrofolate reductase refined at 1.7Å resolution, J. Biol. Chem. **257**:13650 (1982).
13. D. J. Filman, T. J. Bolin, D. J. Filman, D. A. Mathews, R. C. Hamlin, and J. Kraut, Crystal structures of escherichia coli and lactobacillus casei dihydrofolate reductase refined at 1.7Å resolution, J. Biol. Chem. **257**:13663 (1982).
14. C. A. Fierke, K. A. Johnson, and S. J. Benkovic, Construction and evaluation of the kinetic scheme associated with dihydrofolate reductase from escherichia coli, Biochemistry **26**:4085 (1987).
15. C. Bystroff and J. Kraut, Crystal structure of unliganded escherichia coli dihydrofolate reductase, ligand-induced conformational changes and cooperativity in binding, Biochemistry **30**:2227 (1991).
16. V. A. Roberts, P. Dauber-Osguthorpe, D. J. Osguthorpe. J. Wolff, and A. T Hagler, A comparison of the binding of the ligand trimethoprim to bacterial and vertebrate dihydrofolate reductases, Isr. J. Chem. 198-210 (1986).
17. P. Dauber-Osguthorpe, V. A. Roberts, D. J. Osguthorpe, J. Wolff, M. Genest, and A. T. Hagler, Structure and energies of ligand binding to proteins: Escherichia coli dihydrofolate reductase trimethoprim, a drug receptor system, Proteins **4**:31 (1988).
18. U. C. Singh, Probing the salt bridge in the dihydrofolate reductase-methorexate complex by using the coordinate-coupled free-energy perturbation method, Proc. Nat. Acad. Sci. USA **85**:4280 (1988).
19. U. C. Singh and S. J. Benkovic, A free-energy perturbation study of the binding of methotrexate to mutants of dihydrofolate reductase, Proc. Nat. Acad. Sci. **85**:9519 (1988).
20. S. H. Fleishman and C. L. Brooks, Protein drug interactions: Characterization of inhibitor binding in complexes of DHFR with trimethoprim and related derivatives, Proteins: Structure, Function and Genetics **7**:52 (1990).
21. P. L. Cummins, K. Ramnarayan, U. C. Singh, and J. E. Gready, Molecular dynamics/free energy perturbation study on the relative affinities for the binding of reduced and oxidized NADP to dihydrofolate reductase, J. Am. Chem. Soc. **113**:8247 (1991).
22. J. Bajorath, D. H. Kitson, G. Fitzgerald, J. Andzelm, J. Kraut, A. T. Hagler, Electron redistribution on binding of a substrate to an enzyme: Folate and dihydrofolate reductase, Proteins: Structure, Function and Genetics **9**:217 (1991).
23. D. A. Mathews, R. A. Alden, J. T. Bolin, D. J. Filman, S. T. Freer, R. Hamlin, W. G. Hol, R. L. Kisliuk, E. J. Pastore, L. T. Plante, N. Xuong, J. Kraut, Dihydrofolate reductase from lactobacillus casei, J. Biol. Chem. **253**:6946 (1978).
24. J. C. Fontecilla-Camps, C. E. Bugg, C. Temple, J. D Rose, J. A. Montgomery, R. L. Kisliuk, Chemistry and Biology of Pteridines; R. L. Kisluik and G. M. Brown, eds., Elsevier, New York (1979), pp. 235.
25. C. Bystroff, S. J. Oatley, and J. Kraut, Crystal structures of escherichia coli dihydrofolate reductase: The NADP+ holoenzyme and the folate-NADP+ ternary complex duhydrofolate reducatse, Biochemistry **29**:3263 (1990).
26. S. R. Stone and J. F. Morrison, Kinetic mechanism of the reaction catalyzed by dihydrofolate reductase from escherichia coli, Biochemistry **21**:3757 (1982).

27. S. R. Stone and J. R. Morrison, Catalytic mechanism of the dihydrofolate reductase as determined by pH studies, Biochemistry **23**:2753 (1984).
28. E. E. Howell, J. E. Villafranca, M. S. Warren, S. J. Oakley, and J. Kraut, Functional role of aspartic acid-27 in dihydrofolate reductase revealed by mutagenesis, Science **231**:1123 (1986).
29. J. T. Chen, K. Taira, C. P. D. Tu, and S. J. Benkovic, Probing the functional role of phenylalanine-31 of escherichia coli dihydrofolate reductase by site-directed mutagenesis, Biochemistry **26**:4093 (1987).
30. J. Kraut and D. A. Mathews, Biological Macromolecules and Assemblies; F. A. Jurnak, and A. McPherson, eds., John Wiley, New York (1987), pp. 1-72.
31. AMBER (Version 3.3) is a fully vectorized version of AMBER (3.0) with coordinate coupling, intra/inter decomposition and the option to include the polarization energy as part of the total energy. Singh et al., (U. C. Singh, P. K. Weiner, J. W. Caldwell, P. A. Kollman, University of California, San Francisco, 1986).
32. U. C. Singh and P. A. Kollman, An approach to computing electrostatic charges for molecules, J. Comput. Chem. **5**:129 (1984).
33. P. C. Hariharan and J. A. Pople,The influence of polarization functions on molecular orbital hydrogenation energies, Theor. Chim. Acta **28**:213 (1973).
34. R. W. Zwanzig, High-temerature equation of a state by a perturbation method. I. Nonopolar gases, J. Chem. Phys. **22**:1420 (1954).
35. U. C. Singh, F. K. Brown, P. A. Bash, and P. A. Kollman, An approach to the application of free perturbation methods using molecular dynamics: Applications to the transformation of $CH_3OH \rightarrow CH_3CH_3$, $H_3O^+ \rightarrow NH_4^+$, glycine → alanine, and alanine → phenylalanine in aqueous solution and to $H_3O^+(H_2O_3) \rightarrow NH_4^+(H_2O_3)$ in the gas phase, J. Amer. Chem. Soc. **109**:1607 (1987).
36. M. R. Reddy, M. D. Erion, and A. Agarwal, Free energy calculations: Use and limitations in predicting ligand binding affinities, in: Reviews in Computational Chemistry, vol. 16 K. B. Lipkowitz and D. B. Boyd, eds., VCH Publishers, New York, (2000) pp. 217-304.
37. B. G. Rao and U. C. Singh, Hydrophobic hydration: A free energy perturbation theory, J. Am. Chem. Soc. **111**:3125 (1989).
38. B. G. Rao and U. C. Singh, A free energy perturbation study of solvation in methanol and dimethyl sulfoxide, J. Am. Chem. Soc. **112**:3803 (1990).
39. K. Ramnarayan, B. G. Rao, and U. C. Singh, The effect of polarization energy on the free energy perturbation calcultions, J. Chem. Phys. **92**:7057 (1990).
40. W. L. Jorgensen, J. Chandrasekhar, and J. D. Madura, Comparison of simple potential functions for stimulation liquid water, J. Chem. Phys. **79**:926 (1983).
41. J. P. Ryckaert, G. Ciccotti, and H. J. C. Berendsen, Numerical integration of the cartesian equations of motion of a system with constraints: Molecular dynamics of n-alkanes, J. Comput. Phys. **23**:327 (1977).
42. J. H. Freisheim and O. A. Mathews, Folate Antagonists as Therapeutic Agents, vol. 1, F. M. Sirotnak, J. J. Burchall, W. B. Emsiger, and J. A. Montgomery, eds.; Academic Press, New York (1981), pp. 69-131.
43. J. E. Gready, Theoretical studies on the activation of the pterin cofactor in the catalytic mechanism of dihydrofolate reductase, Biochemistry **24**:4761 (1985).

Section Five

Drug Design Case Studies

Chapter 14

Fructose 1,6-Bisphosphatase: Use of Free Energy Calculations in the Design and Optimization of AMP Mimetics

M. Rami Reddy and Mark D. Erion
Metabasis Therapeutics, Inc., San Diego, CA 92121

1. INTRODUCTION

Fructose 1,6-bisphosphatase (FBPase) catalyzes the hydrolysis of fructose 1,6-bisphosphate to fructose 6-phosphate (F6P) and inorganic phosphate and is a key regulatory enzyme in the gluconeogenesis pathway.[1] Inhibitors of FBPase are postulated to be useful for the treatment of diabetes,[1] since excessive flux through the gluconeogenesis pathway is responsible for the hyperglycemia associated with type II diabetes. Discovery of potent and selective inhibitors of FBPase represents a significant challenge for medicinal chemists due to the hydrophilic nature of both the substrate binding site and the allosteric regulatory site. The latter site binds adenosine monophosphate (AMP), which induces a protein conformational change that results in decreased enzymatic activity.

The availability of a high resolution X-ray structure of FBPase,[1, 2] led to our efforts to utilize computer-assisted drug design (CADD) approaches[3, 4] for the discovery of AMP mimetics that bind to FBPase with high affinity and selectivity. CADD has been used successfully for the discovery of several novel enzyme inhibitors, including inhibitors of thymidylate synthase,[5] HIV-1 protease[6, 7] and purine nucleoside phosphorylase.[8] The most accurate computational method for estimating relative binding affinities of structurally similar inhibitors to an enzyme is the free energy perturbation approach using with either molecular dynamics (MD) or Monte Carlo (MC) simulations.[9] Despite its high accuracy, free energy calculations have primarily been used to rationalize experimentally-determined binding affinities[10-13] with few applications focusing on predictions.[14, 15] This chapter describes our efforts to utilize relative solvation and binding free energies for evaluating the binding affinity of AMP mimetics to the AMP binding site prior to synthesis[16, 17] using the iterative process summarized in Figure 1.

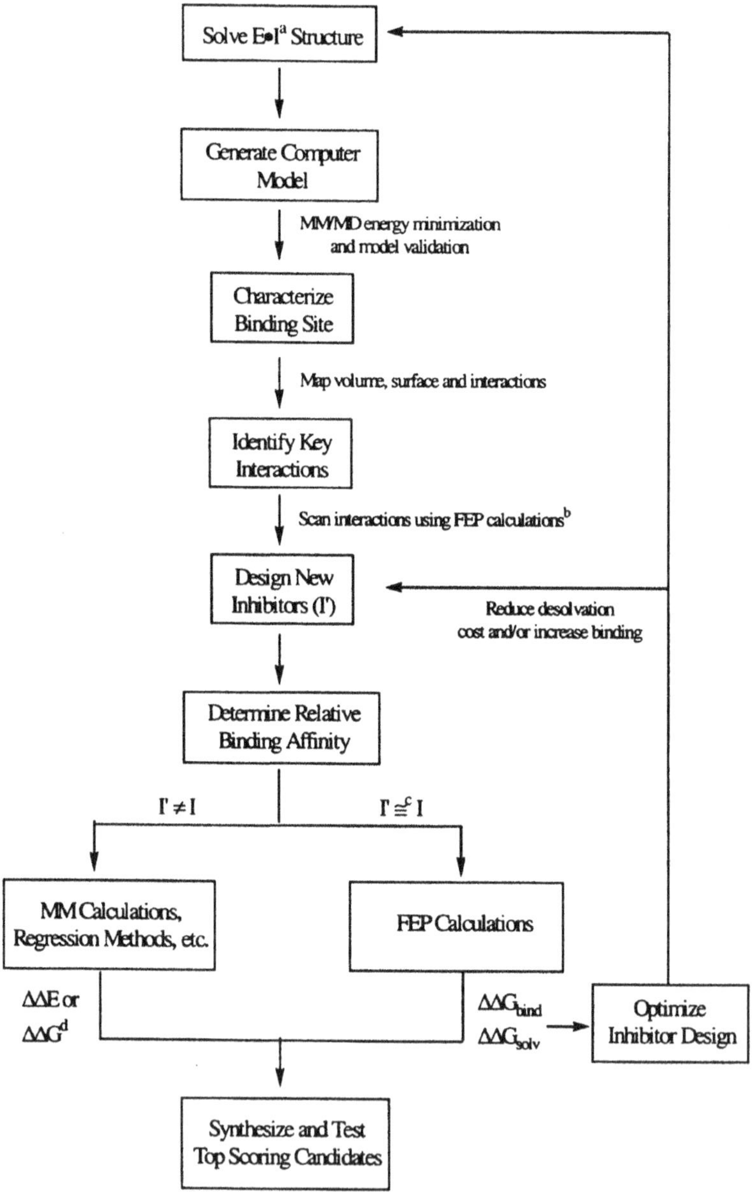

Figure 1. CADD Flowchart. [a]E•I = enzyme-inhibitor complex. [b]scan binding interactions using strategy described in reference 16. [c]I and I' are structurally similar. [d]relative energies or estimated relative free energies.

2. COMPUTATIONAL DETAILS

2.1 Methodology

The thermodynamic cycle-perturbation (TCP) method[9, 18] computes the relative difference in solvation and binding free energies through non-physical paths connecting the desired initial and terminal states (see Figure 2, Chapter 16). This approach enables calculation of the relative change in binding free energy difference between two related compounds by computationally simulating the 'mutation' of one to the other. The relative solvation and binding free energy changes for the two ligands is computed using either single or double topology method.

The single topology method entails changing the appropriate reactant atoms to product atoms. The mutation often results in geometrical changes as well as changes in partial charges and van der Waals parameters. Prior to the mutation, the system was minimized using 500 steps of steepest descent and 2000 steps of conjugate gradient methods and then equilibrated for 20 ps. Each mutation entailed calculations using 101 windows, with each window comprising 1 ps of equilibration and 2 ps of data collection or 303 ps of total MD simulation.

In the double topology or thread method,[19, 20] a single topology is defined for those atoms which are identical in both molecules. For the portion of the molecule that must be transformed, both the starting (reactant) and ending (product) topologies are defined using their associated geometries, with one beginning and the other ending the simulation entirely as dummy atoms. Dummy atoms are identical to real atoms except that their Lennard-Jones parameters and charges are set to zero. At intermediate points during the transformation, all atoms in both topologies have fractional Lennard-Jones parameters and charges. Additionally, molecules with both topologies interact with the environment but not with each other. For each calculation, the system was initially minimized using 500 steps of steepest descent and 2000 steps of conjugate gradient methods and then equilibrated for 20 ps. A two-stage procedure was used to obtain relative free energy differences from the molecular dynamics simulations. During the first stage, the charges of the reactant atoms are turned off while the Lennard-Jones parameters of the product atoms are turned on. During the second stage, the Lennard-Jones parameters of the reactant atoms are turned off while the charges of the product atoms are turned on. This procedure has been used previously[20] to achieve convergence. Each stage of the simulation was performed using 101 windows, each window comprising 1 ps of equilibration and 2 ps of data collection to compute the free energy differences using the thread method because shorter simulations often resulted in convergence problems.[20] Thus, a molecular dynamics simulation of 606 ps run was used for the complete mutation.

2.2 Force Field Parameters

All molecular dynamics, molecular mechanics and FEP calculations were carried out with the AMBER program using an all atom forcefield[21, 22] and the SPC/E model potential[23, 24] to describe water interactions. Electrostatic charges and parameters for the standard residues were taken from the AMBER database. For non-standard solute atoms, partial charges were obtained by fitting wave functions calculated with Gaussian94[25] ab initio 6-31G* basis set level with CHELP[26]. All equilibrium bond lengths, bond angles, and dihedral angles for non-standard residues were taken from ab initio optimized geometries. Missing force field parameters were estimated from similar chemical species within the AMBER database.

2.3 Solvent Simulations

SPC/E explicit water was used for calculating solvation free energies. All the water molecules located greater than 15.0 Å or less than 2.5 Å away from any solute atoms were removed. Aqueous phase dynamics simulations were carried out in a rectangular box using periodic boundary conditions in all directions. Newton's equations of motion for all the atoms were solved using the Verlet algorithm[27] with a 2 fs time step. SHAKE[28] was used for constraining all bond lengths. Constant temperature (N, P & T ensemble) was maintained by velocity scaling all atoms in the system. Non-bonded interaction energies were calculated using a 15.0 Å residue based cutoff.

2.4 Complex Simulations

The EDIT module of AMBER was used to add hydrogens to the protein tetramer and the crystallographic waters. The protonation state of histidine was deduced from analysis of neighboring residues and from hydrogen bonding potential. The total charge on the FBPase tetramer complex was +4 e. No counterions or changes in the customary charge of protein residues were used. The entire system was immersed in a 25.0 Å radius sphere of solvent centered around the mutating group. The water sphere was subjected to a half-harmonic restraint near the boundary to prevent evaporation. During the simulation, all atoms of the protein were fixed beyond 25.0 Å. All non-bonded interactions involving the inhibitors and the charged residues of the protein were computed with an infinite cutoff. A 15.0 Å non-bonded residue based cutoff was used for other residues of the system. The algorithm for the complex simulation was identical to the solvent simulation, except for the absence of periodic boundary conditions in the former.

3. STRUCTURAL ANALYSIS

3.1 AMP Model

FBPase is a tetrameric molecule with four identical polypeptide chains (Cl to C4). Cl and C2 comprise the crystallographic n-symmetric unit. The X-ray structure of the ZMP: human liver FBPase complex[1] was solved at 2.3 Å resolution by Prof. William Lipscomb (Harvard U.). Analysis of the X-ray structure of the FBPase: ZMP complex shows that slight differences exist between each subunit with regard to the atomic positions for several binding site residues (e.g. Arg140 and Lys112 side-chains) as well as the number and position of water molecules. Consequently, we calculated the interaction energy of ZMP in each subunit. The calculated interaction energies after energy minimization indicated that the C4 subunit has the lowest energy (C4<C1<C3<C2). Accordingly, the ZMP binding site of the C4 subunit was used for all molecular modeling calculations.

3.2 Model Validation

A model of adenosine monophosphate (AMP) bound to the AMP binding site was built by first overlaying AMP on ZMP in the enzyme subunit C4 (Figure 2). The model was then energy minimized using 500 steps of

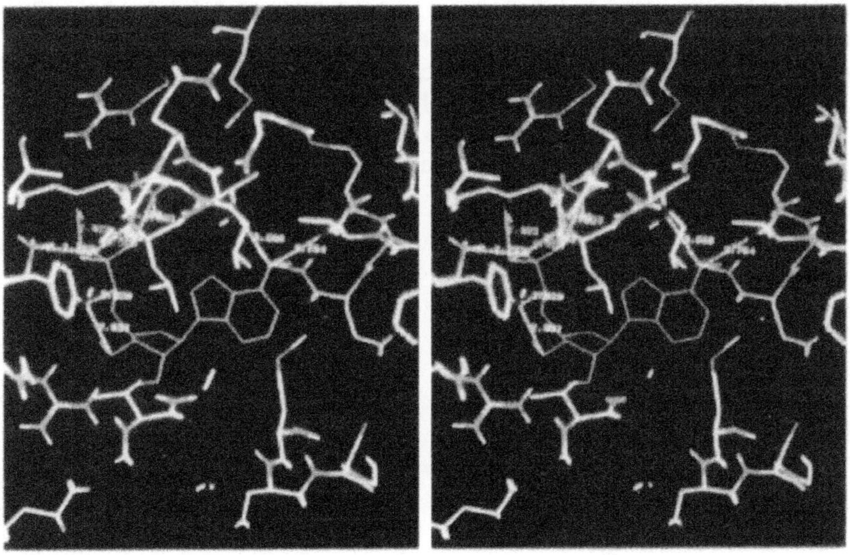

Figure 2. Stereoview of the allosteric binding site of fructose 1,6-bisphosphatase and the binding geometry of AMP (yellow).

steepest descent followed by 2000 steps of conjugate gradient. The complex was then equilibrated with 20 ps MD simulation. The average 'dynamical' structure of the complex was computed from the MD simulation. For time steps of 1 fs and 2 fs, the root mean square (RMS) deviations from the crystal structure were 1.10 Å and 1.17 Å for backbone atoms and 1.55 Å and 1.61 Å for side-chain atoms, respectively. As expected, the largest deviations were observed on the surface (with RMS deviations of 1.28 Å and 1.70 Å for backbone and side chain atoms, respectively) when compared to the rest of the protein. This is primarily due to the flexibility of the protein in the surface region. Nevertheless, the dynamical structure is a good model for calculating relative free energy changes between two similar inhibitors. Since both time steps gave similar structures and are in good agreement with the X-ray structure, we used the larger time step, i.e. 2 fs, for all the free energy calculations reported, to save computer time.

3.3 FEP Methodology Validation

Initially, the relative binding free energy between ZMP (**1**) to AMP (**2**) was performed to test the validity of our protocol. Since this mutation involves significant changes in ligand structure, the 'thread' method[12] was used to accomplish the transformation. For this mutation, the bases of ZMP and AMP are 'threaded' together at C_1' (Figure 3). The calculated difference in binding free energy of 1.70±0.9 kcal/mol is in good agreement with the experimentally measured free energy difference of 1.4 kcal/mol. These results indicate that ZMP binds to FBPase with lower affinity relative to AMP. The lower affinity appears to be due to the higher desolvation free energy of ZMP (1.5 kcal/mol), which results from the increased number of hydrogen bonds formed to solvent water, and the increased conformational freedom (higher entropy) of ZMP relative to AMP.

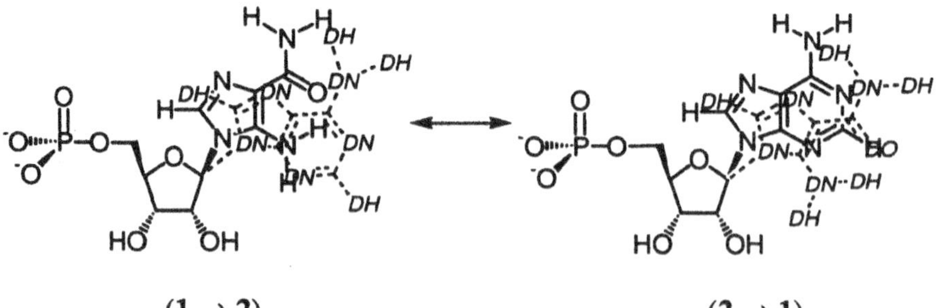

(**1** → **2**) (**2** → **1**)

Figure 3: Dual-topology definition for ZMP (**1**) and AMP (**2**). Dashed structures incorporate "dummy" atoms (D).

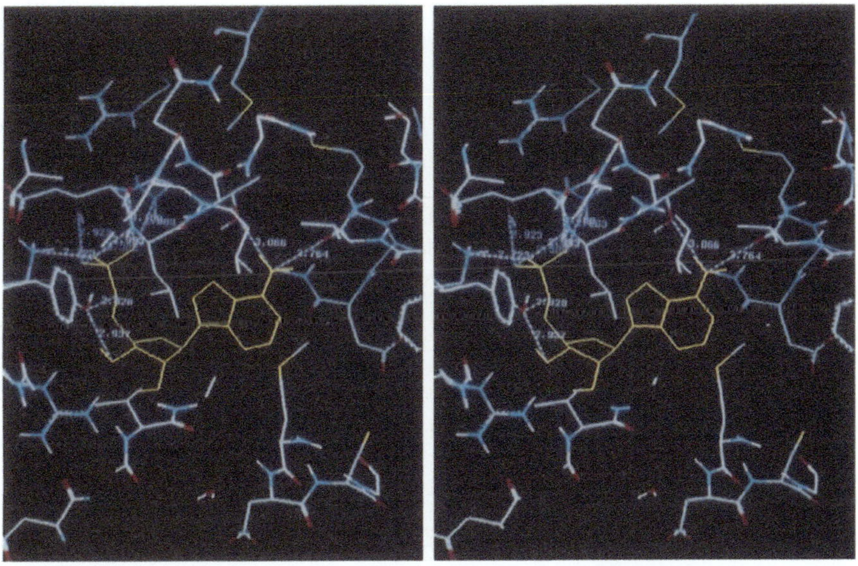

Figure 2. Stereoview of the allosteric binding site of fructose 1,6-bisphosphatase and the binding geometry of AMP (yellow).

4. ANALYSIS OF AMP MIMETICS

Scanning of AMP binding site of FBPase (Figure 2) using the free energy perturbation method[16] indicates that hydrogen bond interactions with the phosphate, 6-amino group, and N7 may be very important to the binding affinity of AMP to FBPase. Accordingly, discovery of more potent and selective analogs required new analogs with groups that provide additional favorable interactions or replacement of these groups with groups that are equally effective. Thus, new AMP analogs were designed using the X-ray structure of FBPase:AMP complex and CADD approaches. Newly designed analogs were examined individually and the most promising analogs selected for further calculations on the basis of their interactions with the binding site residues. The relative binding affinities of selected new analogs were calculated using the free energy perturbation method. Analogs exhibiting significantly enhanced binding affinities were synthesized and evaluated as FBPase inhibitors.

4.1 5'- Substituted AMP Analogs

Phosphates are often poor drug candidates[29] due to their instability in biological fluids and inability to penetrate cell membranes. As a result, we focused our efforts on the discovery of suitable phosphate mimics. In these studies we evaluated various analogs with well-known phosphate replacements (Figure 4), namely phosphonate (**3**), sulfate (**4**), carboxylate (**5**) and dicarboxylate (**6**).

Four mutations, AMP (**2**) → phosphonate (**3**), AMP (**2**) → sulfate (**4**), AMP (**2**) → carboxylate (**5**), and AMP (**2**) → dicarboxylate (**6**) were performed and relative solvation and binding free energies calculated. For the mutations involving replacement of phosphate (**2**) with carboxylate (**5**) and dicarboxylate (**6**), the double topology method was used due to the large structural changes. For the AMP to phosphonate (**3**) and AMP to sulfate (**4**) transformations, the single topology method was used. The free energy differences between AMP and phosphonate (**3**) show that phosphonate costs less to desolvate by 1.0 kcal/mol. The large loss in relative binding affinity is likely due to the loss of a hydrogen bond with the Tyr113 hydroxyl group, and to decreased interaction energies of the phosphonate group with protein residues relative to phosphate. As a result, phosphonate (**3**) is a much weaker inhibitor (3.9 kcal/mol) of FBPase than AMP.

The calculated relative binding free energies for AMP to sulfate (**4**) and AMP to carboxylate (**5**) mutations are 3.5±0.6 kcal/mol and 5.0±1.6 kcal/mol, respectively. These calculated free energies are qualitatively similar to the experimental results of >2.6 kcal/mol and >3.6 kcal/mol for

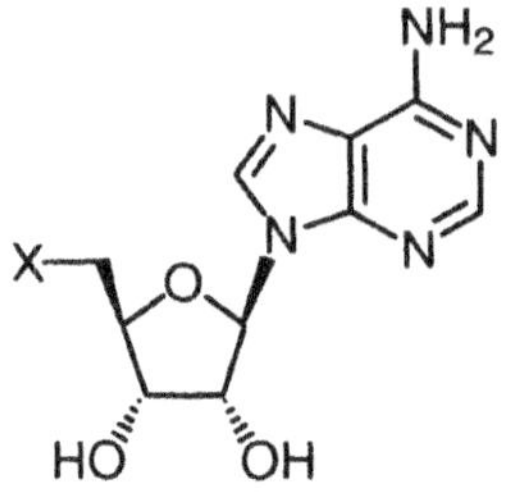

2 X = $OPO_3^{=}$
3 X = $CH_2PO_3^{=}$
4 X = OSO_3^{-}
5 X = CH_2COO^{-}
6 X = $CH(COO^{-})_2$

Figure 4. 5'-Substituted AMP analogs evaluated in FEP calculations

the former and latter transformations. The large change in charge between the phosphate (-2e) and either the carboxylate or sulfate (-1e) is likely to decrease accuracy in these free energy calculations. Nevertheless, the qualitative trend was correctly predicted by the FEP method. Analysis of the calculated relative solvation and binding free energies between dicarboxylate (**6**) and AMP to FBPase showed that the dicarboxylate (**6**) costs 2.3 kcal/mol more to desolvate but gains about 1.1 kcal/mol in the complex as compared to AMP. Accordingly, the dicarboxylate analog (**6**) binds less effectively (1.2 kcal/mol) to FBPase, even though it gained favorable interactions in the binding site residues relative to AMP, because of its much larger desolvation cost. Based on the calculated relative binding free energies none of the four phosphonate mimics were predicted to bind to FBPase as effectively as phosphate.

4.2 AMP Analogs Containing a Modified Pyrimidine Ring

The X-ray structure of the FBPase-AMP complex (Figure 2) revealed unfilled space near N1, C2 and N3. Since the residues in the vicinity were hydrophobic and neither N1 nor N3 participated in a hydrogen bond with the protein, analogs, 2-S-methyl AMP (**7**), 2-methyl AMP (**8**), 2-ethyl AMP (**9**) and 2-chloro AMP (**10**) (Figure 5) were evaluated to determine whether these substituents gained favorable hydrophobic interactions with Met188.

The calculated binding free energies indicate that relative to AMP, 2-methyl AMP (0.20 kcal/mol) and 2-ethyl AMP (0.7 kcal/mol) were predicted to be slightly weaker FBPase inhibitors despite their reduced desolvation costs. The rationale for the decreased binding affinity in the complex was not apparent from analysis of the X-ray structure but the results were consistent with the inhibition potency determined subsequently for 2-ethyl AMP (1.0 kcal/mol). However, consistent with the experimental binding affinity results, 2-S-methyl AMP (0.25 kcal/mol) and 2-chloro AMP (0.8 kcal/mol) also showed weaker FBPase inhibitors than AMP presumably because of the large increase in desolvation costs predicted to occur with these substituents. In addition to the 2-substituted analogs, several deaza AMP analogs; 1-deaza AMP (**11**), 3-deaza AMP (**12**), 1,3-dideaza AMP

(**13**), 1,3-dideaza-1- fluoro AMP (**14**), and 1,3-dideaza-1-ethyl AMP (**15**) were designed (Figure 5) and evaluated to determine whether the expected reduction in desolvation costs produced by the base modification would result in an AMP analog with enhanced binding affinity. The calculated relative solvation and binding free energies for 1-deaza AMP, 3-deaza AMP, 1,3-dideaza AMP, 1,3-dideaza-1-fluoro AMP, 1,3-dideaza-1-ethyl AMP are, 0.7 kcal/mol, -0.6 kcal/mol, 1.1 kcal/mol, -0.5 kcal/mol, 1.3 kcal/mol -0.8 kcal/mol, 1.0 kcal/mol, -1.3 kcal/mol, and 1.2 kcal/mol, -0.5 kcal/mol, respectively to FBPase as compared to AMP. As expected the calculated results indicated decreased solvation and improved binding affinity.

An X-ray structure of the FBPase complexed with AMP (Figure 2) showed the 6-amino group hydrogen bonds with both the hydroxyl group of Tyr31 and the backbone carbonyl oxygen of Val17. In addition, N7 appeared to be within hydrogen bonding distance to the hydroxyl group of Thr31. In order to understand the contribution of the 6-amino hydrogen bonds to the binding affinity of FBPase enzyme, the compounds purine riboside monophosphate (**16**), 6-methyl purine riboside monophosphate (**17**), and 6-chloro purine riboside monophosphate (**18**) (Figure 5) were designed

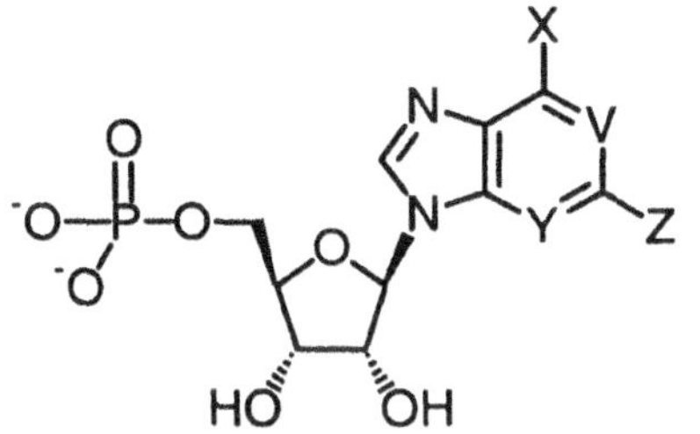

Compounds	X	Y	Z	V
7	NH_2	N	SCH_3	N
8	NH_2	N	CH_3	N
9	NH_2	N	C_2H_6	N
10	NH_2	N	Cl	N
11	NH_2	N	H	CH
12	NH_2	CH	H	N
13	NH_2	CH	H	CH
14	NH_2	CH	H	CF
15	NH_2	CH	H	CC_2H_5
16	H	N	H	N
17	CH_3	N	H	N
18	Cl	N	H	N

Figure 5. Substituted pyrimidine base analogs evaluated in FEP calculations

and evaluated. The relative solvation and binding free energies were calculated between AMP and each of these analogs (**16** to **18**) using the single topology and FEP method. The calculated relative solvation and binding free energies between, AMP and purine riboside monophosphate (**16**), AMP and 6-methyl purine riboside monophosphate (**17**), and AMP and 6-chloro purine riboside monophosphate (**18**), are -4.0 kcal/mol, 2.3 kcal/mol, -3.5 kcal/mol, 2.0 kcal/mol, -4.9 kcal/mol, and 1.2 kcal/mol, respectively and consistent with available experimental data[17]. The observed binding preference of AMP relative to 6-desamino AMP analogs (**16** – **18**) is attributed to the two strong hydrogen bonds formed between the 6-amino group and the carbonyl of Val17 and hydroxyl of Thr31 residues. Apparently, these hydrogen bond interactions dominate the opposing contribution arising from the large desolvation penalty of AMP relative to 6-desamino AMP analogs. The larger desolvation penalty of AMP arises from the two hydrogen bonds formed between water and the 6-amino group compared to similar desamino analogs.

4.3 AMP Analogs Containing a Modified Imidazole Ring

7-deaza AMP (**19**) and formycin monophosphate (**20**) (Figure 6) were evaluated to assess whether AMP analogs lacking N7 or using N7 as a hydrogen bond donor or acceptor were suitable FBPase inhibitors. The mutations between compounds, AMP to 7-deaza AMP (**19**) and AMP to formycin monophosphate (**20**) were completed using the FEP method. Relative to AMP, the desolvation gain for 7-deaza AMP (**19**) is about 0.8 kcal/mol. Apparently, however, the loss of the N7 hydrogen bond to the hydroxyl of Thr31 residue results in a 2.8 kcal/mol loss in binding affinity for 7-deaza AMP (**19**) compared to AMP.

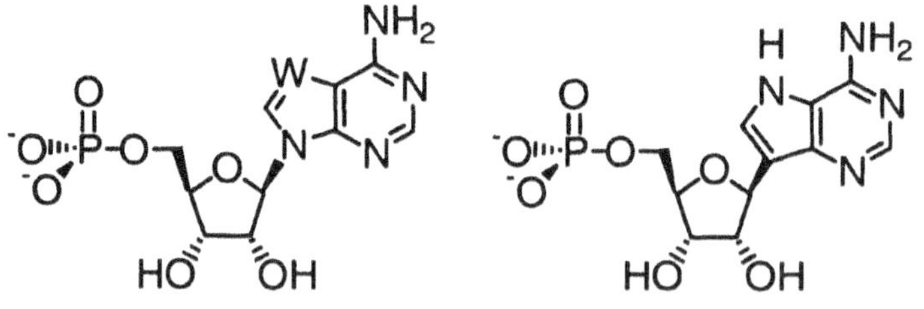

2 W = N
19 W = CH

20

Figure 6. AMP analogs containing modifications in the imidazole ring used in FEP calculations.

The major difference between AMP and formycin monophosphate (**20**) is that in formycin N7 acts as a hydrogen bond donor, whereas N7 in AMP is a hydrogen bond acceptor to the Thr31 hydroxyl group. Despite this significant structural change and the earlier results with 7-deaza AMP showing that N7 is very important, the mutation of AMP to formycin (**20**) led to a relative free energy difference of only 0.6 kcal/mol. The graphical analysis following 20 ps of MD simulation showed that the FBPase binding site accommodates the change in N7 by a slight rearrangement in the binding site residues, especially the Thr31 side chain. Overall, both compounds showed similar number of hydrogen bonds both in complex as well as in solvent states. However, the calculated relative solvation free energy between AMP and formycin monophosphate indicates a small loss (0.5 kcal/mol) of free energy in the solvent state. This was more than compensated in the complex by slightly more favorable interactions between AMP and protein residues as compared to formycin monophosphate (**20**). The changes in free energy, both in the solvent and in the complex, could be due to the loss of an intramolecular hydrogen bond between N7 and the 6-amino group and a change in the geometry of the 6-amino group. The net result is that AMP is a slightly (0.6 kcal/mol) better inhibitor of the FBPase than formycin monophosphate (**20**), which is in agreement with the experimental data (0.3 kcal/mol). These results further indicate that the hydroxyl group of Thr31 forms a hydrogen bond with N7 of the inhibitor, either by donating a proton or accepting a lone pair and maintains similar binding affinity.

5. CONCLUSIONS

An iterative CADD method that combines molecular design, molecular mechanics, molecular dynamics and free energy perturbation calculations with the synthesis, biochemical testing and crystallographic structure determination of protein-inhibitor complexes has been successfully used to predict the rank order of a series of nucleoside monophosphate analogs as fructose 1,6-bisphosphate (FBPase) inhibitors. As part of this paradigm, the free energy calculations were instrumental in identifying a series of inhibitors for the FBPase enzyme. More importantly, once validated, this approach was used in a predictive sense to prioritize design ideas and eliminate the need to synthesize poor inhibitors, thus accelerating the drug design cycle. In all the cases, where experimental data was available, predictions based on this approach were accurate. This study also elucidated the importance of solvation free energy in the binding of inhibitors to an enzyme. If the goal is to design a more potent inhibitor, the energetic cost of desolvation for the addition of polar groups to an inhibitor must be compensated and overcome by stronger ligand-protein interactions. For example, significant desolvation costs are associated with the 6-amino and N7 of AMP. Fortunately, these losses are more than compensated in the complex by forming strong hydrogen bonds with protein residues. While the FEP method is very expensive computationally and can't be used for all proposed analogs of a lead compound, it still represents the only method that routinely generates accurate relative binding free energies. Moreover, the method enables accurate calculation of relative solvation free energies, which, in many cases, represent useful information for drug design.

6. REFERENCES

1. H. M. Ke, Y. P. Zhang, and W. N. Lipscomb, Crystal structure of fructose 1,6-bisphosphatase complexed with fructose 6-phosphate, AMP and magnesium, Proc. Natl. Acad. Sci. USA **87**:5243 (1990).
2. L. Iversen, M. Brzozowski, S. Hastrup, R. Hubbard, J. S. Kastrup, I. K. Larsen, L. Naerum, L. Norskov-Lauridsen, P. B. Rasmussen, L. Thim, F. C. Wiberg, and K. Lundgren, Characterization of the allosteric binding pocket of human liver fructose 1,6-bisphosphate by protein crystallography and inhibitor activity studies, Protein Sci. **6**:971 (1997).
3. M. R. Reddy and A. Parrill, Overview of rational drug design in Rational Drug Design: Novel Methodology and Practical Applications, ACS Symposium Series 719, A. Parrill and M.R. Reddy, eds., Oxford University Press, Washington, DC (1999), pp. 1-11.
4. M. R. Reddy, V. N. Viswanadhan, M. D. Erion, Rapid estimation of relative binding affinities of enzyme inhibitors, in: 3D QSAR in Drug Design, vol. 2, H. Kubinyi, Y. C. Martin, and G. Folker, eds., Kluwer Academic Publishers, Great Britain (1998), pp. 85-98.
5. K. Appelt, R. J. Bacquet, C. A. Bartlett, C. L. Booth, S. T. Freer, M. A. M. Fuhry, M. R. Gehring, S. M. Hermann, E. F. Howland, C. A. Janson, T. R. Jones, C. Kan, V. Kathardekar, K. K. Lewis, G. P. Marzoni, D. A. Matthews, C. Mohr, E. W. Moomaw. C. A. Morse, S. J. Oatley, R. O. Ogden, M. R. Reddy, S. H. Reich, W. S. Schoettlin, W. W. Smith, M. D. Varney, J. E. Villafranca, R. W. Ward, S. Webber, S. E. Webber, K. M. Welsh, and J. White, Design of enzyme inhibitors using iterative protein crystallographic analysis, J. Med. Chem. **34**:1925 (1991).
6. K. Holloway, J. M. Wai, T. A. Halgren, P. M. Fitzgerald, J. P. Vacca, B. D. Dorsey, R. B. Levin, W. J. Thompson, J. L. Chen, J. S. deSolms, N. Gaffin, A. K. Ghosh, E. A. Giuliani, S. L. Graham, J. P. Guare, R. W. Hungate, T. A. Lyle, W. M. Sanders, T. J. Tucker, M. Wiggins, C. M. Wiscount, O. W. Woltersdorf, S.D. Young, P. L. Darke, and J. A. Zugay, A priori prediction of activity for HIV-1 protease inhibitors employing energy minimization in the active site, J. Med. Chem. **38**:305 (1995).
7. M. D. Varney, K. Appelt, V. Kalish, M. R. Reddy, J. Tatlock, C. L. Palmer, W. H. Romies, B. W. Wu, and L. Musick, Crystal-structure based design and synthesis of novel c-terminal inhibitors of HIV protease, J. Med Chem. **37**:2274 (1994).
8. M. D. Erion, J. A. Montgomery, S. Niwas, J. D. Rose, S. Ananthan, M. Allen, J. A. Secrist, S. Y. Babu, C. E. Bugg, W. C. Guida, and S. E. Ealick, Structure-based design of inhibitors of purine nucleoside phosphorylase, 3, 9-arylmethyl derivatives of a 9-deazaguanine substituted on the methylene group, J. Med. Chem. **36**:3771 (1993).
9. M. R. Reddy, M. D. Erion, and A. Agarwal, Free energy calculations: use and limitations in predicting ligand binding affinities, in: Reviews in Computational Chemistry, vol. 6, K. B. Lipkowitz and D. B. Boyd, eds. VCH Publishers, New York (2000), pp. 217-304.
10. M. R. Reddy, V. N Viswanadhan, and J. N. Weinstein, Relative differences in the binding free energies of human immunodeficiency virus 1 protease inhibitors: A thermodynamic cycle-perturbation approach, Proc. Natl. Acad. Sci. USA **88**:10287 (1991).
11. D. M. Ferguson, R. J. Radmer, and P. A. Kollman, Determination of the relative binding free energies of peptide inhibitors to the HIV-1 protease, J. Med. Chem. **34**:2654 (1991).

12. A. J. Tropshaw and J. Hermans, Application of free energy simulations to the binding of a transition state analogue inhibitor to HIV protease, J. Prot. Eng. **5**:29 (1999).
13. B. G. Rao, R. F. Tilton, and U. C. Singh, Free energy perturbation studies on inhibitor binding to HIV-1 protease, J. Am. Chem. Soc. **114**:4447 (1992).
14. K. M. Merz and P. A. Kollman, Free energy perturbation simulation of the inhibition of thermolysin: Prediction of the free energy of binding of a new inhibitor, J. Am. Chem. Soc. **111**:5649 (1989).
15. M. R. Reddy, M. D. Varney, V. Kalish, V. N. Viswanadhan, and K. Appelt, Calculation of relative differences in the binding free energies of HIV-1 protease inhibitors: A thermodynamic cycle perturbation approach, J. Med. Chem. **114**:10117 (1994).
16. M. D. Erion, P. D. van Poelje, and M. R. Reddy, Computer-assisted scanning of ligand interactions: Analysis of the fructose 1,6-bisphosphatase-AMP complex using free energy calculations, J. Am. Chem. Soc. **122**:6114 (2000).
17. M. R. Reddy and M. D. Erion, Calculation of relative binding free energy differences for fructose 1,6-bisphosphatase inhibitors using the thermodynamic cycle perturbation approach, J. Am. Chem. Soc. in press (2001).
18. R. J. Zwanzig, High-temperature equation of state by a perturbation method in nonpolar gases, J. Chem. Phys. **22**:1420 (1954).
19. U.C. Singh, Probing the salt bridge in the dihydrofolate reductase-methotrexate complex by using the coordinate-coupled free-energy perturbation method, Proc. Natl. Acad. Sci. **85**:4280 (1988).
20. M. R. Reddy, R. J. Bacquet, D. Zichi, D. A. Matthews, K. M. Welsh, T. R. Jones, and S. Freer, Calculation of solvation and binding free energy differences for folate-based inhibitors of the enzyme thymidylate synthase, J. Am. Chem. Soc. **114**:10117 (1992).
21. S. J. Weiner, P. A. Kollman, D. A. Case, U. C. Singh, C. Ghio, G. Alagoha, S. Profeta Jr., and P. K. Weiner, A new force field for molecular mechanical stimulation of nucleic acids and proteins, J. Am. Chem. Soc. **106**:765 (1984).
22. U. C. Singh, P. K Weiner, J. K Caldwell, and P. A. Kollman, AMBER version 3.0 University of California at San Francisco, San Francisco, CA (1986).
23. H. J. C. Berendsen, J. R. Grigera, and T. P. Straatsma, The missing term in effective pair potentials, J. Phys. Chem. **91**:6269 (1987).
24. M. R. Reddy and M. Berkowitz, The dielectric constant of SPC/E water, Chem. Phys. Lett. **155**:173 (1989).
25. M. J. Frisch, H. B. Head-Gordon, K. Schlegel, J. S. Raghavachari, C. Binkley, D. J. Gonzalez, D. J Defrees, R. J. Fox, R. Whiteside, C. F. Seeger, J. Melius, R Baker, L. R. Martin, J. J. P. Kahn, E. M. Stewart, S. Fluder, J. A. Topiol, J. A. Pople, Gaussian, Inc.: Pittsburgh, PA. (1994).
26. L. E. Chirlian and M. M. Francl, Atomic charges derived from electrostatic potentials: A detailed study, J. Comput. Chem. **8**:894 (1987).
27. L. Verlet, Computer "experiments" on classical fluids. I. Thermodynamical properties of Lennard-Jones molecules, Phys. Rev. **159**:98 (1967).
28. J. P. Ryckaert, G. Ciccotti, and H. J. C. Berendsen, Numerical integration of the cartesian equations of motion of a system with constraints: Molecular dynamics of n-alkanes, J. Comput. Phys. **23**:327 (1977).
29. J. E. Starrett, D. R. Tortolani, M. J. Hitchcock, J. C. Martin, and M. M. Mansuri, Synthesis and in vitro evaluation of a phosphonate prodrug: Bis(pivaloyloxymethyl) 9-(2-phosphonyl-methoxyethyl)adenine, Antiviral Res. **19**:267 (1992).

Chapter 15

COX–2, SRC SH2 Domain, HIV Reverse Transcriptase, and Thrombin

Computational Approaches to Protein-Ligand Binding

William L. Jorgensen, Melissa L. P. Price, Daniel J. Price, Robert C. Rizzo, Deping Wang, Albert C. Pierce, and Julian Tirado-Rives
Department of Chemistry, Yale University, New Haven, CT, 06520-8107

1. INTRODUCTION

The development of reliable methodology for the computation of relative free energies of binding for ligands with proteins is central to structure-based drug design. Enhanced binding normally correlates with greater drug potency and increased specificity, and can contribute to lower toxicity. Diverse approaches are available ranging from ligand docking and growing procedures with empirical scoring functions to fluid simulations with full atomic detail.[1-3] We have emphasized the latter course owing to its theoretical rigor, the direct connections that it provides between observed and predicted free energies of binding, and the detailed structural information that can be obtained to help understand the variations in binding. Docking programs can be useful for lead generation; however, scoring functions coupled with various levels of neglect of ligand flexibility, protein flexibility, and configurational sampling limit their accuracy. With the full simulation approach, there are several key choices: (a) the sampling procedure, e.g., molecular dynamics (MD) simulations or Monte Carlo (MC) statistical mechanics, (b) representation of the solvent as a continuum or as discrete molecules, (c) the potential-energy evaluation, and (d) the methodology for the free energy calculations. Each of these issues will be discussed in further detail, and as examples of recent successes, free energy perturbation (FEP) results for COX–1, COX–2 and SRC SH2 domain and linear interaction energy (LIE) results for thrombin and HIV–RT are illustrative.

2. COMPUTATIONAL BACKGROUND

Though MD has been the dominant choice of sampling procedure, recent results suggest that Metropolis MC calculations may be particularly efficient for conformational sampling of protein side chains.[4] In view of this and numerous successes of MC simulations for organic host-guest complexation,[5–8] recent work in our laboratory has explored the MC approach for protein-ligand binding. The solvent, water, is represented as discrete molecules with the TIP4P potential[9] the description of specific interactions such as hydrogen bonding with continuum models is a concern for their application along with the loss of detail on variations in water structure. The remaining interactions in our studies have employed the OPLS force fields, which have been parameterized to give correct conformational energetics and properties for organic liquids.[10, 11] Finally, free energy changes can be computed rigorously with FEP or thermodynamic integration methods.[2, 5, 12, 13] In both cases, a free energy change is computed along a reaction path or for converting a system with a molecule A to one with a molecule B, a series of non-physical intermediate states is covered. The two methods are comparably effective and computationally taxing, the latter owing to the need to run the series of calculations for the intermediate states or for many points along the reaction path. Consequently, more efficient, approximate methods are also being evaluated, such as Åqvist's LIE procedure for free energies of binding.[14–20]

2.1 Energetics and Sampling

A classical force field is typically used in which the energy expression consists of harmonic terms for bond stretching and angle bending, a Fourier series for each torsional angle, and Coulomb and Lennard-Jones interactions between atoms separated by three or more bonds (Equations 1– 4). The latter "non-bonded" interactions are also evaluated between intermolecular atom pairs, and they are reduced by a factor of 2 for intramolecular 1,4-interactions. Inhibitors or substrates are represented in an all-atom format with OPLS–AA parameters[11] though sometimes with partial charges obtained from quantum mechanical wavefunctions.

The proteins were described by the OPLS–UA force field with all hydrogens explicit except those on aliphatic carbon[10] in our earlier studies,

$$E_{bond} = \Sigma_i \; k_{r,i} \, (r_i - r_{0,i})^2 \qquad (1)$$

$$E_{bend} = \Sigma_i \; k_{\vartheta,i} \, (\vartheta_i - \vartheta_{0,i})^2 \qquad (2)$$

$$E_{tor} = \Sigma_i\{V_{1,i}(1 + \cos \varphi_i)/2 + V_{2,i}(1 - \cos 2\varphi i)/2 + V_{3,i}(1 + \cos 3\varphi_i)/2\} \quad (3)$$

$$E_{non\text{-}bonded} = \Sigma_i \Sigma_j \{q_i q_j e^2/r_{ij} + 4\,\varepsilon_{ij}\,[(\sigma_{ij}/r_{ij})^{12} - (\sigma_{ij}/r_{ij})^6]\,\} \quad (4)$$

though we now use the all-atom force field, OPLS–AA. The TIP4P water molecules are treated as rigid bodies that only translate and rotate, while the sampling for the inhibitors includes translations, rotations, bond angle variations, and torsional motion in each case. For the proteins, attempted MC moves involve variation of the bond angles and dihedral angles for one randomly picked residue at a time. Variation of bond lengths can also be performed at little additional cost in MC simulations. The protein backbone is sometimes held fixed for efficiency, though our MCPRO program can perform backbone sampling with internal coordinate or Cartesian Monte Carlo.[21, 22] Other, potentially more efficient procedures for backbone sampling are also being explored.[23, 24]

2.2 Free Energy Methods

The FEP calculations use the Zwanzig expression (Equation 5) to compute the free energy change between the reference system X and the perturbed system Y.[2, 5, 12] For binding studies,

$$\Delta G(X \rightarrow Y) = -\, k_B T \ln \langle \exp[-(E_Y - E_X)/k_B T]\rangle_X \quad (5)$$

perturbations are made to convert one ligand to another using the thermodynamic cycle shown below.[25] The conversions use a coupling parameter, λ, that causes one molecule to be smoothly mutated to the other by changing the force field parameters and geometry. The difference in free energies of binding for A and B then comes from Equation 6. Two series of mutations are performed to convert A to B unbound in water and complexed to the protein, which yield ΔG_F and ΔG_C, respectively. The same cycle can be used to compute the effect of a protein mutation on the binding of an inhibitor P with A and B as the two proteins.

The experimental data usually come from inhibition constants or assays of enzyme activity or cell mortality. In general, if a group of inhibitors have the same mechanism of action, the ratios of IC_{50} values should be the same as the ratios of K_is, as long as the assays yielding the IC_{50} values are performed in the same way.[26] The latter condition generally requires only using IC_{50} values from a single source for a given set of inhibitors. Thus, when K_i values are not available, relative binding free energies are approximated from $\Delta\Delta G_b\,(A - B) \approx RT \ln IC_{50}(A)/IC_{50}(B)$.

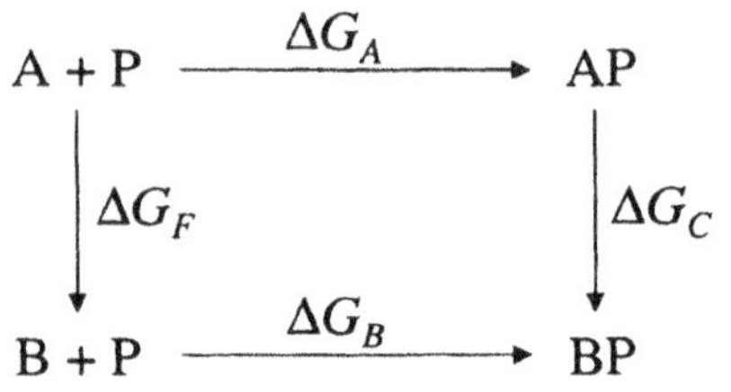

$$\Delta\Delta G_b = \Delta G_A - \Delta G_B = \Delta G_F - \Delta G_C \qquad (6)$$

An approximate approach, LIE, based on linear response theory was introduced by Åqvist et al.[14] In this model, the free energy of interaction of a solute with its environment is given by one-half the electrostatic (Coulombic) energy plus the van der Waals (Lennard-Jones) energy scaled by an empirical parameter, α. For binding a ligand to a protein, the differences in the interactions between the ligand in the unbound state and bound in the complex then provide an estimate of the free energy of binding via Equation 7.

$$\Delta G_b = \beta <\Delta E_{elec}> + \alpha <\Delta E_{vdw}> \qquad (7)$$

The required energy components were obtained from MD simulations for the inhibitors in water and for the protein-inhibitor complexes in water. Key advantages over FEP methods are (a) absolute free energies of binding are readily obtained, and (b) only simulations at the endpoints of a mutation are required. In principle, this might provide a factor of about ten gain in speed; however, we find it necessary to run the LIE calculations several times longer than a typical FEP window to obtain adequate precision for the energy components. Nevertheless, a major plus for the LIE approach is that it is easy to treat structurally diverse ligands, which might be impractical to tackle with FEP calculations. In spite of the approximations in Equation 7 including the neglect of intramolecular energetics, the approach has yielded promising results for several applications.[16] It is expected to be most viable for an analog series of a relatively rigid ligand, a common situation in lead optimization.

Our extension of the LIE approach to calculate free energies of hydration (ΔG_{hyd}) incorporated a third term proportional to the solute's solvent-accessible surface area (SASA), as an index for cavity formation within the solvent.[19, 27] The latter term is needed for cases with positive ΔG_{hyd} such as alkanes and additional improvement occurred when both α and β were allowed to vary. Equation 8 gives the corresponding LIE/SA equation for

ΔG_b. We successfully applied this expression to correlate observed binding affinities for series of inhibitors with thrombin,[18] HIV–RT,[28] and FKBP.[20]

$$\Delta G_b = \beta < \Delta E_{elec} > + \alpha < \Delta E_{vdw} > + \gamma < \Delta SASA > \quad (8)$$

In all cases average errors of ca. 0.8 kcal/mol have been obtained, though the data sets were limited to 35 free energies of hydration, and 7 (thrombin), 12 (HIV-RT), or 11(FKBP) free energies of binding. As described below, results have now been obtained for 40 analogs of nevirapine and thymine analogs with HIV–RT,[29] and for 20 thrombin inhibitors.[30] These studies show that better accuracy can be achieved by considering alternate descriptors. It was also found that convergence of ΔE_{elec} must be carefully validated.

2.3 Typical System Setup and MC Details

The MC simulations for the proteins and inhibitors are carried out with the MCPRO program using water caps with 20–25 Å radius. For the complexes, amino acid residues more than ca. 18 Å from the binding site are removed and only residues within ca. 16 Å are active (sampled). Typically, the ca. 150 residues nearest the inhibitor are retained. The number of water molecules in the calculations is ca. 1500 for the unbound inhibitors and ca. 1000 for the complexes. An attempt to move a protein residue is made every 10 configurations and the period is 20–100 for the inhibitors. The remaining moves are for the water molecules. The protein fragments are normally made neutral, so no counterions are added. Residue-based cutoffs are adopted with truncation of the non-bonded interactions at 12–15 Å. As a typical protocol, the FEP calculations are performed over 10–20 windows. The MC run for each window consists of ca. 10 M configurations of equilibration and 20 M configurations of averaging. The extended LIE calculations typically cover 50 M configurations for equilibration, followed by averaging for 50 M configurations with the unbound inhibitors and for 20 M configurations with the complexes. Initial coordinates are needed for one member of the series of complexes and are usually obtained from the Protein Data Bank. The utility program, pepz, converts the PDB file to a Z–matrix with OPLS atom typing that is suitable for input to MCPRO. Pepz also adds missing hydrogens, and performs the residue truncation and capping. The ligand is typically described by OPLS–AA potential functions except for the charges, which for generality, come from PM3 calculations using the CM1P procedure[31] and are scaled by 1.3 to reflect the condensed-phase environment.[32] The ligands are usually built with XchemEdit[33] and are then positioned in the binding site with our MC docking program, Matador. Matador was derived from MCPRO, but uses only the residues lining the binding site and no explicit water molecules. Its scoring function includes the molecular mechanics energetics and either a distance-dependent dielectric to screen electrostatic interactions or the GB/SA continuum model[34] to account for solvation. The chosen docked structure is input to MCPRO, and the TIP4P water cap is then added. Annealing of this initial hydrated protein-ligand complex can involve conjugate gradient optimizations, MD with the Impact program, [35] and/or elevated-temperature MC.

3. APPLICATIONS

3.1 COX-1 and COX-2

The origins of binding affinity and cyclooxygenase isoform selectivity (COX–2/COX–1) for analogs of celecoxib (**1**, R = Me) have been explored.[36] These inhibitors are COX–2-selective non-steroidal anti-inflammatory drugs (NSAIDs) that are of current interest for their reduced gastrointestinal irritation compared to traditional NSAIDs. Starting from the crystal structure for COX-2 with **1** (R = Br), [37] Matador was used to prepare docked structures for the R = H, methyl, ethyl, hydroxymethyl, hydroxyl, thiomethyl, methoxy, trifluoromethyl, chloro, and fluoro derivatives. MC–FEP calculations were performed and the computed free energies of binding relative to celecoxib are in good accord with the experimental data[38] as illustrated in Figure 1.

The structural information from the simulations also readily explained the experimentally observed binding trends. Furthermore, the docking and FEP results have provided clarification of the binding conformation of the phenylsulfonamide moiety, which is, in fact, rotated nearly 180° from what was originally reported in the 1CX2 crystal structure.

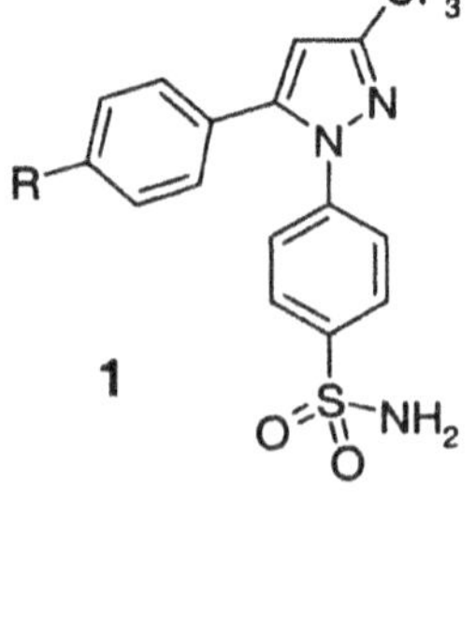

1

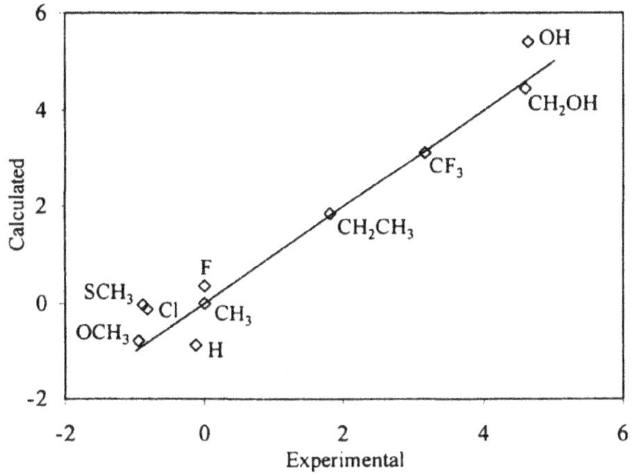

Figure 1. Calculated vs. experimental $\Delta\Delta G_b$ values for celecoxib analogs with COX-2

Convergence of the FEP results was carefully studied. A typical mutation has 5–10 windows each consisting of 15 M configurations of equilibration and 10 M configurations of averaging. From the batch means procedure, the resulting statistical uncertainty in the computed $\Delta\Delta G_b$ values was ca. 0.15 kcal/mol. Two closed perturbation cycles (H → F → Cl → OH → H and Et → Me → OH → OMe → SMe → Et) were also run and gave hystereses of 1.0 and 0.7 kcal/mol. Thus, the current protocol is yielding fine precision. Nevertheless, further detailed analyses have revealed some interesting problems. First, the docking program found two orientations for the thiomethyl group, basically up (CCSC = 90°) and down (CCSC = -90°); note that this is the preferred geometry for thioanisoles as opposed to anisoles which favor CCOC = 0°. Though the barrier to CCSC rotation in the gas phase is only ca. 0.2 kcal/mol,[39] the up and down structures did not interconvert during two MC-FEP calculations for their respective complexes. The tightly packed binding site provides an impeding steric barrier. Though the two FEPs produced the same planar methoxy derivative at the end of the runs, the computed $\Delta\Delta G_b$ values are -1.5 and -0.4 kcal/mol. The latter result came from the structure with the more favorable docking score, and it is closer to the experimental value of -0.07 kcal/mol.

The $\Delta\Delta G_b$ for R = H → F showed the largest error, $\Delta\Delta G_b$ (calc) = 1.24, $\Delta\Delta G_b$ (exptl) = 0.15 kcal/mol. The discrepancy was traced to the existence of a water molecule near the H that is trapped as the FEP progresses to F. If the calculation is run from F to H, the water molecule is not initially placed near the F and it does not appear during the mutation to H; in this case the calculated $\Delta\Delta G_b$(H→F) = 1.52 kcal/mol making the H analog too unfavorable. It is likely that the water molecule should be there for the R = H analog and not for R = F.

We have also performed three FEP calculations for **1** (R = Et → CH_2OH, OCH_3, CH_3) with COX–1. The results are again in excellent agreement with the experimental data and confirm the binding order R = OCH_3 > CH_3 > Et > CH_2OH. Additional studies have been performed that clarify the origin of the 30 to 1000-fold binding preference for COX–2 over COX–1. There are only two residues that differ in the binding site for COX–1 going to COX–2, I523V and H513R. Mutagenesis studies indicate that the I523V change is the dominant factor.[40] We have performed FEP calculations for the celecoxib/COX–2 complex corresponding to the I523V mutant.[41] The energetic and structural results indicate that the substitution of the larger Ile residue indeed makes unfavorable steric interactions with the ligand. This steric contribution stems from repulsive Lennard-Jones interactions between the δ-methyl group on Ile and a sulfonamide oxygen of the ligand. In addition, this steric crunch manifests itself in our simulations by altering the conformation of both the ligand and the Ile residue compared to their gas-phase preferences.

3.2 SRC SH2 Domain

The therapeutic potential of disrupting interactions between the SH2 domain of SRC and its cognate receptor ligands, thereby disabling the associated signaling pathways in hyperproliferative cells and osteoclasts, is an attractive option for the treatment of several cancers and osteoporosis, respectively.[42–45] In order to predict the relative free energies of binding and identify the structural requirements for high affinity binding within a series of ligands, **2**, MC–FEP calculations using the OPLS–AA force field were performed.

Substituents at the 3-position of the central ring target interactions in the glutamic acid binding region of the SH2 domain[46] the amido, methylamido, methyl keto, amino, chloro, methyl, and un-substituted compounds were examined in this study.[47] Agreement with experimental trends in binding affinity is seen, although the computed relative free energy of binding of the amido compound is underestimated relative to the methylamido and keto compounds by ca. 1 kcal/mol.

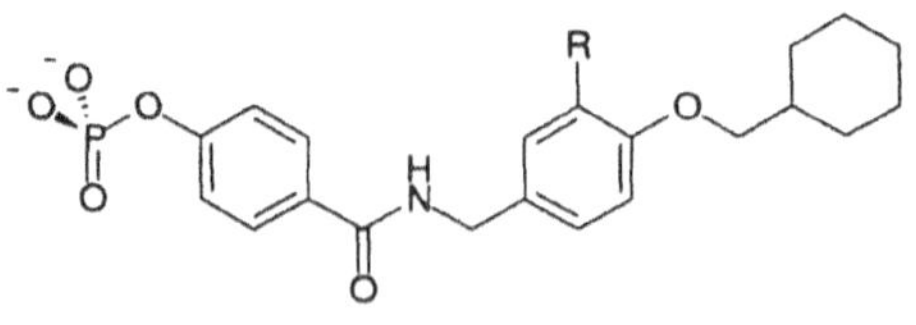

2

These results are reconciled by examination of the hydration energies of model systems, which predict primary amides as too hydrophilic by ca. 1.7 kcal/mol, indicating the amide is too well solvated in the unbound legs. As these compounds are bound at the surface of the protein and have some water contacts, it is unclear how much of the apparent heightened hydrophilicity of the amide is also reflected in the bound mutation, thus cancelling some of the error in the unbound leg. Also, the binding affinity of the unsubstituted compound, was beyond the sensitivity of the experimental assay, consequently only a lower bound for the experimental $\Delta\Delta G_b$ is available, $\Delta\Delta G_b(CONH_2 \rightarrow H) \geq 1.6$ kcal/mol. The calculated relative free energies of binding agree with this result regardless of the pathway used to determine it: $\Delta\Delta G_b(CONH_2 \rightarrow NH_2 \rightarrow Cl \rightarrow H) = 3.5$ kcal/mol and $\Delta\Delta G_b(CONH_2 \rightarrow CH_3 \rightarrow H) = 4.6$ kcal/mol.

The simulations with explicit solvent give valuable detail at the atomic level for the structures of the protein-ligand complexes. The best binder, the amido compound, makes two hydrogen bonds in the glutamic acid binding region, one to the backbone amide of Lys βD6 HN and another to a highly-coordinated water molecule. This water molecule completes its four-coordination with the backbones of Lys βD6 and Ile βE5 and the guanidinium group of Arg βD′1, and is present in every snapshot of every simulation. The next tier of ligands includes the methylamido and keto compounds, which only make the hydrogen bond to the Lys, whereas the poorest binders (amino, methyl, and unsubstituted) also lose this contact. The chloro compound cannot hydrogen-bond in the traditional sense; however, it makes a favorable electrostatic contact with Lys βD6 HN, thus ranking in between the latter two groups.

Conformational searching with the GB/SA continuum solvation model[34] was used to identify the lowest-energy conformer that could be used as the starting geometry for the unbound simulations, thereby minimizing the impact of infrequent exchange between conformers. However, the presence of at least six low-energy minima for the ligands in solution suggests that there is an entropic penalty contributing to the absolute ΔG_b that comes from constraining the flexible ligand upon binding. Furthermore, comparison of low-energy structures to the distributions obtained from the bound simulations establishes that the ligands are not in a low-energy conformation when bound, further contributing an internal-energy penalty to ΔG_b. Several predictions can be made for modifications that lower the relative energy of the binding geometry. In particular, predictions for methylation at the benzyl amide methylene are supported by experimental results for the racemic mixture of the compounds singly methylated at that position.[46] While either methylation reduces the conformational flexibility, Pro(S) methylation provides a favorable contact with Tyr βD5, while a Pro(R) methyl group points towards the solvent.

3.3 HIV-1 Reverse Transcriptase (RT)

Recently, a large effort has been directed at the design of anti-HIV drugs. The focus has been on non-nucleoside inhibitors of HIV-RT (NNRTIs). FDA-approved drugs in this class are nevirapine (**3**), delavirdine (**4**), and Sustiva (efavirenz), while clinical trials are in progress for others including the HEPT analog MKC–442 (**5**, R^1 = *i*-Pr, R^2 = $CH_2OCH_2CH_3$, Y = CH_2), and other candidates such as 9-chloro-TIBO (**6**) have been abandoned. These ligands are non-competitive with the nucleoside analogs such as AZT; they bind in a pocket between the "palm" and "thumb" subdomains of the p66 polymerase chain that is proximal to, but not overlapping with, the polymerase active site.[48, 49]

We reported an extensive LIE study of **6** and 11 analogs that included consideration of the alternative amine epimers and protonation states.[28] The MC simulations were initiated from the crystal structure for the complex of 8–Cl–TIBO and HIV–1 RT.[50] Partial charges came from RHF/6–31G* CHELPG calculations for each inhibitor. The experimental data are IC_{50} values for the effective concentration required to achieve 50% protection of MT–4 cells against the cytopathic effect of HIV–1.[51]

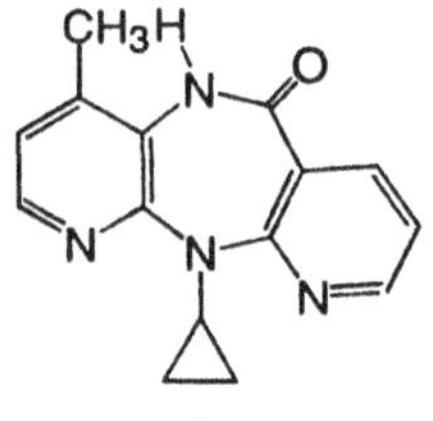

3

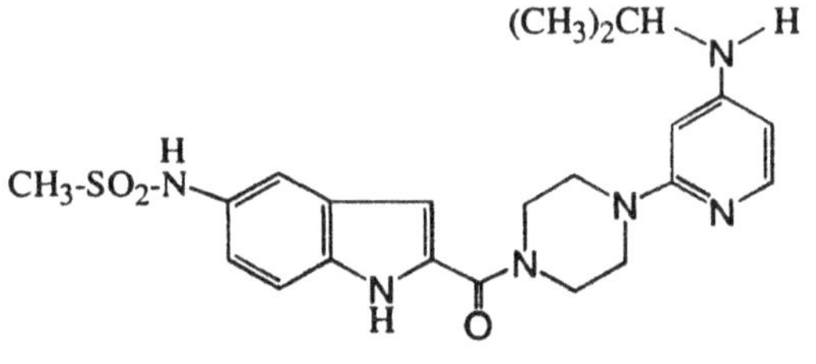

4

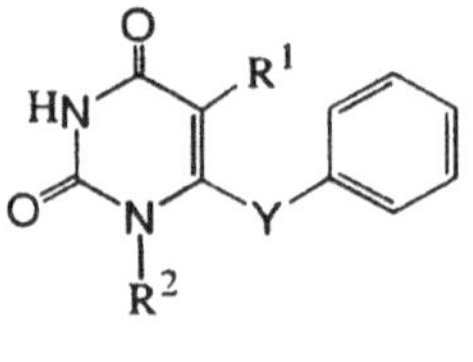

5

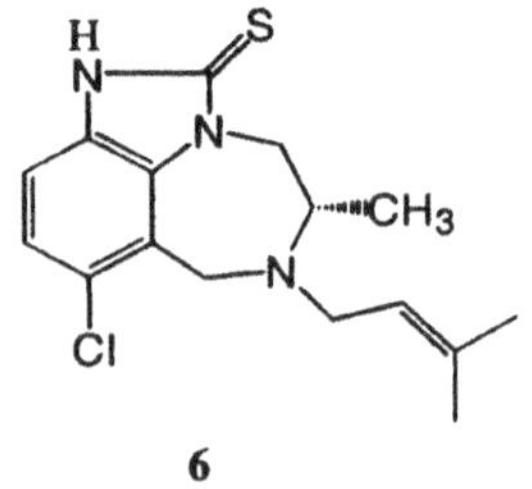

6

Equation 8 yielded a fine correlation with rms error of 0.88 kcal/mol; α, β and γ were -0.150, 0.114, and 0.0286, though the ΔSASA term could be replaced by a constant with little degradation of the fit. Nevertheless, we remained concerned about the size of the data set (it is statistically desirable to have at least 5 data points per descriptor) and slow convergence for the Coulombic energy components.

Thus, we undertook a larger study using 20 HEPT and 20 nevirapine analogs.[29] The variety of R^2 side chains for the HEPTs was particularly substantial ranging from hydrogen to methyl benzyl ether. The convergence issue was carefully addressed for the unbound inhibitors in water, which are generally more problematic than for the bound systems. MC simulations were run for all 20 HEPTs using two slightly different starting structures for the inhibitors in droplets with 1485 TIP4P water molecules. After 5M configurations of equilibration and 10M configurations of averaging, there were discrepancies of up to 10 kcal/mol in the solute-water Coulomb energies for the two starting structures and with total values in the -30 to -60 kcal/mol range. An annealing protocol was then developed that reduced the discrepancies to an average of less than 1 kcal/mol. It consists of repeating five cycles of (1) 5M configurations of equilibration with the solute at 1000 K and the solvent still at 298 K, (2) 5M configurations of equilibration with the entire system at 298 K, and (3) 10M configurations of averaging. During the 1000 K portion, the solute's bond lengths and bond angles are not varied so they do not have to be cooled; the extra energy is targeted to promote conformational changes by focusing the heating on the dihedral angles and total translation and rotation of the solute. Thus, the length of the calculations for the unbound ligands has increased to a total of 100M configurations, which for this system size corresponds to 300-400 ps of MD.[4] The MC simulations for the bound structures, which included 851 water molecules and 42 sampled of 123 total protein residues, covered 20M configurations; this corresponds to about 100 ps of MD.

Without the heating, the fits for the 40 compounds were poor. With the heating, LIE Equation 8 provided a reasonable fit with $r^2 = 0.56$, rms error = 1.24 kcal/mol, and average error = 0.84 kcal/mol over a range of 7 kcal/mol. However, following our work on properties predictions,[52, 53] in addition to the traditional LIE energy components, other terms are also now averaged including the total numbers of solute-water and solute-protein donated and accepted hydrogen bonds, solute dipole moment, and the hydrophobic, hydrophilic, and aromatic components of the solute's SASA. An excellent fit for the 40 data points was obtained (Figure 2), which uses five terms: the protein-ligand Lennard-Jones energy, the reduction in exposed hydrophobic surface area for the ligand, the change in total numbers of hydrogen bonds for the ligand (this is a negative number as no medium forms more hydrogen

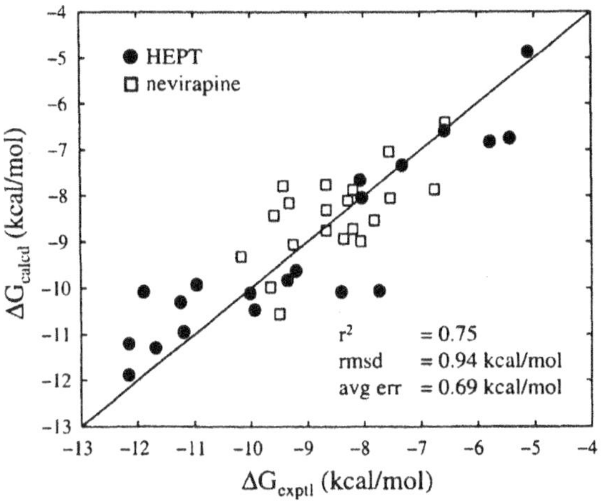

Figure 2. Experimental vs. calculated free energies of binding with HIV-RT

bonds than water), a correction for the nevirapines, if they contain a secondary amide, and a constant:

$$\Delta G_b = 0.30\langle E_{vdw}\rangle + 0.0085\langle \Delta FOSA\rangle - 0.94\langle \Delta HB\rangle - 2.8\langle 2^o\, amides\rangle + 4.6$$

The correction stems from overestimation of the $\Delta\Delta G_{hyd}$ between secondary and tertiary amides with the utilized HF/6-31G* CHELPG charges. The statistical analyses have become far more sophisticated with the use of the JMP program. F ratios (regression model mean/error mean square) established the significance of the descriptors; they all satisfy the condition that the probability of a greater F occurring by chance (Prob>F) is <0.001. They also make perfect physical sense with emphasis on the desolvation penalty for losing hydrogen bonds at a cost in free energy of 1 kcal/mol each, the need for a good protein-ligand fit, and the benefit of burying hydrophobic surface area.

3.4 Mutant Proteins and Drug Resistance

The effect of mutations of HIV-1 RT on the drug binding is also being explored. The principal mutants that confer resistance to the NNRTIs are Y188C/H/L, Y181C, L100I, K103N, and V106A.[54, 55] Interestingly, the mutations are mostly to smaller residues, which likely reduces favorable inhibitor-protein contacts. The only crystal structure of a mutant RT that has been reported is for Y181C-RT with 8–chloro–TIBO at 3.2 Å.[56] Nevertheless, a novel series of MC/FEP calculations has been completed for

Table 1. Inhibition penalties (kcal/mol) for HIV–1 RT mutants

drug	V106A $\Delta\Delta G_{calc}$	V106A ΔG_{expt}	Y181C $\Delta\Delta G_{calc}$	Y181C ΔG_{expt}
Nevirapine	3.3 ± 0.4	2.5, 2.9, 3.5	3.9 ± 0.3	2.6, 2.8, 3.5, 4.9
MKC–442	0.7 ± 0.5	2.9	4.7 ± 0.3	2.9, 5.0
9–Cl–TIBO	1.3 ± 0.5	1.2, 1.8, 2.0	3.0 ± 0.3	1.0, 1.6, 2.7, 2.9
Sustiva	0	0.5, 0.5, 0.7	0	0.1, 0.2, 0.6, 2.1

the effects of the V106A and Y181C mutants on the binding of nevirapine, MKC–442, 9–Cl–TIBO, and efavirenz (Sustiva). This involved FEP calculations for the V106A and Y181C conversions in the presence of the four ligands and for the apoprotein.[57]

There is considerable scatter in the experimental data from different groups, though it is clear that Sustiva has the best resistance profile, and that the computations parallel the experimental data well. It appears that Sustiva benefits from having weaker interactions (looser packing) in the vicinity of V106 and Y181; it makes up for this by better hydrogen bonding with the protein backbone, which is less variable in position.

Shown below are snapshots of structures from our simulations of native RT with Sustiva and nevirapine. No crystal structure had been reported for RT with Sustiva when the calculations were performed; however, confidence in the computed structure comes from the quantitative accord for the inhibition penalties and because Matador yielded docked structures for nevirapine, MKC–442, and 9–Cl–TIBO that are virtually identical to the crystal structures.[57] Sustiva hydrogen bonds with the backbone of K101, as do MKC–442 and 9–chloro–TIBO, and it has hydrophobic interactions in the Y181–Y188 pocket, also typical of NNRTIs.

A crystal structure for the Sustiva/RT complex was subsequently reported and fully confirmed the correctness of the computed structure.[58]

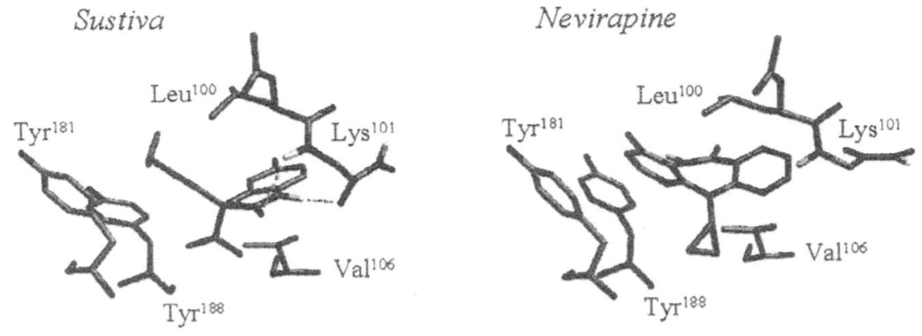

Figure 3. Sustiva (left) and nevirapine (right) shown with select residues in the NNRTI binding site of HIV–RT. The initial structure of Sustiva/RT complex was obtained by docking.

3.5 Thrombin

A similar set of calculations has been performed on a series of 20 active-site-directed thrombin inhibitors with activities ranging from 5 μM to 45 pM.[30] Among these inhibitors, there are two different binding modes and at least seven different functional groups binding in each of the three pockets of the thrombin active site (Figure 4). Though the fits based on the standard LIE equations (Equations 7–8) were poor, alternatives with 3–5 descriptors such as Equation 9 yield good results with rms errors of 1.0 – 1.3 kcal/mol and r^2 values of 0.7-0.8 for the 20 data points that range over 7.0 kcal/mol in ΔG_b (Figure 5).

$$\Delta G_b = 0.15\langle \Delta E_{LJ} \rangle + 0.14\langle \Delta E_{\mathrm{int}} \rangle - 0.74\langle \Delta HB \rangle + 0.44\langle \# RB \rangle - 13.3 \qquad (9)$$

As in the HIV–RT study, terms for Lennard-Jones interactions and hydrogen bonding prove to be important for estimating binding affinities. In this case, descriptors for the number of rotatable bonds in the ligand (*#RB*) and for the change in internal energy of the ligand upon binding (ΔE_{int}) are also important. The latter term accounts for the fact that a ligand is less likely to bind to the protein if conformational strain is induced upon binding, while the former serves as an estimate of the loss of conformational freedom in the bound inhibitor. Both of these descriptors have been used successfully in earlier ligand binding studies, but they are likely to be particularly important for more flexible ligands such as those considered in this work.[59] It is interesting to note that none of the descriptors used here or in the HIV–RT study depend on long range-interactions or slowly converging quantities,

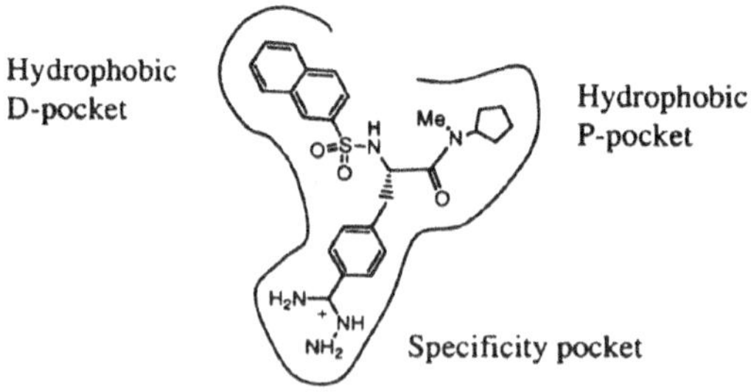

Figure 4. Inhibitor LB30057 in the binding site of thrombin

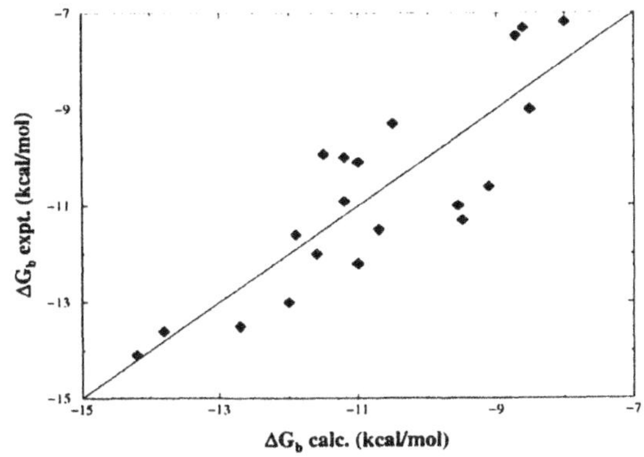

Figure 5. Experimental vs. calculated free energies of binding

suggesting that similar results should be possible with considerably reduced computational effort in future projects.

4. REFERENCES

1. C. L .M. J. Verlinde and W. G. H. Hol, Structure-based drug design: Progress, results and challenges, Structure **2**:577 (1994).
2. M. L. Lamb and W. L. Jorgensen, Computational approaches to molecular recognition, Curr. Op. Chem. Biol. **1**:449 (1997).
3. T. J. Marrone, J. M Briggs, and J. A. McCammon, Structure-based drug design: Computational advances, Ann. Rev. Pharmacol. Toxicol. **37**:71 (1997).
4. W. L. Jorgensen and J. Tirado-Rives, Monte Carlo vs. molecular dynamics for conformational sampling, J. Phys. Chem. **100**:14508 (1996).
5. W. L. Jorgensen, Computational insights on intermolecular interactions and binding in solution, Chemtracts - Organic Chemistry **4**:91 (1991).
6. W. L. Jorgensen and T. B. Nguyen, Modeling the complexation of substituted benzenes by a cyclophane host in water, Proc. Nat. Acad. Sci. USA **90**:1194 (1993).
7. E. M. Duffy and W. L. Jorgensen, Structure and binding complexes of Rebek's acridine diacid with pyrazine, quinoxaline, and pyridine from Monte Carlo simulations with an all-atom force field, J. Am. Chem. Soc. **116**:6337 (1994).
8. N. A. McDonald, E. M. Duffy, and W. L. Jorgensen, Monte Carlo investigations of selective anion complexation by a bis(phenylurea) p-tert-butylcalix[4]arene, J. Am. Chem. Soc. **120**:5104 (1998).
9. W. L. Jorgensen, J. Chandrasekhar, J. D Madura, R. W. Impey, and M. L. Klein, Comparison of simple potential functions for simulating liquid water, J. Phys. Chem. **79**:926 (1983).
10. W. L. Jorgensen and J. Tirado-Rives, The OPLS force field for proteins. Energy minimizations for crystals of cyclic peptides and crambin, J. Am. Chem. Soc. **110**:1657 (1988).

11. W. L. Jorgensen, D. S. Maxwell, and J. Tirado-Rives, Development and testing of the OPLS all-atom force field on conformational energetics and properties of organic liquids, J. Am. Chem. Soc. **118**:11225 (1996).
12. P. A. Kollman, Free energy calculations: Applications to chemical and biochemical phenomena, Chem. Rev. **93**:2395 (1993).
13. W.L. Jorgensen, Computation of free energy changes in solution, in: Encyclopaedia of Computational Chemistry, P. v. R. Schleyer, ed., Wiley, New York (1998), pp. 1061–1070.
14. J. Åqvist, C. Medina, and J. E. Sammuelsson, A new method for predicting binding affinities in computer-aided drug design, Protein Eng. **7**:385 (1994).
15. T. Hansson, and J. Åqvist, Estimation of binding free energies for HIV–1 protease inhibitors by molecular dynamics simulations, Protein Eng. **8**:1137 (1995).
16. T. Hansson, J. Marelius, and J. Åqvist, Ligand binding affinity predictions by linear interaction energy methods, J. Comput. Aided Mol. Design **12**:27 (1998).
17. M. D. Paulsen, and R. L. Ornstein, Binding free energy calculations for P450cam-substrate, Protein Eng. **9**:567 (1996).
18. D. K. Jones-Hertzog, and W. L. Jorgensen, Binding affinities for sulfonamide inhibitors with human thrombin using Monte Carlo simulations with a linear response method, J. Med. Chem. **40**:1539 (1997).
19. N. A. McDonald, H. A. Carlson, and W. L. Jorgensen, Free energies of solvation in chloroform and water from a linear response approach, J. Phys. Org. Chem. **10**:563 (1997).
20. M. L. Lamb, J. Tirado-Rives, and W. L. Jorgensen, Estimation of the binding affinities of FKBP12 inhibitors using a linear response method, Bioorg. Med. Chem. **7**:851 (1999).
21. W. L. Jorgensen, Monte Carlo simulations for liquids, in: Encyclopaedia of Computational Chemistry, P. v. R. Schleyer, ed., Wiley, New York (1998), pp. 1754–1763.
22. W. L. Jorgensen and J. Tirado-Rives, MCPRO, Version 1.65, Yale University, New Haven, CT. (2000).
23. M. W. Deem, M. W. and J. S. Bader, A configurational bias Monte Carlo method for linear and cyclic peptides, Mol. Phys. **87**:1245 (1996).
24. D. Hoffmann and E. W. Knapp, Polypeptide folding with off-lattice Monte Carlo dynamics: The method, Eur. Biophys. J. **24**:387 (1996).
25. T. P. Straatsma and J. A. McCammon, Computational alchemy, Ann. Rev. Phys. Chem. **43**:407 (1992).
26. Y. Cheng and W. H. Prusoff, Relationship between the inhibition constant and the concentration of inhibitor which causes 50 percent inhibition of an enzymatic reaction, Biochem. Pharmacol. **22**:3099 (1973).
27. H. A. Carlson and W. L. Jorgensen, An extended linear response method for determining free energies of hydration, J. Phys. Chem. **99**:10667 (1995).
28. R. H. Smith Jr., W. L. Jorgensen, J. Tirado-Rives, M. L. Lamb, P. A. Janssen, C. J. Michejda, and M. B. Kroeger Smith, Prediction of binding affinities for TIBO inhibitors of HIV-1 reverse transcriptase using Monte Carlo simulations in a linear response method, J. Med. Chem. **41**:5272 (1998).
29. R. C. Rizzo, J. Tirado-Rives, and W. L. Jorgensen, Estimation of binding affinities for HEPT and nevirapine analogs with HIV–1 reverse transcriptase via Monte Carlo simulations, J. Med. Chem. **44**:145 (2001).
30. A. C. Pierce and W. L. Jorgensen, Estimation of binding affinities for selective thrombin inhibitors via Monte Carlo simulations, J. Med. Chem. **44**:1043 (2001).

31. J. W. Storer, D. J. Giesen, C.J. Cramer, and D. G. Truhlar, Class IV charge models: A new semiempirical approach in quantum chemistry, J. Comput.-Aided Mol. Design **9**:87 (1995).
32. G. A. Kaminski and W. L. Jorgensen, A QM/MM method based on CM1A charges: Applications to solvent effects on organic equilibria and reactions, J. Phys. Chem. B **102**:1787 (1998).
33. D. Lim and W. L. Jorgensen, XChemEdit, Version 0.91, Yale University, New Haven, CT (2000).
34. W. C. Still, A. Tempczyk, R. C. Hawley, and T. A. Hendrickson, Semianalytical treatment of solvation for molecular mechanics and dynamics, J. Am. Chem. Soc. **112**:6127 (1990).
35. J. Banks, R. Friesner, E. Gallicchio, A. Ghosh, R. Levy, R. Murphy, A. Wallqvist, and R. Zhou, Integrated Modeling Program using Applied Chemical Theory, Version 1.0, Schrodinger, Inc., Jersey City (1990).
36. M. .L. P. Price and W. L. Jorgensen, Origin of the selectivity of celecoxib analogs with COX–1 and COX–2 from docking and Monte Carlo simulations, J. Am. Chem. Soc. **122**:9455 (2000).
37. R. G. Kurumbail, A. M. Stevens, J. K. Gierse, J. J. McDonald, R. A. Stegeman, J. Y. Pak, D. Gildehaus, J. M. Miyashiro, T. D. Penning, K. Seibert, P. C. Isakson, and W. C. Stallings, Structural basis for selective inhibition of cyclooxygenase-2 by anti-inflammatory agents, Nature **384**:644 (1996).
38. T. D. Penning, J. J. Talley, S. R. Bertenshaw, J. S. Carter, P. W. Collins, S. Docter, M. J. Graneto, L. F. Lee, J. W. Malecha, J. M. Miyashiro, R. S. Rogers, D. J. Rogier, S. S. Yu, G. D. Anderson, E. G. Burton, J. N. Cogburn, S. A. Gregory, C. M. Koboldt, W. E. Perkins, K. Seibert, A. W. Veenhuizen, Y. Y. Zhang, and P. C. Isakson, Synthesis and biological evaluation of the 1,5-diarylpyrazole class of cyclooxygenase-2 inhibitors: Identification of 4-[5-(4-Methylphenyl)-3-(trifluoromethyl)-1H-pyrazol-1-yl] benzenesulfonamide (SC-58635, Celecoxib), J. Med. Chem. **40**:1347 (1997).
39. T. A. Halgren and R. B. Nachbar, Merck molecular force field. IV. Conformational energies and geometries for MMF94, J. Comput. Chem. **17**:587 (1996).
40. J. K. Gierse, J. J. McDonald, S. D. Hauser, S. H. Rangwala, C. M. Koboldt, and K. Seibert, A single amino acid difference between cyclooxygenase-1 (COX-1) and -2 (COX-2) reverses the selectivity of COX-2 specific inhibitors, J. Biol. Chem. **271**:15810 (1996).
41. M. L. P. Price and W. .L. Jorgensen, Rationale for the observed COX-2/COX-1 selectivity of celecoxib from Monte Carlo simulations, Bioorg. Med. Chem. Lett. **11**:in press (2001).
42. W. Mao, R. Irby, D. Coppola, L. Fu, M. Wloch, J. Turner, H. Yu, R. Garcia, R. Jove, and T. J. Yeatman, Activation of c-Src by receptor tyrosine kinases in human colon cancer cells with high metastatic potential, Oncogene **15**:3083 (1997).
43. B. S. Verbeek, T. M. Vroom, S. S. Adriaansen-Slot, A. E. Ottenhoff-Kalff, J. G. N. Geertzema, A. Hennipman, and G. Rijksen, c-Src protein expression is increased in human breast cancer. A immunohistochemical and biochemical analysis, J. Pathol. **180**:383 (1996).
44. B. F. Boyce, T. Yoneda, C. Lowe, P. Soriano, and G. R. Mundy, Requirement of $pp60^{c\text{-}src}$ expression for osteoclasts to form ruffled borders and resorb bone in mice, J. Clin. Invest. **90**:1622 (1992).
45. C. Lowe, T. Yoneda, B. F. Boyce, H. Chen, G. R. Mundy, and P. Soriano, Osteopetrosis in src-deficient mice is due to an autonomous defect of osteoclasts, Proc. Natl. Acad. Sci. USA **90**:4485 (1993).

46. E. A. Lunney, K. S. Para, J. R. Rubin, C. Humblet, J. H. Fergus, J. S. Marks, and T. K. Sawyer, Structure-based design of a novel series of nonpeptide ligands that bind to the pp60^{src} SH2 domain, J. Am. Chem. Soc. **119**:12471 (1997).
47. D. J. Price, and W. L. Jorgensen, Computational binding studies of human pp60$^{c\text{-}src}$ SH2 domain with a series of nonpeptide, phosphophenyl-containing ligands. Bioorg. Med. Chem. Lett. **10**:2067 (2000).
48. L. A. Kohlstaedt, J. Wang, J. M. Friedman, P. A. Rice, and T. A. Steitz, Crystal structure at 3.5 Å resolution of HIV–1 reverse transcriptase complexed with an inhibitor, Science **256**:1783 (1992).
49. B. G. Turner, and M.F. Summers, Structural biology of HIV, J. Mol. Biol. **285**:1 (1999).
50. J. Ding, K. Das, H. Moereels, L. Koymans, K. Andries, P. A. Janssen, S.H. Hughes, and E. Arnold, Structure of HIV-1 RT/TIBO R 86183 complex reveals similarity in the binding of diverse nonnucleoside inhibitors, Nature Struct. Biol. **2**:407 (1995).
51. W. Ho, M. J. Kukla, H. J. Breslin, D. W. Ludovici, P. P. Grous, C. J. Diamond, M. Miranda, J. D. Rodgers, C. Y. Ho, E. De Clercq, R. Pauwels, K. Andries, M. A. C. Janssen, and P. A. Janssen, Synthesis and anti-HIV-1 activity of 4,5,6,7-tetrahydro-5-methylimidazo[4,5,1-jk][1,4]benzodiazepin-2(1H)-one (TIBO) derivatives, J. Med. Chem. **38**:794 (1995).
52. E. M. Duffy, and W. L. Jorgensen, Prediction of properties from simulations: Free energies of solvation in hexadecane, octanol, and water, J. Am. Chem. Soc. **122**:2878 (2000).
53. W. L. Jorgensen, and E. M. Duffy, Prediction of drug solubility from Monte Carlo simulations, Bioorg. Med. Chem. Lett. **10**:1155 (2000).
54. D. D. Richman, D. Havlir, J. Corbeil, D. Looney, C. Ignacio, S. A. Spector, J. Sullivan, S. Cheeseman, K. Barringer, D. Pauletti, C. K. Shih, M. Myers, and J. Griffin, Nevirapine resistance mutations of immunodeficiency virus type 1 selected during therapy, J. Virol. **68**:1660 (1994).
55. R. Schinazi, B. A. Larder, and J. Mellors, Mutations in HIV-1 reverse transcriptase and protease associated with drug resistance, Int. Antiviral News **2**:72 (1994).
56. K. Das, J. Ding, Y. Hsiuo, A. D. Clark, H. Moreels, L. Koymand, R. Pauwels, P. A. J. Janssen, P. L Boyer, P. Clark, R. H. Smith, M. B. Kroeger Smith, C. J. Michejda, S. . Hughes, and E. Arnold, Crystal structures of 8-Cl and 9-Cl TIBO complexed with wild-type HIV-1 RT and 8-Cl TIBO with the Tyr181Cys HIV-1 RT drug-resistant mutant, J. Mol. Biol. **264**:1085 (1996).
57. R. C. Rizzo, D. Wang, J. Tirado-Rives, and W. L. Jorgensen, Validation of a model for the complex of HIV–1 reverse transcriptase with Sustiva through computation of resistance profiles, J. Am. Chem. Soc. **122**:12898 (2000).
58. J. Ren, J. Milton, K. L. Weaver, S. A. Short, D. I. Stuart, and D. K. Stammers, Structure **8**:1089 (2000).
59. J. Boström, P. O. Norrby, and T. Liljefors, Conformational energy penalties of protein-bound ligands, J. Comput.-Aided Mol. Design **12**:383 (1998).

Chapter 16

HIV-1 Protease: Structure-Based Drug Design Using the Free Energy Perturbation Approach

M. Rami Reddy[†] and Krzysztof Appelt[‡]
[†]*Metabasis Therapeutics, Inc., San Diego, CA 92121*
[‡]*Agouron Pharmaceuticals, Inc., San Diego, CA 92121*

1. INTRODUCTION

The human immunodeficiency virus 1 (HIV-1) is a member of the retroviruse family and is the cause of the debilitating and fatal disease acquired immune deficiency syndrome (AIDS).[1, 2] As part of an overall effort to develop an effective treatment for AIDS, researchers mounted an intense campaign to understand and exploit the critical pathways in the life-cycle of HIV-1.[3-4] Inhibition of HIV-1 protease emerged as one of the most promising drug targets identified to date.[5-10] A wide range of approaches were used to discover HIV-1 protease inhibitors. These included the screening of chemical libraries,[11] designing substrate analogs,[12] and using the X-ray structure of HIV-1 protease in structure-based drug design approaches.[13, 14] To complement the structural information, a variety of computational tools[15-18] were developed and incorporated in a computer-assisted drug design (CADD) strategy.

Advances in protein crystallography and molecular simulations greatly aided CADD paradigms and the accuracy of binding affinity predictions.[19, 20] Inhibitor design used methods ranging from graphical visualization of the ligand in the binding site to calculation of relative binding affinities using molecular dynamics simulations in conjunction with the FEP approach.[14, 21-24] Structure-based drug design typically follows the steps outlined in the flowchart shown in Figure 1. The process begins by generating a working computational model from crystallographic data. This step usually entails developing molecular mechanics force field parameters for non-standard residues, assigning the protonation states of histidines, and orientating carbonyl and amide groups of asparagine and glutamine amino acid residues based upon neighboring donor/acceptor groups. After generation of the computer model, the active site is characterized using a variety of visualization tools. For example, hydrophobic and hydrophilic regions of the active site are readily identified by calculating the electrostatic potential at different surface grid points. This information is then used to design and optimize lead compounds.

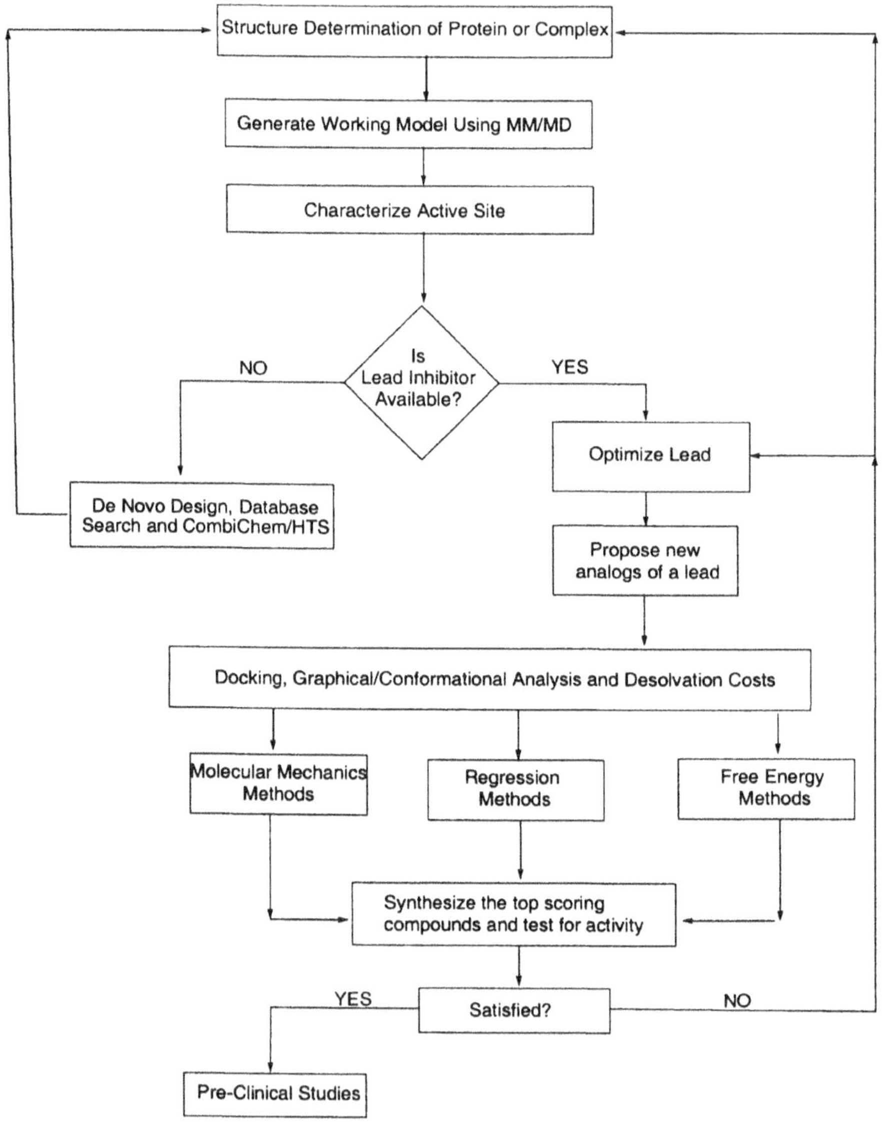

Figure 1. Flowchart for ligand optimization

Discovery of a lead inhibitor often requires considerable experimentation. In most cases the lead is discovered through screening of compound libraries[25] or combinatorial chemistry libraries,[26] although there have been a few reports suggesting that de novo molecular design methods were used to design new structures by sequentially adding molecular fragments to a growing structure, e.g. adding functionality to an appropriately-sized molecular scaffold or adding fragments building toward the center of a molecule starting from distant sites thought to interact with the target.[27, 28]

The last step is to optimize the lead compounds. The computational methods in combination with SAR information has proved to be a powerful approach to determine areas on the molecule to expand, contract, or modify. Analogs of the lead compound are ranked using one or all of the following methods: 1) qualitative predictions using molecular mechanics calculations;[29-32] 2) semi-quantitative predictions[33, 34] based on regression methods that incorporate interaction variables (intra-and intermolecular interaction energies) and ligand properties (desolvation, log P, etc.), and 3) quantitative predictions[35-37] based on relative free energy calculations. The later method is a computationally demanding and sophisticated procedure to compute highly accurate binding free energy differences. This chapter reviews the use of free energy calculations for the design of HIV-1 protease inhibitors.

2. METHODOLOGY

2.1 Thermodynamic-Cycle Perturbation Approach

The thermodynamic cycle-perturbation (TCP) approach[38, 39] is a method for computing the relative changes of binding free energy using non-physical paths connecting the desired initial and terminal states. This approach enables calculation of the relative change in binding free energy difference

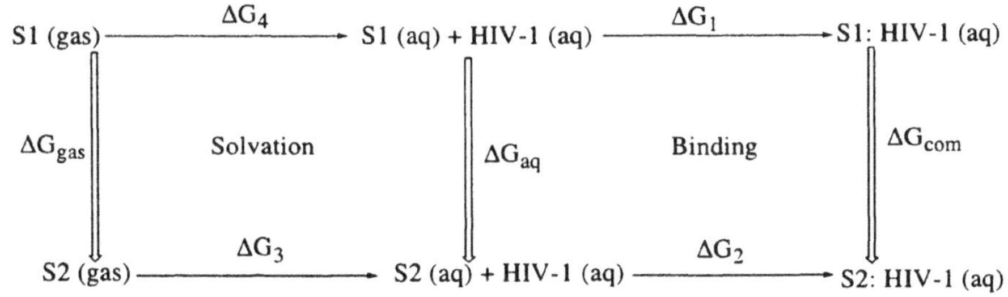

Figure 2. TCP cycle for calculating binding affinities of inhibitors to HIV-1 protease

($\Delta\Delta G_{bind}$) between two related compounds, by computationally simulating the 'mutation' of one to the other. The relative solvation free energy change for two substrates is computed using the first cycle shown in Figure 2, as represented in the following equation:

$$\Delta G_3 - \Delta G_4 = \Delta G_{aq} - \Delta G_{gas} = \Delta\Delta G_{sol} \quad (1)$$

The relative binding free energy change for the two substrates is computed using the second cycle (Figure 2), which is represented by the following equation:

$$-k_BT \ln (k2/k1) = \Delta G_2 - \Delta G_1 = \Delta G_{com} - \Delta G_{aq} = \Delta\Delta G_{bind} \quad (2)$$

where k_B is the Boltzmann constant, T is the absolute temperature, and k1 and k2 refer to the experimentally measured binding constants for S1 and S2 inhibitors respectively, whereas ΔG_1 and ΔG_2 are the corresponding free energy differences. The free energy change for converting S1 into S2 is computed by perturbing the Hamiltonian of reactant (initial) state S1 into that of the product (final) state S2. This transformation is accomplished through a parameterization of terms comprising the interaction potentials of the system with a change of state variable that maps onto reactant and product states when that variable is 0 and 1, respectively. The total free energy change for the mutation from the initial to the final state is computed by summing 'incremental' free energy changes over several windows visited by the state variable changing from 0 to 1.

2.2 Single Topology Method

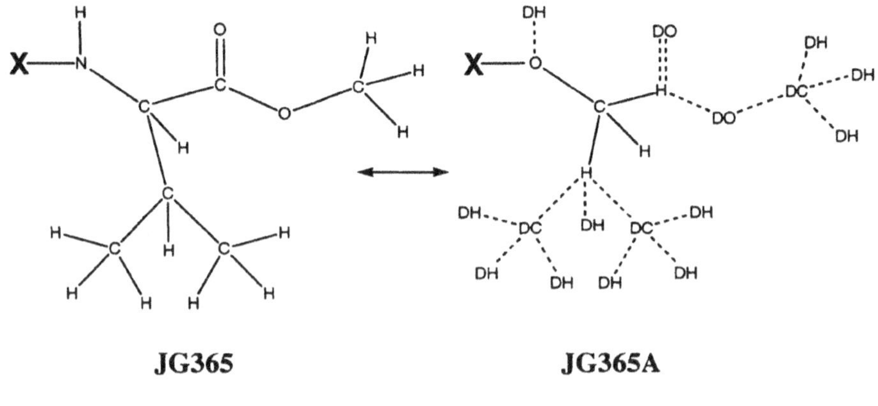

JG365 **JG365A**

Figure 3. Single topology definition for compound **JG365** and **JG365A**. The chemical symbols with the "D" prefix indicate dummy atoms and **X**=Ac-Ser-Leu-Asn-(Phe-Hea-Pro)

In this commonly used method, the appropriate reactant atoms are changed to product atoms. These mutations include changes in the geometry, partial charges, and van der Waals parameters. Figure 3 shows a typical example of single topology method for the mutation of **JG365** inhibitor to **JG365A** shown in Figure 5 (a).

2.3 Double Topology Method

In the double topology method[40, 41] a single topology is defined for those atoms that are identical in both molecules in the sense that force constants and equilibrium geometries are the same (partial charges can vary). For the portion of the molecule which must be transformed, both the starting (reactant) and ending (product) topologies are defined with their correct geometries, one beginning and the other ending the simulation entirely as dummy atoms. Dummy atoms are identical to real atoms except for their Lennard-Jones parameters and charges which are set to zero. At intermediate points during the transformation, all atoms in both topologies have fractional Lennard-Jones parameters and charges. Molecules with both topologies interact with the environment, but not with each other. A typical example of the thread method is shown in Figure 4 for the mutation of the HIV-1 protease inhibitors **JG365** and its analog **JG365A** which are shown in Figure 5 (a).

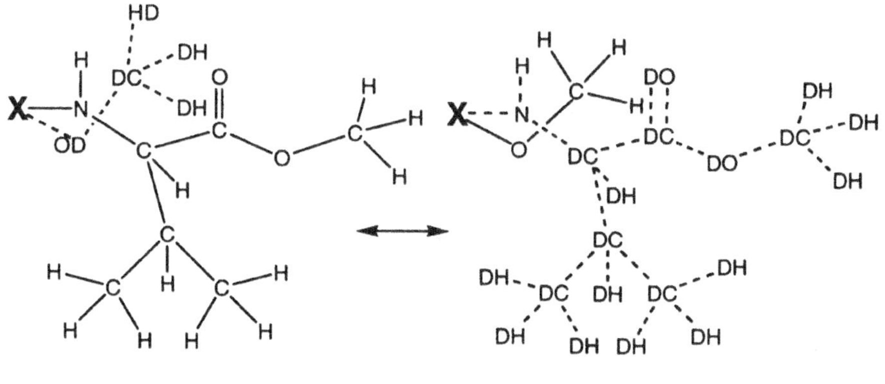

JG365 **JG365A**

Figure 4. Single topology definition for compound **JG365** and **JG365A**. The chemical symbols with the "D" prefix indicate dummy atoms and **X**=Ac-Ser-Leu-Asn-(Phe-Hea-Pro)

3. VALIDATION OF FEP METHODOLOGY

Validation of free energy perturbation methodology for estimating relative binding affinities of known HIV-1 protease inhibitors is very important before using this methodology for predicting relative binding affinities of unknown HIV-1 protease inhibitors. As discussed below, three studies using AMBER and the FEP methodology provided support for the protocol and its subsequent use in the design of HIV-1 protease inhibitors.

Reddy et al.[13] reported the first validation study for the mutation of **JG365** into **JG365A** (Figure 5(a)) using the X-ray structure of HIV-1 protease complexed with **JG365**.[6] All the computational details are described elsewhere.[13] The calculated relative solvation free energy difference ($\Delta\Delta G_{sol}$) between inhibitors **JG365** and **JG365A** indicated that removing the valine residue results in a difference of about 8 kcal/mol. This large relative difference was attributed to three good hydrogen bonds formed between the valine backbone atoms of **JG365** inhibitor and solvent water. The calculated relative solvation free energy (8.0 kcal/mol) agreed with the experimental result obtained for isolated valine (8.2 kcal/mol).[42] This agreement suggested that the force field parameters and FEP methodology are very good for calculating relative solvation free energies between these inhibitors.

(a)

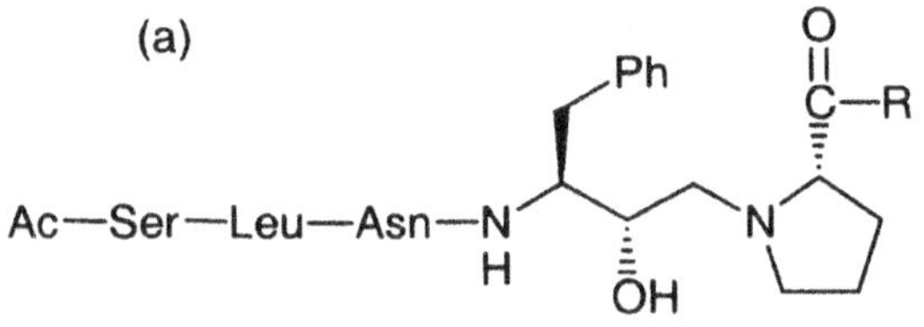

R = Ile-Val-OMe (**JG365**)
R = Ile-OMe (**JG365A**)

(b)

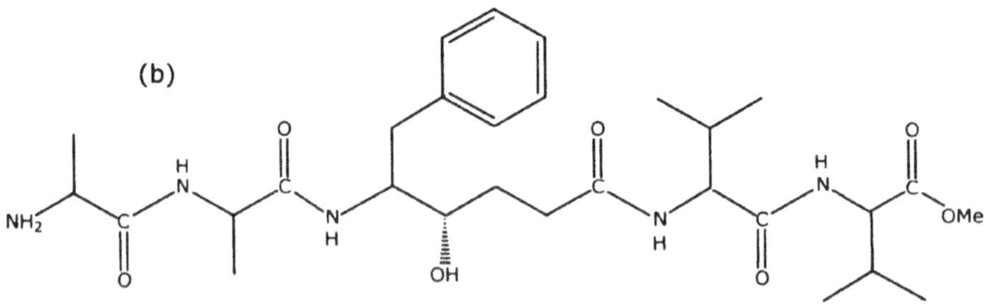

Figure 5. HIV-1 Protease inhibitors. (a) **JG365** and **JG365A**, (b) Hydroxyethlene isostere inhibitor.

The relative binding free energy difference for **JG365** and **JG365A** inhibitors to HIV-1 protease was calculated using the second cycle shown in Figure 2. The X-ray structure of HIV-1 protease complexed with Ac-Ser-Leu-Asn-(Phe-Hea-Pro)-Ile-Val-OMe, where Hea is hydroxyethylamine, was used as the starting configuration. The Val residue of the inhibitor was mutated to nothing (**JG365A**) with Hea in both the (R) and (S) configurations. The (R) configuration of Hea was generated starting from the (S) configuration of the complex crystal structure. The relative differences in the binding affinity for the mutation of Val to nothing with (S) and (R) Hea are 2.95 ± 1.02 kcal/mol and 3.55 ± 1.10 kcal/mol, respectively. Therefore, the average calculated relative binding difference ($\Delta\Delta G_{bind}$ (cal)) for the mutation with (R) and (S) configurations of Hea was 3.25 ± 1.06 kcal/mol and was in good agreement with experimentally measured (as found by binding of a racemic mixture) value of 3.8 ± 1.3 kcal/mol. The calculated results indicate that, even though **JG365** costs about 8 kcal/mol more to desolvate, it is a better inhibitor of HIV-1 protease than **JG365A**. A comparison of the HIV-1 protease-inhibitor complexes suggests that the high binding preference for **JG365** is due to a good hydrogen bond and strong electrostatic interactions between the carbonyl oxygen of valine and Arg8 as well as good hydrophobic interactions between the valine side-chain (**JG365**) and other protein residues. These interactions dominate over an opposing contribution arising from the larger desolvation penalty of **JG365** compared to **JG365A**.

In the second validation study, Ferguson et al.[36] used the same structure of HIV-1 protease complexed with **JG365** and computed the relative binding free energies of the S-(OH) and R-(OH) diastereomers of the peptide inhibitor, Ac-Ser-Leu-Asn-(Phe-Hea-Pro)-Ile-Val-OMe (**JG365**)[6] to the HIV-1 protease. The calculations were carried out using the AMBER 3.0 revision A with united-atom force field. During the free energy calculation in the reverse direction [R-(OH) back to S-(OH)], the structure of the S-OH diastereomer did not return to the original conformation. To compensate for this, the reverse simulation was corrected by constraining the backbone atoms of the protein. A close agreement with the experimental value was found for the protonated Asp125 model where calculated relative binding free energy and experimental binding free energy were 2.8 ± 0.2 kcal/mol and 2.6 kcal/mol, respectively.

Rao et al.[37] reported the third validation study on HIV-1 protease inhibitors. In this study the hydroxyethlene isostere inhibitor (Figure 5(b)) was modeled beginning from the X-ray structure of HIV-1 complex to the inhibitor MVT-101. The calculated binding free energy differences for the mutations of (S)-OH to hydrogen, and (R)-OH to hydrogen in hydroxyethlene isostere were 3.92 ± 1.03 kcal/mol, and 3.07 ± 0.26 kcal/mol, respectively. The calculated (Asp diad unconstrained and constrained) and experimental binding free energy differences for the (S)-OH to (R)-OH mutation were 3.37 ± 0.64 kcal/mol, 2.16 ± 0.65 kcal/mol and

2.6 kcal/mol, respectively. The calculated binding free energy difference using constraints on the Asp diad more closely matched the experimental result. However, two experimental studies[43, 44] have reported that the R-diastereomers of a few inhibitors bind to HIV-1 PR better than the corresponding S-diastereomer, a result that is confirmed by the calculations using no constraints.

In summary all the relative free energy calculations on HIV-1 protease inbitors indicated that FEP methodology and force field parameters in the AMBER database would be useful for estimating relative binding affinities of inhibitors to HIV-1 protease and therefore suggested that this method could be used for screening proposed analogs of potential HIV-1 protease inhibitors prior to synthesis.

4. DESIGN OF NON-PEPTIDIC INHIBITORS

At the beginning of this project, several peptide-based inhibitors for HIV-1 protease were available, but all were peptides and therefore relatively poor drug candidates.[45, 46] Thus, efforts were focused on structural modifications that would lessen the peptide nature of these leads. A model of the unsubstituted hydroxyethylene-based inhibitor, created by simply overlaying an AM1[47] minimized structure of an initial compound (Figure 5(b)) on the MVT-101:HIV-1 protease complex reported by Wlodawer and coworkers[48] (pdb:HVP4) was used as a guide. The C-terminal Val-Val-methyl ester was

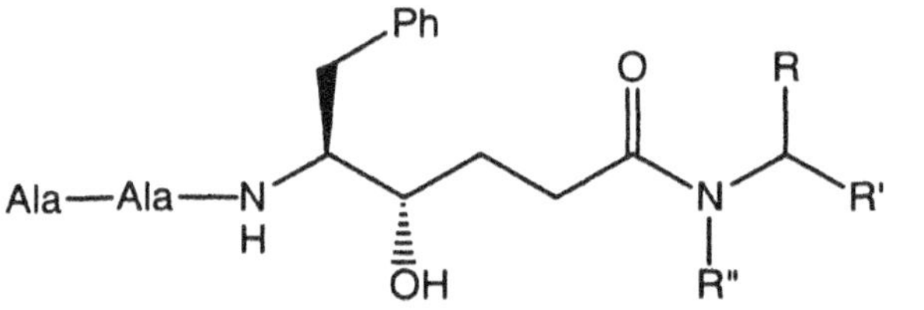

Compound	R	R'	R"
1	Ph	Ph	H
2	2-Indole	Ph	H
3	2-Benzimidazole	Ph	H
4	2-Indole	m-CF_3Ph	H
5	2-Indole	m-CH_3Ph	H
6	2-Indole	p-NH_2Ph	H
7	2-Indole	4-pyridyl	H
8	2-Indole	cyclohexyl	H
9	2-Indole	Ph	Benzyl

Figure 6. HIV-1 inhibitors evaluated by FEP calculations.

then removed and replaced with a diphenhydramine moiety (**1** in Figure 6). This simple scaffold served to fill the P2' and P3' sites in the protein with phenyl groups. Later, compound **1** was synthesized, the binding constant (K_i = 1.67μM) measured and the X-ray structure of the complex solved. Concurrently, compound **2** was designed and selected for synthesis based on space-filling considerations and molecular mechanics calculations that indicated that it would have better enthalpic interactions with the protein than **1**. After synthesis, biochemical testing revealed that compound **2** indeed a more potent inhibitor than **1** by a factor of eight (K_i = 0.2 μM).

4.1 X-ray Structures

The X-ray structures of HIV-1 protease complexes with inhibitor **1** and **2** were solved.[49] Both exist in the orthorhombic space group $P2_12_12_1$ with the unit cell dimensions of a = 66.4 Å, b = 92.4 Å, and c = 28.8 Å. The resolutions and crystallographic R factors for HIV-1 complexed with compound **1** were 2.6 Å and 0.163, respectively, and for HIV-1 complexed with compound **2** were 2.5 Å and 0.158, respectively. Crystallographic analysis of HIV-1 protease complexed with compound **2** indicated that the increased binding, relative to compound **1**, was probably due to a combination of both increased hydrophobic interactions with active site protein residues as well as an additional hydrogen bond from the indole NH to the Gly27 carbonyl oxygen via an intervening water molecule.[50] Analysis of the crystal structures of the complexes with **1** and **2** indicated that there was some empty space in the vicinity of the phenyl ring. Therefore, there was a possibility of enhancing hydrophobic interactions by suitable substitutions in the meta position of the phenyl ring.

4.2 Computational Details

All molecular dynamics, molecular mechanics and FEP calculations were carried out with AMBER 3.3 using an all-atom force field[51, 52] and with the SPC/E water potential[53, 54] to describe water interactions. Electrostatic charges and parameters for the standard residues were taken from the AMBER database. For nonstandard solute atoms, partial charges were obtained by electrostatic fitting with CHELP[55] from ab initio 6-31G*//3-21G* basis set level wave functions calculated with Gaussian 88.[56] All equilibrium bond lengths, bond angles, and dihedral angles for nonstandard residues were taken from optimized geometries. Missing force field parameters were estimated from similar chemical species within the AMBER database.

To compute relative solvation free energies, the solute was first solvated with SPC/E water using the AMBER box option and all solvent molecules located more than 10 Å from any of the solute atoms were removed. Water molecules located less than 2.5 Å from the solute atom were also removed. Aqueous phase MD simulations were carried out in a rectangular box using periodic boundary conditions in all directions. Newton's equations of motion for all the atoms were solved using the Verlet algorithm[57] with a 2 fs time step and SHAKE[58] was used for constraining all bond lengths. A constant temperature (NPT) ensemble was maintained by velocity scaling of all-atoms in the system. Nonbonded interaction energies were calculated using a 10 Å residue based cutoff. Molecular dynamics simulations in conjunction with the thermodynamic cycle approach shown in the Figure 2 was used to calculate relative solvation free energies between ligands given in Figure 6.

For the protein complex simulations (second cycle in Figure 2), it was necessary to generate all the hydrogen atom coordinates in order to use the all-atom force field. The EDIT module of AMBER was used to add hydrogens to the protein dimer and the crystallographic waters. One of the aspartic acids in the catalytic dyad (Asp124) was protonated for the simulations. The histidine protonation at one or both ring nitrogens was deduced from hydrogen bonding and other features of its environment. The total charge on HIV-1 protease was +5 e. No counter ions or changes in the customary charge of protein residues were used. While such an electrostatic model is far from ideal, alternatives sometimes adopted have their own drawbacks. The entire system was immersed in a 25 Å radius sphere of solvent from the center of mutating groups, which was subjected to a half-harmonic restraint near the boundary to prevent evaporation. During the simulation, all atoms of the protein beyond 25 Å were fixed. All nonbonded interactions involving the inhibitors and the charged residues of the protein were computed with no cutoff. A 10 Å nonbonded residue-based cutoff was used for other residues of the system. The algorithm for the complex simulation was identical to the solvent simulation, except for the absence of periodic boundary conditions in the former.

4.3 Structural Comparison

A comparison of structural differences between the X-ray and the energy minimized structure can provide information about the quality of the force field parameters. Assuming the force field is satisfactory, a comparison between the X-ray structure and the MD-averaged structure can provide information about the quality of the X-ray structure as well as determine the time step required in the simulations. An energy minimization (500 steps of steepest descent followed by 2000 steps of conjugate gradient optimization) of the HIV-1-protease:**2** complex was performed. Using the minimized coordinates, two separate 20 ps MD simulations were carried out using 1 fs and 2 fs time steps. The average "dynamical" structure of the complex was computed for each simulation. The root mean square (RMS) deviations between the X-ray structure and energy minimized structure were 0.4 Å and

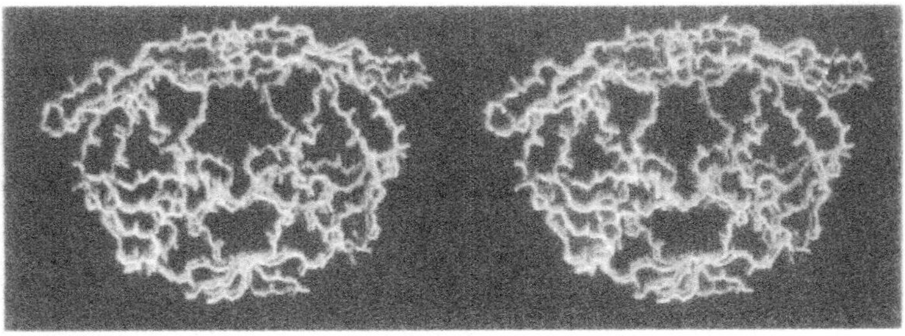

(a)

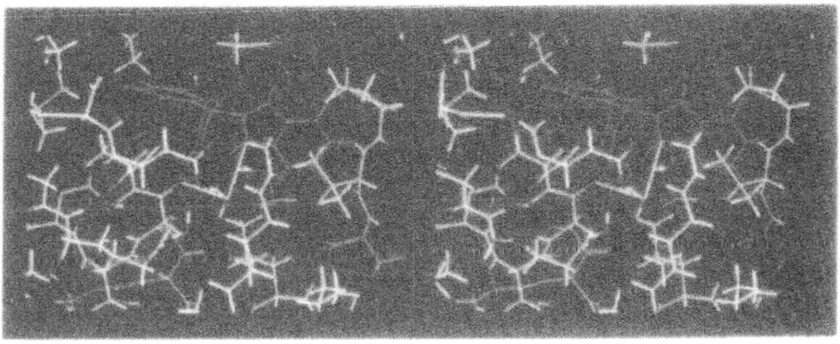

(b)

Figure 7. (a) Stereoview of comparison of the main chain of the X-ray structure of the HIV-1 protease complex with compound **2** (red) with the main chains of the minimized complex (yellow) and a 20ps average dynamical structure of the same complex of HIV-1 protease (green). (b) Stereoview of the active-site geometry of the crystal structure (in half-bond color) of the HIV-1 protease complexed with the compound **2** (with the indole and phenyl groups shown in red) as revealed by X-ray crystallography.

0.6 Å for backbone and side-chain atoms, respectively. For the MD-averaged structure using time steps of 1 fs and 2 fs, the RMS deviations from the crystal structure were 1.01 Å and 1.05 Å for backbone atoms and 1.5 Å and 1.55 Å for side-chain atoms, respectively. Larger deviations were observed in the flap region (RMS deviations of 1.30 Å and 1.75 Å for backbone and side-chain atoms, respectively). Figure 7a show the structure comparisons of the average dynamical structure from 20ps dynamic trajectory (with a time step of 2 fs) with the X-ray structure of the HIV-1 protease dimmer. This is due primarily to the flexibility of the protein in the flap region. Nevertheless, the dynamical structure was a good model for calculating relative free energy changes between two similar inhibitors. These comparisons indicate that the quality of the X-ray structure, and the force field parameters were adequate for performing MD simulations. Since both time steps yielded good agreement with the X-ray structure, the longer time step of 2 fs was used for all free energy calculations reported, in the interest of saving computer time.

4.4 Analysis of Lead Inhibitors Using FEP Methodology

Mutation of compound **1** to compound **2** was performed to validate the FEP method for HIV-1 protease inhibitors. Since this mutation involves significant changes in the ligand structure (phenyl to indole), the commonly used "single topology method" was unsuitable for this type of mutation. Therefore, the "thread" method,[40-41] which was developed for large structural changes between two ligands, was used to accomplish this nonphysical transformation. For the mutation **1** $\rightarrow$ **2**, the phenyl group of molecule **1** and the indole group of molecule **2** are "threaded" together at C_α positions similar to compounds shown in Figure 3 was used, where one is real and the other is dummy. In all free energy simulations, the system was initially minimized (using 500 steps of steepest descent and 2000 steps of conjugate gradient methods) and then equilibrated for 20 ps. A two-stage procedure was used to obtain relative free energy differences from the molecular dynamics simulations. During the first stage, the charges of the reactant atoms are turned off while the Lennard-Jones parameters of the product atoms are turned on. During the second stage, the Lennard-Jones parameters of the reactant atoms are turned off while the charges of the product atoms are turned on. This procedure was used previously [29-32] and shown to give better convergence. Each stage of the simulation was performed using 101 windows. Each window consisted of 1 ps equilibration and 2 ps of data collection except for the last 10 windows which required longer times for equilibration (3 ps) and data collection (6 ps). Trying to compute the free energy differences using the thread method with shorter simulations can lead to problems in convergence. Thus, a molecular dynamics simulation of 726 ps was needed for the complete mutation.

Each mutation used the doublewide sampling procedure. The results reported are based on the averages of the backward and forward simulations of the mutation. Table 1 lists the results of the FEP calculations. Although compound **2** was found to have a greater desolvation penalty than compound **1**, it binds more tightly to HIV-1 protease by forming more favorable interactions in the complex. In particular, it forms additional hydrogen bonds and hydrophobic interactions, as shown in Figure 7b, with protein residues. In summary, FEP calculations suggested that compound **2** showed 1.9 kcal/mol greater binding affinity relative to compound **1**, which is consistent with the experimental results.

The convergence of the calculated free energy results was tested by comparing the free energy differences between compound **2** and **1** to HIV-1 protease by varying the molecular dynamics simulation length (363 ps and 726 ps). The calculated results for 363 ps and 726 ps were 2.2 $\pm$ 0.8 kcal/mol and 1.9 $\pm$ 0.6 kcal/mol, respectively whereas the experimental result was 1.3 $\pm$ 0.3 kcal/mol. Therefore, the calculated result obtained with longer simulation more closely matches the experimental result. Error bars were estimated for each window by dividing the window statistics into four

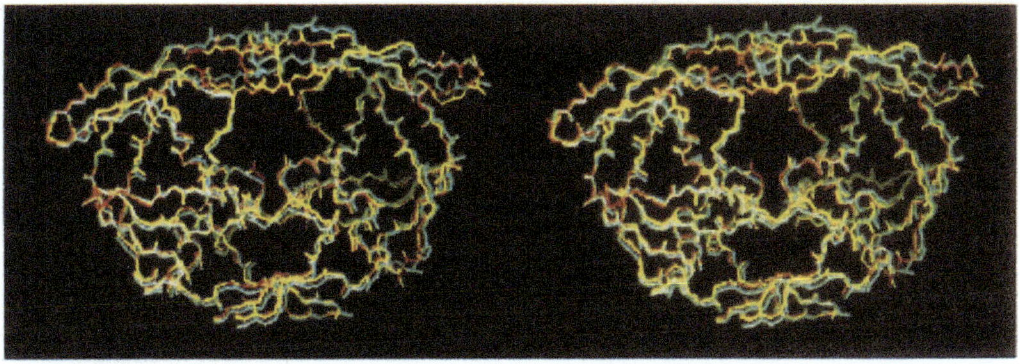

(a)

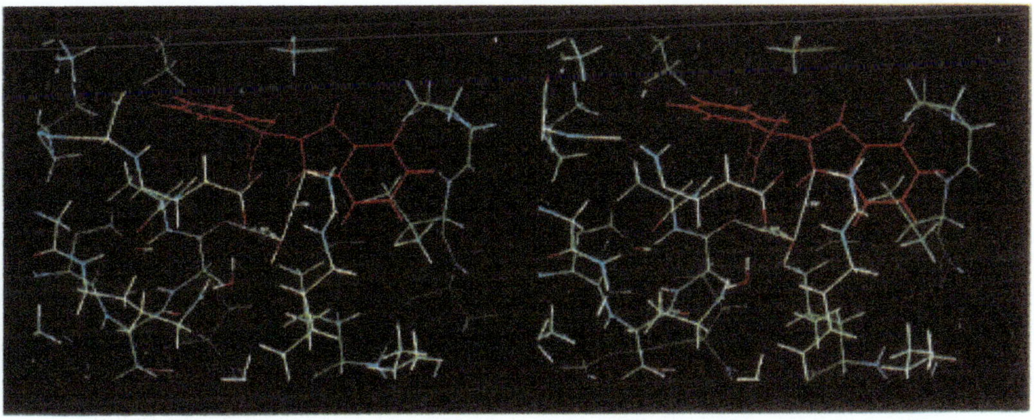

(b)

Figure 7. (a) Stereoview of comparison of the main chain of the X-ray structure of the HIV-1 protease complex with compound **2** (red) with the main chains of the minimized complex (yellow) and a 20ps average dynamical structure of the same complex of HIV-1 protease (green). (b) Stereoview of the active-site geometry of the crystal structure (in half-bond color) of the HIV-1 protease complexed with the compound **2** (with the indole and phenyl groups shown in red) as revealed by X-ray crystallography.

groups (in both forward and backward directions) and computing the standard deviation for the indicated free energy change. The root mean square of these window errors is reported in Table 1 as a measure of the statistical uncertainty in the result for each mutation.

4.5 Lead Optimization Using FEP Calculations

Compound **3** was evaluated using the free energy perturbation method based on the possibility that replacement of a carbon with nitrogen in the indole ring of compound **2** could enhance the electrostatic interaction with the main chain Gly48 of HIV-1 protease. The desolvation penalty for compound **3,** however, was expected to be greater than that of compound **2** because of this additional hydrophilic nitrogen. Two mutations (**1**→ **3** and **2** → **3**) were performed prior to the synthesis and testing of compound **3**. For the mutation **1**→ **3**, the thread method was used again because of the large structural change involved. For the other mutation **2** → **3**, a single topology method, was used. The calculated relative binding free energy results (Table 1) indicated that the compound **1** binds better (by 1.2 kcal/mol) to HIV-1 protease than the compound **3,** whereas compound **3** binds weaker to HIV-1 protease as compared to compound **2** by 1.3 kcal/mol.

In another attempt to improve on our lead compound **2**, we focused on the other phenyl ring in the active site. The crystal structures of HIV-1 protease complexes with **1** and **2** showed that there existed some unfilled space in the vicinity of the phenyl ring. Based on graphical analysis, substituents at the meta position of the phenyl ring could be used to enhance hydrophobic interactions whereas substituents at the para position could be used to form hydrogen bonds or gain electrostatic interactions with Asp129. Initially, more than 20 analogs were proposed, but based on graphical and conformational analyses and assessment of desolvation costs, six molecule (**4** to **9**) were identified for further computational study using FEP.

Table 1. Relative binding free energies (kcal/mol) of inhibitors to HIV-1 protease.

"Mutations"	$\Delta\Delta G_{sol}$(calc)	$\Delta\Delta G_{bind}$(calc)	$\Delta\Delta G_{bind}$(expt)
1→**2**	-3.00±0.40	-2.30±0.60	-1.30±0.30
2→**1**	3.50±0.40	1.90±0.60	1.30±0.30
1→**3**	-5.00±0.60	1.20±0.80	0.70±0.20
2→**3**	-2.00±0.50	1.30±0.60	1.95±0.31
2→**4**	-1.00±0.40	0.20±0.50	-0.16[a]±0.24
2→**5**	0.04±0.20	0.40±0.50	-0.06[a]±0.26
2→**6**	-3.20±0.40	1.10±0.60	-
2→**7**	-0.90±0.30	0.80±0.50	-
2→**8**	0.90±0.50	1.80±0.70	2.03[a]±0.40
2→**9**	-1.00±0.50	0.70±0.60	0.86[a] ±0.30

[a]Experimental values for these molecules are based on a different N-terminal group, an asparagine-quinoline moiety replacing NH_2-Ala-Ala in the compounds **2**, **4**, **5**, **8** and **9**.

Results in Table 1 show that the desolvation penalty relative to compound **2** is higher for compounds **6** (by -3.2 kcal/mol), **4** (by -1.0 kcal/mol), **9** (by -1.0 kcal/mol) and **7** (by -0.9 kcal/mol), whereas it is slightly lower for compounds **5** (by 0.04 kcal/mol) and **8** (by 0.90 kcal/mol) because of the hydrophobic substituents. While it is important to take the desolvation penalty into consideration in the design of new analogs of a lead compound, it should be noted that in some cases the design modification that increases the desolvation penalty might improve binding if the additional polar groups form good hydrogen bonds and/or electrostatic interactions with the protein residues, as were the cases of compounds **1** → **2** and **JG365** → **JG365A**.

Compounds **4** and **5** were predicted to have a relative binding free energy near zero. The compound **4** gained free energy in the complex but this was not enough to overcome the desolvation penalty. We still felt that there is some possibility of improving binding with these compounds, because of the error bounds on these free energy estimates. We therefore, made some design modifications at the N-terminal position involving the replacement of NH_2-Ala-Ala with the aparagine-quinaloyl moiety. These modified structures for compound **2, 4, 5, 8** and **9** were synthesized and tested. No relative free energy calculations were performed for these N-terminal modified variants. Experimental results (Table 1) for modified compounds show slight improvement in binding. Though free energy predictions (**2** → **4**, **2** → **5**) are based on the original structures (reported as free energy difference between **2** and **4** or **5** in Table 1), these are comparable to the experimental results of modified **2, 4, 5, 8** and **9** structures, because each member of a given pair has the same N-terminal moiety as the other member of that pair. As may be seen from Table 1, relative free energy results indicate a slight net loss of binding when **2** is replaced with **4** and **5**. Discrepancy between relative free energy calculations and experimental results could be due to the modifications at the N-terminal site as well as errors both in the experimental measurements and in calculations. Overall, relative free energy results are in good agreement with experiments. In the case of compounds **6** and **7**, relative free energy calculations indicated weaker binding (by 1 kcal or more) relative to compound **2**, primarily because of the larger desolvation penalty on these ligands, and hence these compounds were not synthesized. The free energy calculations predicted lower binding affinity to HIV-1 protease for compounds, **8** (by 1.80 kcal/mol) and, **9** (by 0.70 kcal/mol) as compared to compound, **2** which was later confirmed experimentally.

5. CONCLUSIONS

In summary, successful application of a computer-assisted drug design paradigm is described in this report. As part of this paradigm, the free energy perturbation approach has been instrumental in screening a series of promising inhibitors for the HIV-1 protease. Once validated, this approach was used in a predictive sense to prioritize design ideas and eliminate the need to synthesize poor inhibitors, thus accelerating the drug design cycle. In all cases where experimental data is available, predictions based on this approach were shown to be correct with an error margin of less than 1 kcal/mol. This study also brought to light the role played by solvation free energy in binding. The energetic cost of desolvation for the addition of polar groups to an inhibitor has to be compensated and overcome by stronger ligand-protein interactions, if the goal is to design a stronger inhibitor. Careful design, modeling and FEP calculations enable the achievement of that goal.

6. REFERENCES

1. F. Barre-Sinoussi, J. C. Chermann, F. Rey, M. T. Nugeyre, S. Chamaret, J. Gruest, C. Dauguest, C. Axler-Blin, F. Vezinet-Brun, C. Rouzioux, W. Rozenbaum, and L. Montagnier, Isolation of a T-lymphotropic retrovirus from a patient at risk for acquired immune deficiency syndrome (AIDS), Science **220**:868 (1983).
2. R. C. Gallo, P. S. Sarin, E. P. Gelman, M. Robert-Guroff, E. Richardson, V. S. Kalyanaraman, D. Mann, G. D. Sidhu, R. E. Stahl, S. Zolla-Prazner, J. Leibowitch, and M. Popovic, Isolation of human T-cell leukemia virus in acquired immune deficiency syndrome (AIDS), Science **220**:865 (1983).
3. A. Wlodawer, J. W. Erickson, Structure-based inhibitors of HIV-1 protease, Annu. Rev. Biochem. **62**:543 (1993).
4. K. Appelt, Crystal Structures of HIV-1 protease inhibitor complexes, Prospect. Drug Discovery Des. **1**:23 (1993).
5. A. Wlodawer, M., Miller, M. Jaskolski, B. K. Sathyanarayana, E. Baldwin, I. Weber, L. Selk, L. Clawson, J. Schneider, and S. Kent, Conserved folding in retroviral Protease: Crystal structure of a synthetic HIV-1 protease, Science **245**:616 (1989).
6. A. L. Swain, M. Miller, J. Green, D. H. Rich, J. Schneider, S. B. Kent, and A. Wlodawer, X-ray crystallographic structure of a complex between a synthetic protease of human immunodeficiency virus 1 and a substrate based hydroxyethyl-amine inhibitor, Proc. Natl. Acad. Sci. **87**:8805 (1990).
7. J. W. Erickson, D. J. Neidhart, J. VanDrie, D. Kempf, X. C. Wang, D. W. Norbeck, J. J. Plattner, J. W. Rittenhouse, M. Turon, N. Wideburg, W. E. Kohlbrenner, R. Simmer, R. Helfrich, D. A. Paul, and M. Knigge, Design, activity, and 2.8 Å crystal structure of a C_2 symmetric inhibitor complexed to HIV-1 protease, Science **249**:527 (1990).
8. J. R. Huff, HIV Protease: A novel chemotherapeutic target for AIDS, J. Med. Chem. **34**:2305 (1991).
9. T. D. Meek, Inhibitors of HIV-1 protease, J. Enzyme Inhibition **6**:65 (1992).
10. D. H. Rich, C. Q. Sun, J. V. N. Varaprasad, A. Pathiasseril, M. V. Toth, G. R. Marshall, M. Clare, R. A. Mueller, and K. Houseman, Effect of hydroxyl group configuration in hydroxyethlamine dipeptide isosteres on HIV protease inhibition, J. Med. Chem. **34**:1222 (1991).
11. B. C. M. Potts, D. J. Faulkner, J. A. Chan, G. C. Simolike, P. Offen, M. E. Hemling, and T. A. Francis, Didemnaketals A and B, HIV-1 protease inhibitors from the Ascidian Didemnum sp., J. Am. Chem. Soc. **113**:6321 (1991).
12. M. L. Dreyer, B. W. Metcalf, T. A. Tomaszek, T. J. Carr, A. C. Chandler III, L. Hyland, S. A. Fakhoury, V. W. Magaard, M. L. Moore, J. E. Strickler, C. Debouck, and T. D.

Meek, Human immunodeficiency virus 1 protease expressed in Escherichia coli behaves as a dimeric aspartic protease, Proc. Natl. Acad. Sci. **86**:9752 (1989).
13. M. R. Reddy, V. N. Viswanadhan, and J. N. Weinstein, Relative differences in the binding free energies of human immunodeficiency virus 1 protease inhibitors: A thermodynamic cycle-perturbation approach, Proc. Natl. Acad. Sci. USA **88**:10287 (1991).
14. M. R. Reddy, M. D. Erion, and A. Agarwal, Free energy calculations: Use and limitations in predicting ligand binding affinities, in: Reviews in Computational Chemistry, K. B. Lipkowitz and D. B. Boyd eds, VCH Publishers, New York, (2000), vol. 16, pp. 217-304.
15. I. D. Kuntz, Structure based strategies for drug design and discovery, Science **257**:1078 (1992).
16. P. J. Goodford, A computational procedure for determining energetically favourable binding sites on biologically important macromolecules, J. Med. Chem. **28**:849 (1985).
17. J. A. McCammon, Computer-aided molecular design, Science **238**:486 (1987).
18. W. E. Harte Jr. and D. L. Beveridge, Prediction of the protonation state of the active site aspartyl residues in HIV-1 protease-inhibitor complexes via molecular dynamics simulation, J. Am. Chem. Soc. **115**:3883 (1993).
19. K. Appelt, R. J. Bacquet, C. A. Bartlett, C. L. Booth, S. T. Freer, M. A. M. Fuhry, M. R. Gehring, S. M. Hermann, E. F. Howland, C. A. Janson, T. R. Jones, C. Kan, V. Kathardekar, K. K. Lewis, G. P. Marzoni, D. A. Matthews, C. Mohr, E. W. Moomaw. C. A. Morse, S. J. Oatley, R. O. Ogden, M. R. Reddy, S. H. Reich, W. S. Schoettlin, W. W. Smith, M. D. Varney, J. E. Villafranca, R. W. Ward, S. Webber, S. E. Webber, K. M. Welsh, and J. White, Design of enzyme inhibitors using iterative protein crystallographic analysis, J. Med. Chem. **34**:1925 (1991).
20. J. A. Montgomery, S. Niwas, J. D. Rose, J. A. Secrist, Y. S. Babu, C. E. Bugg, M. E. Erion, W. C. Guida, and S. E. Ealick, Structure-based design of inhibitors of purine nucleoside phosphorylase 1: 9-(arylmethyl) derivatives of 9-deazaguanine J. Med. Chem. **36**:55-69 (1993).
21. M. D. Erion, P. D. van Poelje, and M. R. Reddy, Computer-assisted scanning of ligand interactions: Analysis of the fructose 1,6-bisphosphatase-amp complex using free energy calculations, J. Am. Chem. Soc. **122**:6114 (2000).
22. J. A. McCammon, Theory of biomolecular recognition, Curr. Opin. Struct. Biol. **8**:245 (1991).
23. J. Hermans, ed. Molecular Dynamics & Protein Structure Polycrystal, West Springs, IL (1985).
24. J. Gao, K. Kuczera, B. Tidor, and M. Karplus, Hidden thermodynamics of mutant proteins: A molecular dynamics analysis, Science **244**:1069 (1989).
25. J. H. van Drie, Strategies for the determination of pharmacophoric 3D database queries, J.Comput.-Aided Mol. Design **11**:39 (1997).
26. R. D. Brown and Y. C. Martin, Designing combinational library mixtures using algorithm, J. Med. Chem. **40**:2304 (1997).
27. H. J. Bohm, Site directed structure generation by fragment joining, Perspect. Drug Disc. Des. **3**:21 (1995).
28. D. K. Gehlhaar, K. E. Moerder, D. Zichi, C. V. J. Sherman, R. C. Ogden, and S. T. Freer, De novo design of enzyme inhibitors by Monte Carlo ligand generation, J. Med. Chem. **38**:466 (1995).
29. C. E. Sansom, J. Wu, and I. T. Weber, Molecular mechanics analysis of inhibitor binding to HIV-1 protease, Protein Eng. **5**:659 (1992).
30. M. D. Erion, J. A. Montgomery S. Niwas, J. D. Rose, S. Ananthan, M. Allen, J. A. Secrist, S. Y. Babu, C. E. Bugg, W. C. Guida, and S. E. Ealick, Structure-based design of inhibitors of purine nucleoside phosphorylase 3: 9-arylmethyl derivatives of 9-deazaguanine substituted on the methylene group, J. Med. Chem. **36**:3771 (1993).
31. V. N. Viswanadhan, M. R. Reddy, A. Wlodawer, M. D. Varney, and J. N. Weinstein, An approach to rapid estimation of relative binding affinities of enzyme inhibitors:

Application to peptidomimetic inhibitors of the human immunodeficiency virus type 1 protease, J. Med. Chem. **39**:705 (1996).
32. K. Holloway, J. M. Wai, T. A. Halgren, P. M. Fitzgerald, J. P. Vacca, B. D. Dorsey, R. B. Levin, W. J. Thompson, J. L. Chen, J. S. deSolms, N. Gaffin, A. K. Ghosh, E. A. Giuliani, S. L. Graham, J. P. Guare, R. W. Hungate, T. A. Lyle, W. M. Sanders, T. J. Tucker, M. Wiggins, C. M. Wiscount, O. W. Woltersdorf, S. D. Young, P. L. Darke, and J. A. Zugay, A priori prediction of activity for HIV-1 protease inhibitors employing energy minimization in the active site, J. Med. Chem. **38**:305 (1995).
33. M. R. Reddy and A. Parrill, Overview of rational drug design, in: Rational Drug Design: Novel Methodology and Practical Applications, ACS Symposium Series 719, A. Parrill and M. R. Reddy, eds., Oxford University Press, Washington, DC (1999), pp. 1-11.
34. M. R. Reddy, V. N. Viswanadhan, M. D. Erion, Rapid estimation of relative binding affinities of enzyme inhibitors, in: 3D QSAR in Drug Design, vol. 2, H. Kubinyi, Y. C. Martin, and G. Folker, eds., Kluwer Academic Publishers, Great Britain (1998), pp. 85-98.
35. M. R. Reddy, M. D. Varney, V. Kalish, V. N. Viswanadhan, and K. Appelt, Calculation of relative differences in the binding free energies of HIV1 protease inhibitors: a thermodynamic cycle perturbation approach, J. Med. Chem. **114**:10117 (1994).
36. D. M. Ferguson, R. J. Radmer, and P. A. Kollman, Determination of the relative binding free energies of peptide inhibitors to the HIV-1 protease, J. Med. Chem. **34**:2654 (1991).
37. B. G. Rao, R. F. Tilton, and U. C. Singh, Free energy perturbation studies on inhibitor binding to HIV-1 protease, J. Am. Chem. Soc. **114**:4447 (1992).
38. R. J. Zwanzig, High-temperature equation of state by a perturbation method in nonpolar gases, J. Chem. Phys. **22**:1420 (1954).
39. D. L. Beveridge and F. M. DiCapua, Free energy via molecular simulation applications to chemical and biomolecular systems, Ann. Rev. Biophys. Biophys. Chem. **18**:431 (1989).
40. U. C. Singh, Probing the salt bridge in the dihydrofolate reductase-methotrexate complex by using the coordinate-coupled free-energy perturbation method, Proc. Natl. Acad. Sci. **85**:4280 (1988).
41. M. R. Reddy, R. J. Bacquet, D. Zichi, D. A. Matthews, K. M. Welsh, T. R. Jones, and S. Freer, Calculation of solvation and binding free energy differences for folate-based inhibitors of the enzyme thymidylate synthase, J. Am. Chem. Soc. **114**:10117 (1992).
42. R. Wolfenden, Affinities of amino acid side-chains for solvent water, Biochemistry **20**:849 (1981).
43. A. L. Swain, M. M. Miller, J. Green, D. H. Rich, J. Schneider, S. B. H. Kent, and A. Wlodawer, X-ray crystallographic structure of a complex between a synthetic protease of human immunodeficiency virus 1 and a substrate-based hydroxyethyl-amine inhibitor, Proc. Natl. Acad. Sci. **87**:8805 (1990).
44. N. A. Roberts, J. A. Martin, D. Kinchington, A. V. Broadhurst, J. C. Craig, I. B. Duncan, S. A. Galpin, B. K. Handa, J. Kay, A. Krohn, R. W. Lambert, J. H. Merrett, J. S. Mills, K. E. B. Parkes, S. Redshaw, A. J. Ritchie, D. L. Taylor, G. I. Thomas, and P. J. Machin, Rational design of peptide-based HIV protease inhibitors, Science **248**:358 (1990).
45. D. H. Rich, C. Q. Sun, J. V. N. V. Prasad, A. Pathiasseril, M. V. Toth, G. R. Marshall, M. Clare, R. A. Mueller, and K. Houseman, Effect of hydroxyl group configuration in hydroxyethylamine dipeptide isosteres on HIV protease inhibition: Evidence for multiple binding modes, J. Med. Chem. **34**:1222 (1991).
46. J. R. Damewood, Peptide mimetic design with the aid of computational chemistry in Reviews in Computational Chemistry, vol. 9, K. B. Lipkowitz and D. B. Boyd, eds., VCH Publishers, New York (1996) pp. 1-79.
47. M. J. S. Dewar, E. G. Zoebhisch, E. F. Healy, and J. J. P. Stewart, AM1: A new general purpose quantum mechanical molecular model, J. Am. Chem. Soc. **107**:3902 (1985).
48. M. Miller, J. Schneider, B. K. Sathyanarayana, M. V. Selk, S. B. B. H. Kent, and A. Wlodawer, Structure of complex of synthetic HIV-1 protease with a substrate-based inhibitor at 2.3 Å resolution, Science **246**:1149 (1989).

49. K. Appelt, Crystal structures of HIV-1 protease inhibitor complexes: Prospects, Drug Disc. Design **1**:23 (1993).
50. M. D. Varney, K. Appelt, V. Kalish, M. R. Reddy, J. Tatlock, C. L. Palmer, W. H. Romies, B. W. Wu, and L. Musick, Crystal-structure based design and synthesis of novel c-terminal inhibitors of HIV protease, J. Med Chem. **37**:2274 (1994).
51. S. J. Weiner, P. A. Kollman, D. A. Case, U. C. Singh, C. Ghio, G. Alagoha, S. Profeta Jr., and P. K. Weiner, A new force field for molecular mechanical stimulation of nucleic acids and proteins, J. Am. Chem. Soc. **106**:765 (1984).
52. U. C. Singh, P. K Weiner, J. K. Caldwell, and P. A. Kollman, AMBER version 3.0 University of California at San Francisco, San Francisco, CA (1986).
53. H. J. C. Berendsen, J. R. Grigera, and T. P. Straatsma, The missing term in effective pair potentials, J. Phys. Chem. **91**:6269 (1987).
54. M. R. Reddy, M. Berkowitz, The dielectric constant of SPC/E water, Chem. Phys. Lett. **155**:173 (1989).
55. L. E. Chirlian and M. M. Francl, Atomic charges derived from electrostatic potentials: A detailed study, J. Comput. Chem. **8**:894 (1987).
56. M. J. Gaussian, M. Frisch, H. B. Head-Gordon, K. Schlegel, J. S. Raghavachari, C. Binkley, D. J. Gonzalez, D. J Defrees, R. J. Fox, R. Whiteside, C. F. Seeger, J. Melius, R Baker, L. R. Martin, J. J. P. Kahn, E. M. Stewart, S. Fluder, J. A. Topiol, and J. A. Pople, Gaussian, Inc., Pittsburgh, PA. (1988).
57. L. Verlet, Computer "experiments" on classical fluids. I. Thermodynamical properties of Lennard-Jones molecules, Phys. Rev. **159**:98 (1967).
58. J. P. Ryckaert, G. Ciccotti, and H. J. C. Berendsen, Numerical integration of the cartesian equations of motion of a system with constraints: Molecular dynamics of n-alkanes, J. Comput. Phys. **23**:327 (1977).

Chapter 17

Thymidylate Synthase: Free Energy Calculations for Estimating Inhibitor Binding Affinities

Tai-Sung Lee[†] and Peter A. Kollman[‡*]
[†]*Molecular Simulations, Inc., San Diego, CA 92121*
[‡*]*Dept. of Pharmaceutical Chemistry, University of California, San Francisco, CA 94143*

1. INTRODUCTION

Thymidylate Synthase (TS) is a 70 kDa dimeric protein that catalyzes the conversion of 2'-deoxyuridine 5'-monophosphate (dUMP) into 2'-deoxythymidine 5'-monophosphate (dTMP) using 5,10-methylene-5,6,7,8-tetrahydrofolate as cofactor. Inhibitors of TS represent potential chemotherapy agents since DNA synthesis requires dTMP and TS catalyzes an important step in the only known pathway for dTMP biosynthesis. Given the importance of TS as a drug target, free energy calculations have been applied to the design and evaluation of potential TS inhibitors with the goal of finding inhibitors that are selective for pathogen or cancer isozymes.[1, 2] Calculations have also been used to study TS specificity [3] and the relative binding free energies of the inhibitor methotrexate (MTX) to wild type and mutant forms of TS.[4] In all of the above cases, the results have tied nicely into available experimental data on these systems.

2. METHODS

The methods used in these calculations are described in detail in the original references. In brief, the various studies of TS used different versions of AMBER[5] and carried out free energy perturbation or thermodynamic integration calculations on free ligand or ligand bound to either wild type TS[1-3] or mutant protein.[4] Simulations used either the Weiner et al.[6] or Cornell et al.[7] force fields and electrostatic potential or RESP[8] derived charges for the inhibitors. The free energy calculations ranged in length from ~50 to ~300 ps and, in every case, multiple runs were conducted to test for statistical accuracy. Calculations on the free ligand used periodic boundary conditions. Calculations on the protein complex used a solvated active site and a spherical mobile zone around the bound ligand where protein residues outside this mobile zone were kept rigid.

3. VALIDATION STUDIES

Reddy et al.[1] calculated the first relative binding free energy of two TS inhibitors (Figure 1), namely 10-propargyl-5,8-dideazafolic acid (PDDF) and 10-formyl-5,8 dideazafolic acid (FDDF). The calculations evaluated binding to the binary complex consisting of the *E. coli* TS and 5-fluoro-2'-deoxyuridylate monophosphate (FdUMP) with the X-ray structure of the TS: FdUMP: PDDF complex[9, 10] providing the starting coordinates.

The two inhibitors, PDDF and FDDF, differed in the ^{10}N-substitution, i.e. a propargyl (–C≡C–H) group for a formyl (–C(O)H) group. Interestingly, PDDF was calculated to have a 3.6 kcal/mol more favorable solvation free energy than FDDF suggesting that the propargyl group was more difficult to desolvate than the formyl group in aqueous solution. Relative solvation free energy calculations on but-1-yne vs. acetaldehyde supported these results, since acetaldehyde was more solvated ($\Delta\Delta G$ = -3.4 kcal/mol (calculated) and -3.3 kcal/mol (experimental)). In the protein, the propargyl group interacts with a backbone carbonyl of the synthase and makes good hydrophobic contacts with side-chain atoms, whereas the formyl oxygen of FDDF, which does not interact with the side-chain atoms, accepts a hydrogen bond from a bound water molecule and has a repulsive interaction with the carbonyl oxygen of 5-fluoro-2'-deoxyuridylate. Thus, PDDF gains very strong favorable interactions in the complex as compared to FDDF, which more

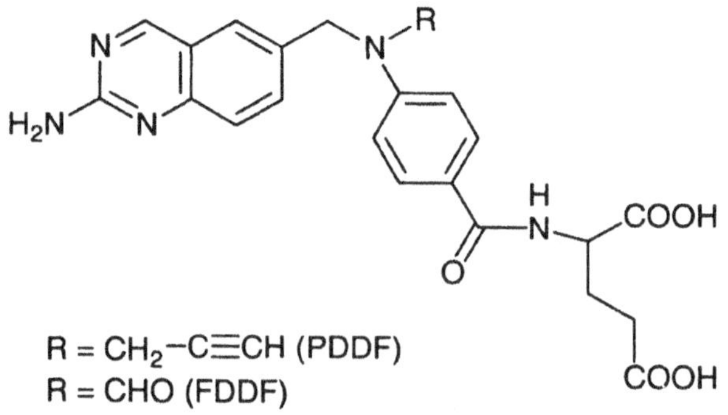

Figure 1. TS inhibitors used for free energy calculations.

than compensates for the higher desolvation costs associated with PDDF. As a result the calculated relative binding free energy is 2.9 kcal/mol, favoring PDDF, compared to the experimental relative binding free energy of 3.8 kcal/mol.[1]

The second calculation reported for TS[3] was focused on the origin of the enzyme substrate specificity. Wild type TS methylates 2'-deoxyuridine 5'-monophosphate (dUMP) and not 2'-deoxycytidine 5'-monophosphate (dCMP). Rastelli et al. calculated relative binding free energies of dUMP and dCMP to TS (Figure 2) and two Asn229 mutants using the X-ray structure of the TS:dUMP complex[11] and the free energy perturbation method. The calculated relative binding free energy of dUMP and dCMP with TS was analyzed as the sum of two components, the relative free energy difference of these two ligands in solvent water and the relative free energy difference in the protein complex. The calculated solvation free energy (-9.4 kcal/mol) indicates that dCMP is better solvated than dUMP in water despite the fact that cytosine and uracil each have five potential hydrogen bonding sites and therefore would have been predicted to interact with bulk solvent to approximately the same extent. The difference was attributed in part to the larger dipole moment of dCMP, which would be expected to increase the strength of the interactions with bulk solvent. In contrast, dCMP gained only -6.56 kcal/mol in the TS complex as compared to dUMP. Thus, the calculated relative binding free energy between dCMP and dUMP to TS (-6.56 +9.4 =2.84 kcal/mol) indicated that TS prefers to bind dUMP by 2.84 kcal/mol as compared to dCMP, which is consistent with the experimental value of 3.6 kcal/mol. Moreover, the simulations support the key role of Asn229 in the TS substrate preference for dUMP over dCMP. Repulsion between the base of dCMP and the Asn229 side-chain (Figure 2) reduces its free energy of binding to the protein and causes the displacement of this nucleotide into a position unsuitable for reaction. Similar relative binding free energy calculations on the Asn229Asp and Asn229Val mutants led to much smaller relative binding free energies, which is consistent with the experimental findings showing that the mutant enzymes exhibit little preference for dUMP relative to dCMP methylation.

In another study, Reddy and Villafranca[4] carried out free energy calculations analyzing TS mutants and their binding of methotrexate (MTX). The relative binding free energies of MTX with wild type TS relative to the two TS mutants Ile79Val and Ile79Ala ($\Delta\Delta G_{bind}$ (calc) = 0.4 kcal/mol and 1.9 kcal/mol for Ile79Val and Ile79Ala, respectively; whereas $\Delta\Delta G_{bind}$ (exp) = 0.6 kcal/mol and 1.4 kcal/mol, respectively) revealed the importance of large non-polar groups at position 79 to MTX binding. The $\Delta\Delta G_{bind}$ for Phe176Ile (calculated 1.8 kcal/mol; experimental 1.2 kcal/mol) indicated that a portion of the MTX binding affinity could be attributed to aromatic–

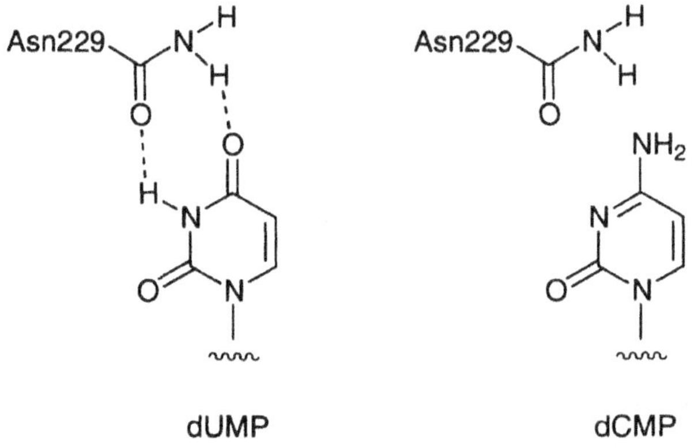

Figure 2. Hydrogen bond potential of dUMP vs. dCMP with the side-chain amide of Asn229.

MTX interactions, since, despite the more favorable desolvation of Ile compared to Phe, MTX bound more strongly to TS with Phe176.

4. NON-ADDITIVITY IN TS INHIBITION

Jones et al.[12] synthesized a total of 31 propargyl lipophilic quinazoline analogs in an effort to develop a potent antifolate inhibitor of TS. These compounds were quite promising in that they bound to TS almost as strongly as the tight binding polyglutamic acid inhibitors while possessing a molecular structure expected to exhibit superior pharmacological properties.

Inspired by these results, we attempted to reproduce the interesting non-additivity observed by Jones using free energy calculations (Figure 3). The experimental results indicated that the free energy change for the di-substituted compound was less than the sum of the free energy gains associated with each mono-substituted compound relative to the unsubstituted molecule. The hope from our studies was that accurate predictions of the non-additivity would lead to accurate predictions of the inhibitory potential of related analogs and ultimately to the discovery of an even more potent inhibitor. Our results showed that we could reproduce the non-additivity with reasonable accuracy. Moreover, using our software PROFEC,[10] we suggested a new inhibitor, which we calculated to bind even better to TS than any synthesized by Jones et al.[12]

The relative free energies for the molecules shown in Figure 3 (a–d) were calculated using the PROFEC program and the free energy program in the AMBER package.[5, 13] The calculated relative binding free energy for adding

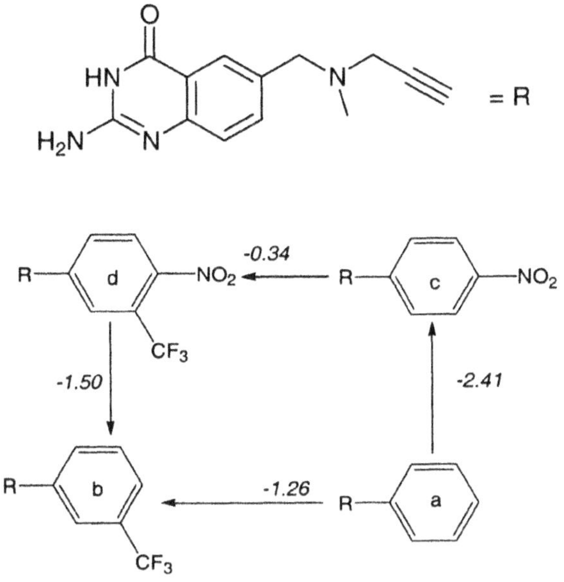

Figure 3. Four compounds designed by Jones et al[12]. The upper is the common structure and the lower shows different substitutions for these compounds. Each compound is named as the letter appearing on the aniline ring.

a $-CF_3$ group to the unsubstituted inhibitor (a→b) is -1.92 kcal/mol, compared to the experimental result of -1.26 kcal/mol. The calculated relative binding free energy for adding a $-NO_2$ group is -3.12 kcal/mol vs. the experimental (a→c) value of -2.41 kcal/mol. However, when one considers the relative binding free energy changes ($\Delta\Delta G$) for the double substitution, the experimental value (a → d) is -2.76 kcal/mol, whereas one would expect it to be -1.26 + (-2.41), or -3.67 kcal/mol, i.e. a non-additivity of 0.9 kcal/mol. The calculated relative binding free energy for mutating a→d via b is -4.59 kcal/mol and a→d via c is -3.95 kcal/mol, with the average relative binding free energy of -4.27 kcal/mol for a →d. This is 0.8 kcal/mol larger than the sum of the relative binding free energies for a→b and a→c mutations (-3.12 + (-1.92)), i.e. -5.04 kcal/mol. Thus, the calculations reasonably reproduce the non-additive effect of double substitution. By carrying out MD trajectories on the various inhibitors, we found[2] that residues Val262 and Ile79 move on either $3-CF_3$ or $4-NO_2$ substitution and that it appears that both substitution sites "compete" for interaction with these residues. This observation may provide a reasonable rationalization for the non-additivity observed for the $3-CF_3$ and $4-NO_2$ substituted compounds.

5. DESIGN OF POTENT TS INHIBITORS

The PROFEC calculations and similar qualitative analyses suggested that the 2,3-difluoro substitution of the 4-NO_2 analog (compound g in Figure 4) should increase binding strength, whereas adding 5,6-difluro to the 3-CF_3 analog (compound h in Figure 4) should decrease binding. As one can see from Figure 4, the addition of the 2,3-difluoro substitution to compound a (a→f) resulted in a calculated relative binding free energy of -2.98 kcal/mol. The effect of the 2,3-difluoro substitution on the 4-NO_2 analog (a→c→g) gave a relative free energy of (-3.12 +(-2.38)) -5.50 kcal/mol, whereas the effect of adding both the 5,6-difluro substitutions and a 3-CF_3 group (a→f→h) on the relative binding free energy was approximately -2 kcal/mol (the average of -2.22 and -1.66 kcal/mol). Thus, the calculated results predicted that the 2,3-difluoro, 4-NO_2 analog would bind about 1 order of magnitude more tightly than the 3-CF_3, 4-NO_2 disubstituted analog, which is the tightest binding inhibitor reported in this series.[13]

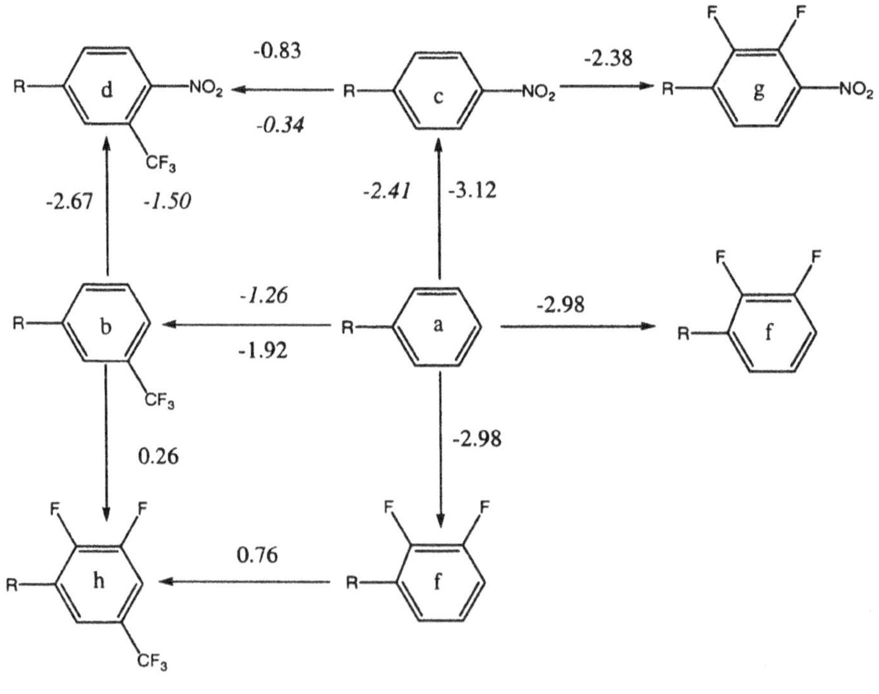

Figure 4. The free energy calculation results for all mutations. The numbers in italics are the relative binding free energies calculated from experimental K_i values. Other numbers are the relative binding free energies from the simulations. Units are kcal/mol.

6. CONCLUSIONS

Free energy calculations provided important insight into ligand binding to TS. The calculated relative binding free energies were in good agreement with experimental results.[1-4] In the most recent study, a combination of free energy calculations and molecular dynamics simulations was used to analyze an inhibitor series and provide a qualitative analysis of the free energy changes resulting from various aromatic substitutions. The results reproduced experimental values and rationalized the non-additivity in TS inhibitor binding. Moreover, the studies were extended to the design of new TS inhibitors with one of the compounds predicted to be an even more potent TS inhibitor. It is hoped that this work will inspire an interest in synthesizing this new compound.

7. REFERENCES

1. M. R. Reddy, R. J. Bacquet, D. Zichi, D. A. Matthews, K. M. Welsh, T. R. Jones, and S. Freer, Calculation of solvation and binding free energy differences for folate-based inhibitors of the enzyme thymidylate synthase, J. Am. Chem. Soc. **114**:10117 (1992).
2. T. Lee and P. A. Kollman, Theoretical studies suggest a new antifolate as a more potent inhibitor of thymidylate synthase, J. Am. Chem. **122**:4385 (2000).
3. G. Restelli, B. Thomas, P. A. Kollman, D. V. Santi, Insight into the specificity of thymidylate synthase from molecular dynamics and free energy perturbation calculations, J. Am. Chem. Soc. **117**:7213 (1995).
4. M. R. Reddy, M. D. Erion, and A. Agarwal, Free energy calculations: Use and limitations in predicting ligand binding affinities, in: Reviews in Computational Chemistry, vol. 16, K. B. Lipkowitz and D. B. Boyd, eds. VCH Publishers, New York (2000), pp. 217-304.
5. D. A. Pearlman, D. A. Case, J. W. Caldwell, W. S. Ross, T. E. Cheatham, S. Debolt, D. Ferguson, G. Seibel, and P. Kollman, AMBER, a package of computer programs for applying molecular mechanics, normal-mode analysis, molecular dynamics and free energy calculations to simulate the structural and energetic properties of molecules. Comput. Phys. Commun. **91**:1 (1995).
6. S. J. Weiner, P. A. Kollman, D. A. Case, C. Singh, C. Ghio, G. Alagona, S. Profeta, and P. K. A. Wiener, New force field for molecular mechanical simulation of nucleic acids and proteins, J. Amer. Chem. Soc. **106**:765 (1984).
7. W. Cornell, P. Cieplak, C. Bayly, I. Gould, K. Merz, D. Ferguson, D. Spellmeyer, T. Fox, J. Caldwell, and P. A. Kollman, Second generation force field for the simulation of proteins, nucleic acids, and organic molecules, J. Am. Chem. Soc. **117**:5179 (1995).
8. C. Bayly, P. Cieplak, W. Cornell, and P. A. Kollman, A well-behaved electrostatic potential based method using charge restraints for deriving atomic point changes: The RESP model, J. Phys Chem. **97**:10269 (1993).
9. D. A. Matthews, K. Appelt, S. J. Oatley, and N. H. Xuong, Stereochemical mechanism of action for thymidylate synthase based on the X-ray structure of the covalent inhibitory

ternary complex with 5-fluoro-2'-deoxyuridylate and 5,10-methylenetetrahydrofolate, J. Mol. Biol. **214**:923 (1990).

10. D. A. Matthews, J. E. Villafranca, P. A. Janson, W. W. Smith, K. Welsh, and S. Freer, Crystal structure of Escherichia coli thymidylate synthase containing bound 5-fluoro-2'-deoxyuridylate and 10-propargyl-5,8-dideazafolate, J. Mol. Biol. **214**:937 (1990).
11. J. S. Finer-Moore, E. B. Fauman, P. G. Foster, K. M. Perry, D. V. Santi, and R. J. Stroud, Refined structures of substrate-bound and phosphate-bound thymidylate synthase from lactobacillus casei, J. Mol. Biol. **232**:1101 (1993).
12. T. R. Jones, M. D. Varney, S. E. Webber, K. K. Lewis, G. P. Marzoni, C. L. Palmer, V. Kathardekar, K. M. Welsh, S. Webber, D. A. Matthews, K. Appelt, W. W. Smith, C. A. Janson, J. E. Villafranca, R. J. Bacquet, E. F. Howland, C. L. Booth, S. M. Herrmann, R. W. Ward, J. White, E. W. Moomaw, C. A. Bartlett, and C. A. Morse, Structure-based design of lipophilic quinazoline inhibitors of thymidylate synthase, J. Med. Chem. **39**:904 (1996).
13. R. Radmer and P. Kollman, The application of three approximate free energy calculations methods to structure based ligand design: Trypsin and its complex with inhibitors, J. Comput.-Aided Mol. Des. **12**:215 (1998).

Chapter 18

Dihydrofolate Reductase: Free Energy Calculations for the Design of Mechanism-Based Inhibitors

Jill E. Gready and Peter L. Cummins
Computational Molecular Biology and Drug Design Group, Division of Biochemistry and Molecular Biology, John Curtin School of Medical Research, Australian National University, Canberra, ACT 2601, Australia

1. INTRODUCTION

The enzyme dihydrofolate reductase (DHFR) catalyses the nicotinamide adenine dinucleotide phosphate (NADPH) dependent reduction of folate to tetrahydrofolate, and dihydrofolate (DHF) to tetrahydrofolate, and is a target for various antineoplastic and antibacterial drugs.[1, 2] DHFR has become a popular test case for computer-aided drug design. Consequently several molecular dynamics and free energy perturbation (MD/FEP) calculations on the DHFR-binding of the anti-cancer drug methotrexate and the antibacterial drug trimethoprim have been reported in the literature.[3-10] The majority of these studied small polar to nonpolar transformations of the 3-, 4- and 5-substituents of the phenyl ring in trimethoprim.[5-9] Fewer studies have been carried out on methotrexate[3, 4] or methotrexate derivatives[10] binding to DHFR. Due to the availability of experimental binding constants and the small structural differences involved, the trimethoprim derivatives appear to have been the dominant test case for new free energy methodologies as applied to DHFR-binding ligands. Gerber et al.[9] developed an efficient strategy based on free energy derivatives to estimate differences in ligand binding and applied it to the trimethoprim derivatives. Although orders of magnitude more efficient than the more rigorous FEP approach, there is some loss of accuracy. In addition to the MD/FEP and related calculations on relative binding free energies, some other methods not based on MD simulation have also been employed. An efficient empirical scoring function approach for estimating binding constants has been developed and used to predict the binding constants of DHFR-inhibitor complexes of trimethoprim derivatives,[11] while the Poisson-Boltzmann equation has been applied to the study of the pH-dependence of methotrexate inhibition of DHFR.[12]

As part of our strategy for the rational design of biologically active molecules, for a number of years we have been engaged in the use of MD simulation together with standard FEP and thermodynamic integration (TI) calculations to study the relative affinities of the novel mechanism-based 8-substituted pterins and N5-deazapterins for DHFR. Underpinning the rational design of these new types of substrates and inhibitors of DHFR is a detailed understanding of enzymic mechanism, i.e. how DHFR catalyses the reduction of folate and DHF. To address this question, we are currently using combined quantum mechanical and molecular mechanical (QM/MM) methodologies together with MD simulation and free energy (FEP and TI) methods. In the present chapter, we review our approach to the use of free energy calculations in the design of mechanism-based DHFR-binding ligands.

2. MECHANISM-BASED SUBSTRATES AND INHIBITORS OF DHFR

The design rationales we used in the development of mechanism-based 8-substituted-pterin (8-R-pterin) substrates of DHFR[13-19] suggested that 8-R-N5-deazapterins might well be inhibitors of DHFR. Central to the pterin-substrate design is substitution at the N8 position, which produces a more basic compound. Experimental work carried out in our laboratory has shown that 8-substituted-N5-deazapterins are indeed inhibitors of DHFR and display strong binding in ternary complexes with NADPH cofactor.[20] The non-8-R-substituted compound, N5-deazafolate, is also known to be an inhibitor and binds more tightly than folate to DHFR from *E. coli* and chicken liver.[21] Quantum chemical calculations of protonation behaviour and electronic spectra[17, 20] in solution indicate that N5-deazapterins (i.e.analogues of N5-deazafolate) protonate at N8, whereas N3 is the protonation site for the 8-R-substituted compounds,[13, 14, 17] findings also supported by experiment.[17, 19, 22] Spectral studies are also consistent with binding of the N3 protonated forms of 8-substituted N5-deazapterins in the active site of DHFR.[24] Structures of the N8-protonated and neutral forms of 6-substituted-N5-deazapterins (e.g. 6-methyl-N5-deazapterin and N5-deazafolate) are shown in Figure 1. The structures of N3-protonated and neutral 6-R'-8-R-N5-deazapterins (e.g. 6,8-dimethyl-N5-deazapterin) are shown in Figure 2. Both the 8-substituted (N3-protonated) and non-8-substituted (N8-protonated) cations are stabilized by the resonance-delocalized extended-guanidinium group.[13, 16, 17]

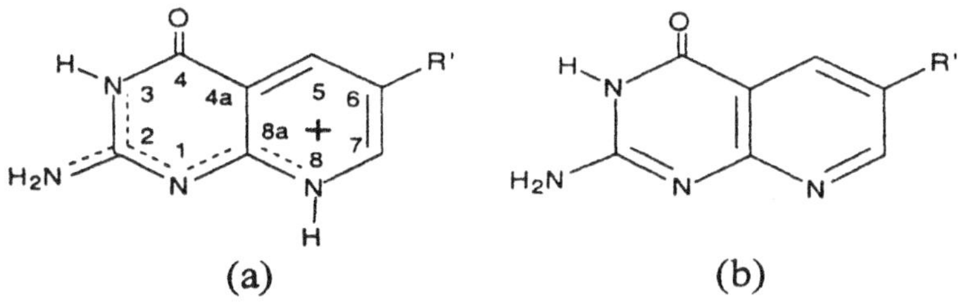

Figure 1. Structures for (a) the stable cation showing the extended guanidinium resonance, and (b) the neutral form of 6-R'-N5-deazapterins. In the inhibitor N5-deazafolate, the side chain is R' = methylene(*p*-aminobenzoyl)-L-glutamate.

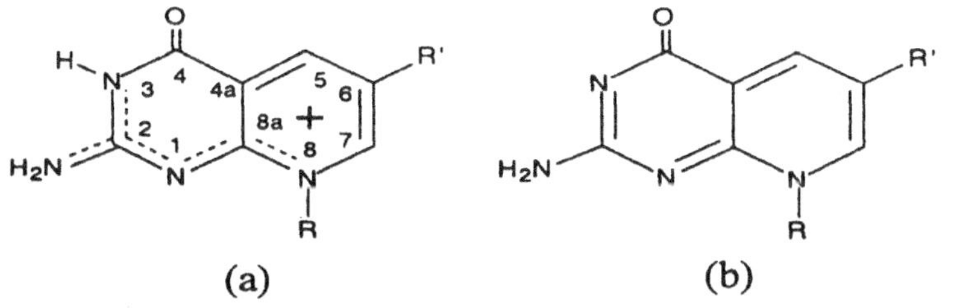

Figure 2. Structures for (a) the stable cation showing the extended guanidinium resonance, and (b) the neutral form of 6-R'-8-R-N5-deazapterins.

3. FREE ENERGY METHODS

The difference in binding affinity between two forms of a ligand A and B reduces to solving the equation for the free energy difference (Chapter 16) given by

$$\Delta\Delta G_{bind} = \Delta G_{com} - \Delta G_{aq} \tag{1}$$

where ΔG_{aq} is the free energy for the mutation of the ligand from A to B in aqueous solution and ΔG_{com} is for the corresponding mutation for the ligand bound in the solvated protein complex. The MM force field used in the simulations to compute ΔG_{com} and ΔG_{aq} has the general form

$$V = V_{bad} + V_{ele} + V_{vdW} \tag{2}$$

where V_{bad} includes all bond, angle and dihedral terms, V_{ele} is the electrostatic term arising from nonbonded coulomb interactions between atomic charges, and V_{vdW} is the nonbonded van der Waals (vdW) 6-12 Lennard-Jones potential interaction term. The free energy changes ΔG_{com} and ΔG_{aq} may be computed using either FEP or thermodynamic integration (TI) methods. A detailed account of these methodologies can be found in Chapter 2. Following the procedure outlined by Rao and Singh,[25] we usually evaluate the ΔG's in two steps. In the first step the free energy change due to mutation of the electrostatic terms (V_{ele}) only in the force field is calculated, followed by the free energy change due to mutation of the vdW terms (V_{vdW}) in the second step.

The practical difficulties inherent in the estimation of free energy differences by molecular simulation are well known and have been discussed in various review articles,[26-29] including this volume. Many of these difficulties are associated with the choice of initial conditions, subsequent MD simulation conditions, simulation times and the efficiency of configuration-space sampling. Note that actual X-ray structures of complexes formed between DHFR and 8-substituted-N5-deazapterins have not been determined. Consequently we must use starting coordinates for DHFR from available X-ray studies on complexes where other substrates or inhibitors are bound. These substrate or inhibitor molecules are then deleted from the structure and the 8-substituted-N5-deazapterin modeled into the binding site using the known binding-geometry information.[15-18] If the initial X-ray structure turns out to be a poor guess to the structure of the true complex, lengthy equilibration may be required, and a priori it is not clear how long the simulation must be run in order to obtain adequate statistics for a free energy determination. A number of free energy determinations should be made by varying the simulation time in order to test the validity of the results. The success of free energy simulations, therefore, depends critically on the development of protocols which are appropriate for a given problem. The other major concern for accurate free energy determination is the quality of the potential energy function and the inclusion of solvent. We have examined many of these issues involved in the calculation of ΔG_{com} and ΔG_{aq} for the binding of 8-substituted pterins and N5-deazapterins to DHFR.

4. FREE ENERGY OF SOLVATION

For molecules and molecular ions, such as the cations of 8-methyl-N5-deazapterin and 8-methyl-pterin, the charge distribution (which is represented in MD simulations by a set of discrete atomic charges) will be dependent on the chosen quantum chemical model. Differences in the charge distributions of these cations may influence both the relative binding and solvation thermodynamics. Consequently, we studied the relative solvation thermodynamics of similar DHFR-binding molecular ions.[30] Atomic charges were obtained from a fit to the molecular electrostatic potential (MEP) calculated using quantum chemical methods. Due to the importance of electrostatics in the binding and solvation of ions, we restricted our attention in this study to free energy changes arising from mutation of the intermolecular (i.e. cation-solvent) electrostatic terms (V_{ele}) in the MD force field.

In order to examine the effect of truncating the non-bonded interactions, simulations were performed using a box containing 804 water molecules. The free energies in Table 1 were computed using TI carried out over 80 ps simulation times. We found that TI gives essentially the same results as the FEP method. Note the cutoff radii are determined by the box size according to $R_c < d_{min}/2$ where d_{min} is the minimum box dimension. Calculations have also been carried out using a smaller number of solvent waters (approximately 400 molecules). If the interaction radius in the 804 water molecule system is kept the same as for the simulations performed using the smaller 400 water molecule system, i.e. 8 Å for both the cation-solvent and solvent-solvent interactions, the magnitude of the free energy change is increased by a small amount (ca. 0.1 kcal/mol). Water molecules which interact with the cation will also interact with water images due to the boundary condition. Many of these water images will in turn have been influenced by interactions with the cation and may, therefore, occupy unrealistic configurational states.[31] Consequently, in the simulations using the smaller box the dielectric relaxation of the solvent is effectively incomplete and the free energy change can be artificially raised, i.e. more positive, although in this particular case the effect appears to be almost negligible. This effect is known to be larger for the mutation of neutral to charged atoms in solution, i.e. the hydration of simple ions.[31]

Increasing the cation-solvent interaction radius from 8 to 10 Å while the solvent-solvent interactions remain at 8 Å results in only a small change (< 0.1 kcal/mol) in the free energy. However, if the solvent-solvent interaction radius is also increased to 10 Å the change in the free energy becomes +0.4 kcal/mol which is much larger than would be expected from the water-image effect. It appears that for this mutation the choice of cutoff radius for the solvent-solvent interactions is more important than the radius for the cation-

Table 1. Dependence of free energies (kcal/mol) on the residue-based cutoff radii R_c (Å) for generation of cation-solvent (c-s) and solvent-solvent (s-s) interaction lists. Free energy changes are given for the forward ΔG_f (i.e. 8-methyl-N5-deazapterin → 8-methyl-pterin) and reverse ΔG_r mutations of the electrostatic terms.

Simulation[a] R_c(c-s)	R_c(s-s)	ΔG_f	ΔG_r	$\Delta G_f + \Delta G_r$
8	8	-4.205	4.129	-0.076
10	8	-4.102	4.133	0.031
10	10	-3.730	3.778	0.048

[a] 804 TIP3P water molecules; AM1 optimized geometry; AM1 atomic charges; simulation time of 80 ps.

solvent interactions. The hydration free energies of simple ions are also found to be sensitive to the solvent-solvent interaction cutoff, but also very dependent on the ion-solvent cutoff.[31] In the present mutation, charge is conserved and the effective Hamiltonian for the change is dipolar (R^{-3}) whereas in the hydration studies[31] the charge state of the solute is being changed (with R^{-1} dependence). Also the ions considered here are much larger and, therefore, have a much smaller charge density compared with Na^+ or Cl^-. Consequently, we would expect the dependence of the free energy on the cation-solvent cutoff to be very much reduced.

The electrostatic free energy changes using the TIP3P or SPC water models and several levels of ab initio SCF charges were computed by performing integrations over 80 ps simulation times (Table 2). The 6-31G* and 6-311G** charges computed at AM1 optimized geometries (AM1/6-31G* and AM1/6-311G**) yield similar free energy changes. In all cases the ab initio charges give a free energy change which is more negative than that for AM1 charges computed using the same simulation conditions. The AM1 and AM1/6-31G free energy changes differ by 1 kcal/mol, while the free energies from simulations in which AM1/3-21G, AM1/6-31G* or AM1/6-311G** charges were used differ by only 5-10% from those obtained using the AM1 charges. Also, some variation in the sampling becomes apparent with the use of different solvent/cation models, as the hystereses in the SPC-AM1/6-31G and TIP3P-AM1/6-31G* calculations are notably larger than for the other 80 ps electrostatic mutations. Free energies computed using 3-21G geometries and 3-21G or 6-31G* charges (3-21G/3-21G and 3-21G/6-31G*) are 0.5 to 0.8 kcal/mol larger than those obtained at the AM1 geometries (AM1/3-21G and AM1/6-31G*). Other studies[32, 33] had reported more significant differences between free energies calculated using AM1 and 6-31G* charges. As we have used AM1 geometries and charges

Table 2. Dependence of free energies (kcal/mol) on the atomic charges computed at the AM1 and ab initio SCF level for both AM1 and 3-21G optimized geometries. Free energy changes are given for the forward ΔG_f (i.e. 8-methyl-N5-deazapterin → 8-methyl-pterin) and reverse ΔG_r mutations of the electrostatic terms.

Solvent[a]	Cation geometry/charges	ΔG_f	ΔG_r	$\Delta G_f + \Delta G_r$
TIP3P	AM1/AM1	-4.066	4.070	0.004
	AM1/3-21G	-4.557	4.531	-0.026
	3-21G/3-21G	-5.347	5.169	-0.178
	AM1/6-31G	-5.161	5.139	-0.022
	AM1/6-31G*	-4.204	4.400	0.196
	3-21G/6-31G*	-4.964	4.928	-0.036
	AM1/6-311G**	-4.335	4.306	-0.029
SPC	AM1/6-31G	-4.988	5.235	0.247
	AM1/6-311G**	-4.043	3.946	-0.097

[a] 389 water molecules; 8 Å residue-based cutoff for non-bonded interactions; simulation time of 80 ps.

extensively in studies of pterin and N5-deazapterin binding to DHFR, it is encouraging that for the present mutations between cations the results indicate the AM1 model gives free energies reasonably close (within ca. 1 kcal/mol) to those from simulations based on ab initio SCF geometries and charges.

The free energy was also calculated for the change 6,8-dimethyl-N5-deazapterin → 6-methyl-N5-deazapterin in aqueous solution using the FEP method.[34] In this simulation the 8-methyl substituent is changed to hydrogen. Like the CH → N mutation between pterin and N5-deazapterin discussed above, a non-polar hydrophobic group is changed to a polar one that can form H bonds with the solvent water molecules. However, the N-CH_3 → N-H mutation in the 8-position involves a more substantial change in volume. In this case the free energy change may well be largely dependent on the vdW parameters of the hydrogen at N8, rather than on the partial atomic charges. Thus, we calculated the free energy change for two values of the atomic hardness, $\varepsilon = 0.02$ and $\varepsilon = 0.0$ for the N8 hydrogen. The total free energy change, ΔG_{aq}, has been obtained as the sum of electrostatic and vdW contributions. The results in Table 3 show a relatively small free energy change for mutation of the coulomb energy terms V_{ele}. This is to be expected since the total ligand charge is conserved and the atomic partial charges[34] indicate only minor differences in the molecular charge distributions and, hence, MEPs of the two ligands. In contrast, mutation of the vdW terms gives a substantial free energy difference, which is very dependent on the choice for ε, i.e. the strength of the H bonding between N8-H and water. Care needs to be taken here in the interpretation of this result and the meaning of "electrostatic" and "vdW" free energy components as obtained by partial mutation of the potential energy function. Although the free energy change is obtained by mutation of vdW terms only, the origin of the H bonding remains largely electrostatic. The magnitude of the repulsive

Table 3. Free energy changes (kcal/mol) calculated for the mutation 6,8-dimethyl-N5-deazapterin → 6-methyl-N5-deazapterin in solution. Results for 80 ps simulation, with 200 ps simulation in parentheses.

ε^a	$\Delta G_{aq}(ele)^b$	$\Delta G_{aq}(vdw)^b$	$\Delta G_{aq}(total)^c$
0.02	-0.25 ± 0.01	-3.83 ± 0.07	-4.08 ± 0.08
	-0.25 ± 0.01	3.38 ± 0.02[d]	-3.63 ± 0.03
	-0.25 ± 0.01	(-3.48 ± 0.05)	(-3.73 ± 0.06)
0.00	-0.25 ± 0.01	-6.63 ± 0.02	-6.88 ± 0.03
	-0.25 ± 0.01	-5.56 ± 0.10[d]	-5.81 ± 0.11
	-0.25 ± 0.01	(-6.11 ± 0.06)	(-6.36 ± 0.07)

[a] vdW parameters ε = 0.0 and 0.02 kcal/mol for H(N8) of 6-methyl-N5-deazapterin. [b] Mean value and standard error from the FEP calculations (ref. 34). [c] Total free energy change is the sum of the electrostatic and vdW components. [d] Reverse mutation.

vdW energy depends on the strength of the coulomb energy. Strongly interacting molecules will have a significant overlap of electron density and, hence, also a large repulsive vdW energy term. Thus, the magnitude of electrostatic binding energy is reflected simply in the mutation of the vdW terms.

Brooks and Freischman[5, 6] have also carried out extensive FEP studies on the relative solution thermodynamics of a DHFR-binding drug, but involving multiple polar to nonpolar type mutations. They mutated the 3',4',5'-methoxybenzyl group of trimethoprim (TMP) to 4' (para)-ethyl (PET) or 3',4',5'-triethyl (TET) moieties. This O → CH_2 mutation is comparable with the N → CH mutation between pterin and N5-deazapterin. As expected the nonpolar ethyl derivatives were found to be less stable in solution than TMP. However, the calculated ΔG_{aq} are not additive with respect to the number of ethyl substituents as the $\Delta G_{aq} = 3.8 \pm 0.5$ kcal/mol for TMP → TET (three ethyl substituents) is only twice the $\Delta G_{aq} = 1.9 \pm 0.2$ kcal/mol for TMP → PET (one ethyl substituent).[6] In addition to the free energy, the enthalpy and entropy changes were also computed. There is a large contribution from the entropy in the mutation from TMP → PET, but not in the TMP → TET mutation, which is consistent with the 4'-methoxy group being more hydrophilic than the methoxy groups at the 3' or 5' positions.

5. FREE ENERGY OF BINDING

We have examined the convergence of the computed free energies and structures obtained by MD simulation by performing calculations under different conditions, i.e. initial X-ray coordinates, simulation times and force-field parameters for the mutation 6,8-dimethyl-N5-deazapterin → 6-methyl-N5-deazapterin for the enzyme-bound cation.[34] The MD/FEP simulations were carried out on several examples of DHFR from bacterial[35-37] and vertebrate[38-41] sources. Many of these X-ray crystal structures have the cofactor $NADP^+$ or NADPH bound, and all have either a known substrate or inhibitor molecule bound in the active site. For some of the complexes studied, the mutations were carried out in both directions, i.e. the free energy was calculated for the change 6,8-dimethyl-N5-deazapterin → 6-methyl-N5-deazapterin and also for the change 6-methyl-N5-deazapterin → 6,8-dimethyl-N5-deazapterin. As an additional source of error in the calculations derives from the approximations used to obtain potential energy functions in MD simulations, the sensitivity of the computed free energies and structures to the possible H-bond interaction was tested by carrying out simulations using two potential energy functions which differ only in the assignment of the atomic hardness parameter (ε) for the description of interactions between the enzyme and H(N8) of the 6-methyl-N5-deazapterin ligand.

The free energy change, ΔG_{com}, was obtained as the sum of electrostatic and vdW contributions, according to the respective mutations of electrostatic and vdW potential energy terms. The vdW contribution and total free energy terms are given in Table 4, where the change for a total mutation is given by ΔG_{com}(total) = ΔG_{com}(ele) + ΔG_{com}(vdW). The results in parentheses are for the vdW mutation performed over a 200 ps simulation time. Note that for all systems the electrostatic components, i.e. difference between ΔG_{com}(total) and ΔG_{com}(vdW) are quite small compared with the vdW components, as was found for the corresponding mutation for the cation in aqueous solution (Table 3).

The vdW component depends heavily on the enzyme source of the initial coordinates for the complexes and also on the ε parameter for H(N8). With the exception of the lcDHFR.ligand complex, the parameter $\varepsilon = 0$ kcal/mol consistently gives lower ΔG_{com}(vdW) values than $\varepsilon = 0.02$ kcal/mol. However, no definite trend emerges when comparing ΔG_{com}(vdW) for the different initial structures or when considering the presence or absence of NADPH cofactor. The relative thermodynamic stabilities of ligand binding, $\Delta\Delta G_{bind}$, also vary markedly, giving a range of values depending on initial coordinates and ε. The simulation performed on lcDHFR(ε=0) yields $\Delta\Delta G_{bind} = 1.37 \pm 0.15$ kcal/mol, i.e. the 6,8-dimethyl-N5-deazapterin complex is a factor of ca. 10 more stable than the 6-methyl-N5-deazapterin complex. In contrast, rhDHFR(ε=0) yields $\Delta\Delta G_{bind} = -3.66 \pm 0.21$ kcal/mol, i.e. the 6-methyl-N5-deazapterin complex is the more stable by a factor greater than 10^2.

The vdW mutation was repeated for several complexes using the longer simulation time of 200 ps. The results are included in Table 4 for comparison with the 80 ps simulations. There are only small (< 0.5 kcal/mol) differences between the 80 and 200 ps estimates of $\Delta\Delta G_{bind}$ for the two ecDHFR(ε = 0) simulations. Thus, regardless of whether the simulation time for the vdW mutation is 80 ps or 200 ps, values for $\Delta\Delta G_{bind}$ differ by almost 2 kcal/mol for the two initial ecDHFR structures (ε = 0). A small (< 0.5 kcal/mol) difference between the 80 and 200 ps simulations is also obtained for clDHFR.NADPH(ε=0). By contrast, on increasing the simulation time changes in $\Delta\Delta G_{bind}$ that are greater than 1 kcal/mol are obtained for lcDHFR(ε = 0), clDHFR(ε = 0.02) and rhDHFR(ε = 0). For lcDHFR the change on going from 80 ps to 200 ps is as high as 3 kcal/mol. The reverse 200 ps mutation for rhDHFR(ε = 0) gives a hysteresis of 1.8 kcal/mol which is approximately the same absolute magnitude as for the corresponding 80 ps mutation.[34] These results indicate that the free energies are generally still not converged after 200 ps.

Table 4. Free energy changes (kcal/mol) for the mutation 6,8-dimethyl-N5-deazapterin → 6-methyl-N5-deazapterin for the ligands bound to various dihydrofolate reductases with and without NADPH cofactor. Results are for 80 ps simulation, with 200 ps simulation in parentheses.

System	ε^a	$\Delta G_{com}(vdw)^b$	$\Delta G_{com}(total)^c$	$\Delta\Delta G_{bind}{}^d$
lcDHFR[e]	0.02	-6.05 ± 0.15	-6.68 ± 0.17	-2.60 ± 0.25
	0.00	-4.88 ± 0.10	-5.51 ± 0.12	1.37 ± 0.15
		(-7.42 ± 0.02)	(-8.05 ± 0.04)	(-1.69 ± 0.11)
lcDHFR.NADPH[e]	0.02	-6.53 ± 0.08	-6.96 ± 0.09	-2.88 ± 0.17
	0.00	-9.33 ± 0.05	-9.76 ± 0.06	-2.88 ± 0.09
ecDHFR[f]	0.02	-3.82 ± 0.10	-4.34 ± 0.11	-0.26 ± 0.19
	0.00	-7.90 ± 0.09	-8.42 ± 0.10	-1.54 ± 0.13
		(-7.34 ± 0.01)	(-7.86 ± 0.02)	(-1.50 ± 0.09)
ecDHFR[g]	0.02	-4.40 ± 0.01	-5.09 ± 0.03	-1.01 ± 0.11
	0.00	-5.82 ± 0.03	-6.51 ± 0.05	0.37 ± 0.08
		(-5.57 ± 0.07)	(-6.26 ± 0.09)	(0.10 ± 0.16)
ecDHFR.NADPH[g]	0.02	-3.79 ± 0.02	-4.28 ± 0.03	-0.20 ± 0.11
	0.00	-7.02 ± 0.11	-7.51 ± 0.12	-0.63 ± 0.15
clDHFR[h]	0.02	-2.79 ± 0.03	-3.33 ± 0.04	0.85 ± 0.12
		(-3.62 ± 0.07)	(-4.16 ± 0.08)	(-0.43 ± 0.14)
	0.00	-5.39 ± 0.02	-5.93 ± 0.03	0.95 ± 0.06
clDHFR.NADPH[h]	0.02	-3.60 ± 0.02	-4.03 ± 0.03	0.05 ± 0.11
	0.00	-5.35 ± 0.03	-5.78 ± 0.04	1.10 ± 0.07
		(-4.41 ± 0.01)	(-4.84 ± 0.02)	(1.52 ± 0.09)
rhDHFR[i]	0.02	-6.29 ± 0.07	-6.54 ± 0.08	-2.46 ± 0.16
	0.00	-10.29 ± 0.17	-10.54 ± 0.10	-3.66 ± 0.21
		(-7.34 ± 0.02)	(-7.59 ± 0.03)	(-1.23 ± 0.10)

[a] vdW parameters e = 0.0 and 0.02 kcal/mol for H(N8) of 6-methyl-N5-deazapterin. [b] Mean value and standard error from the FEP calculations (ref. 34). [c] Total free energy change is the sum of the electro-static and vdW components. [d] From ΔG_{aq}(total) in Table 3 and Equation 1. [e] DHFR.NADPH.methotrexate coordinates (ref. 35). [f] DHFR.-methotrexate coordinates (ref. 36). [g] DHFR.NADP$^+$.folate coordinates (ref. 37). [h] DHFR.NADP$^+$.biopterin coordinates (ref. 40). [i] DHFR.folate coordinates (ref. 41).

Figure 3 shows the correlation between thermodynamic stability ($\Delta\Delta G_{bind}$ results from Table 4) and the N8-oxygen distance in the 6-methyl-N5-deazapterin complexes obtained after mutation. This distance has a value of 2.7 Å in the X-ray structure of the rhDHFR.N5-deazafolate complex,[38] from which ligand protonation and H bonding are inferred, compared with a distance of 3.4 Å in the ecDHFR.N5-deazafolate X-ray structure,[42] suggesting binding of unprotonated ligand. A range of distances is obtained from the simulations, although the majority lie between 2.5 and 3.2 Å, which may be regarded as the H-bonded range.

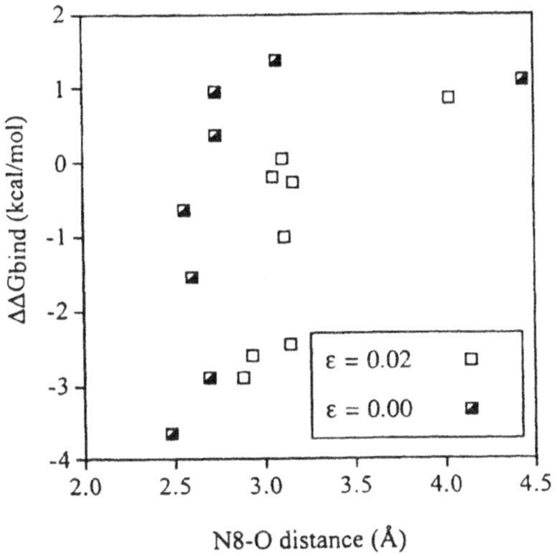

Figure 3. Correlation between the calculated $\Delta\Delta G_{bind}$ values (Table 4) and the N8-O distances (where O is the carbonyl oxygen of the conserved Ile/Leu residue in the active site of DHFR) in the 6-methyl-N5-deazapterin MD complexes.

These distances for $\varepsilon = 0$ are within ±0.2 Å of the experimentally observed distance (2.7 Å) in the rhDHFR.N5-deazafolate complex, whereas the parameter $\varepsilon = 0.02$ kcal/mol yields H bonds which are longer by 0.2 to 0.5 Å, consistent with previous experience that $\varepsilon(H) > 0$ leads to an underestimation of H-bond strength.[43] The majority of simulations with a N8-O distance within the H-bonded range yield $\Delta\Delta G_{bind} < 0$, suggesting that complexes involving H-bonded 6-methyl-N5-deazapterin are the thermodynamically more stable ones. Nevertheless, several of the H-bonded structures at the higher end of the H-bonded range are found to give $\Delta\Delta G_{bind} > 0$, namely lcDHFR($\varepsilon = 0$), ecDHFR($\varepsilon = 0$) and clDHFR($\varepsilon = 0$), while clDHFR.NADPH ($\varepsilon = 0.02$) gives $\Delta\Delta G_{bind} > 0$. Two of the structures [clDHFR($\varepsilon = 0.02$) and clDHFR.NADPH($\varepsilon = 0$)] are well outside the H-bonded range at 4.0 and 4.5 Å respectively, i.e. distances which are typical of the N8-O separations found in the initial 6,8-dimethyl-N5-deazapterin complexes before mutation. Both of these structures correspond to $\Delta\Delta G_{bind} > 0$. The 6-methyl-N5-deazapterin structures exhibit[34] a higher degree of overlap with the initial 6,8-dimethyl-N5-deazapterin structures than those where H-bonding at N8 takes place. Not surprisingly, therefore, H-bonding appears to involve quite a substantial displacement of the binding geometry from the initial non-H-bonded state. This fact is also reflected in the small difference between the free energies for the forward and reverse vdW mutations in the calculations on clDHFR.NADPH(ε = 0).[34] As no large variations in structure occur during the vdW mutation (in either direction), the free energies are not expected to be very dependent on the mutation path, unlike the H-bonded structures which often give rise to a much larger hysteresis. The H bond does, however, have time to form over the longer simulation of 200 ps. For the clDHFR.NADPH(ε = 0) complex, the 200 ps simulation gave an MD structure with an N8-O distance of 2.75 Å compared with 4.44 Å obtained from the 80 ps simulation (Figure 3). The simulations performed for clDHFR(ε = 0.02) yielded 3.21 Å for 200 ps and 4.03 Å for 80 ps.

6. LINEAR RESPONSE APPROXIMATION

The linear response approximation (LRA) allows for the estimate of a free energy change without the need to mutate potential energy terms as in the FEP or TI methods. Consequently, the LRA is more generally applicable and well suited for obtaining the absolute binding free energy of a ligand, i.e. the free energy relative to the unbound ligand in solution, rather than just the free energy difference $\Delta\Delta G_{bind}$ between chemically similar ligands. However, its accuracy depends on the availability of known binding constants. A free energy of binding ΔG_{bind} incorporating the ligand desolvation term is given within the LRA by[44]

$$\Delta G_{bind} = \frac{1}{2} < \Delta V_{l-s}^{ele} > + \ \alpha < \Delta V_{l-s}^{vdw} > \qquad (3)$$

where ΔV_{l-s}^{ele} and V_{l-s}^{vdw} are, respectively, the electrostatic and van der Waals energy differences between the ligand in the solvated protein and the unbound ligand in solution. The parameter α is determined by fitting a set of calculated free energies of binding with experimental data. To determine α for DHFR-binding ligands it was most suitable to choose compounds whose binding modes are expected to be close to that of the biopterin. The 8-methyl-, 5,8-dimethyl-, 6,8-dimethyl- and 7,8-dimethyl-N5-deazapterin cations were chosen as good candidates for this purpose as the substituents are small enough to avoid significant changes in the enzyme structure and they cover a range of free energy values from -6.4 to -8.7 kcal/mol. A value of α = -0.32 gives a good correlation with the experimental data, resulting in an average absolute error of 0.31 kcal/mol for the set of ligands used in the calibration.[45] This value was then used to estimate the binding free energy of a series of larger 8-substituted-N5-deazapterins in different binding pockets in the active site of DHFR that had been generated by simulated annealing, and, thus, predict the most likely binding geometry by correlating with experimentally determined binding constants.[45]

7. HYDROPHOBIC HYDRATION

The results of the calculations on the binding of trimethoprim to DHFR[5-8] suggest that entropic contributions and desolvation effects including solvent structural changes may play an essential role in the binding to DHFR. In particular, solvent structural changes may be significant in enhancing protein-ligand binding. As changes in the solvent structure surrounding a ligand may be influenced by the presence of hydrophobic groups, we have studied the relative binding free energies of methyl-substituted 8-methylpterins and 8-methyl-N5-deazapterins.[46]

The relative free energies of the various methylated ligands in Figure 4 were computed as the sum of electrostatic and vdW mutations using the FEP method and over different mutation pathways.[46] The free energies computed over the various pathways were then averaged to obtain the values given in Table 5, with the corresponding standard deviation for the error estimate. The combined effects of sampling errors in ΔG_{aq} and ΔG_{com} yielded uncertainties larger than 1 kcal/mol in the relative thermodynamic stabilities of ligand binding of the different methylated forms.

Errors of this magnitude make the useful prediction of free energies a difficult task, when differences of only one to three kcal/mol are involved. Nevertheless, within the error limits of the computed free energy differences, the trend is that relative to 8-methyl-N5-deazapterin or 8-methyl-pterin, the compounds methyl substituted in the 5, 6 or 7 positions are thermodynamically more stable when bound to DHFR largely by virtue of a hydrophobic effect, i.e. methyl substitution reduces the affinity of the ligand for the solvent more than it reduces affinity for the DHFR active-site. The stability of ligand binding to DHFR appears to be optimal with a 6-methyl substituent: additional 5-methyl and/or 7-methyl substitution has little effect

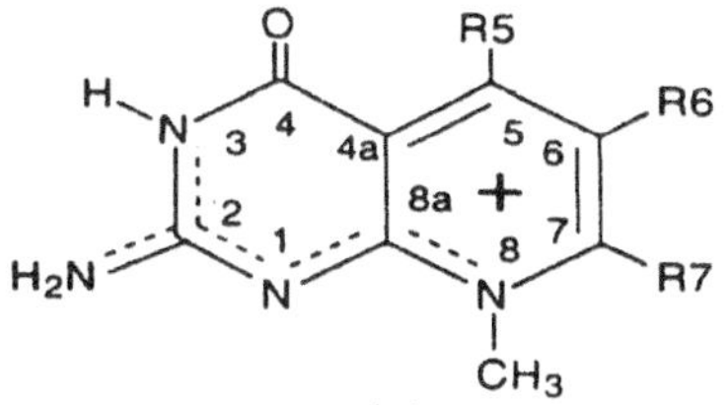

(a)

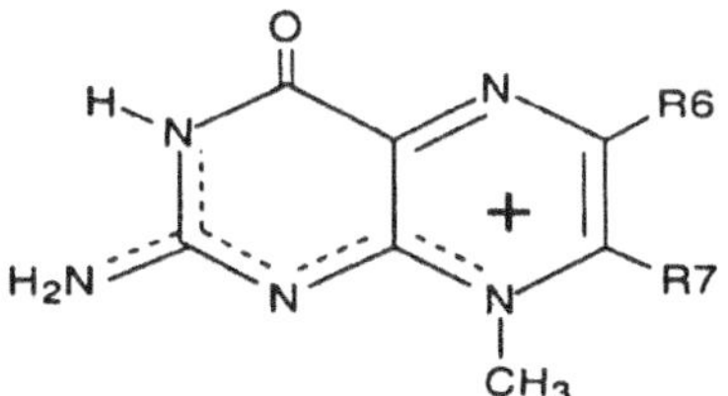

(b)

Figure 4. Structures for cations (i.e. N3 protonated forms) of the 8-methyl substituted (a) N5-deazapterins and (b) pterins: R5, R6 and R7 are either H or CH_3.

Table 5. Free energies (kcal/mol) relative to 8-methyl-N5-deazapterin

Ligand	ΔG_{aq} [a]	ΔG_{com} [a]	$\Delta\Delta G_{bind}$
5,6,7,8-tetramethyl-N5-deazapterin	3.56±0.24	1.28±0.36	-2.28±0.60
6,7,8-trimethyl-N5-deazapterin	2.43±0.47	0.43±0.68	-2.00±1.15
5,7,8-trimethyl-N5-deazapterin	2.63±0.23	1.02±0.57	-1.61±0.80
5,6,8-trimethyl-N5-deazapterin	3.14±0.24	1.14±0.36	-2.00±0.60
7,8-dimethyl-N5-deazapterin	1.56±0.34	-0.05±0.63	-1.61±0.97
6,8-dimethyl-N5-deazapterin	1.57±0.32	-0.65±0.55	-2.23±0.87
5,8-dimethyl-N5-deazapterin	2.01±0.24	1.18±0.52	-0.83±0.76
6,7,8-dimethyl-pterin	-1.75±0.26	-3.94±0.63	-2.19±0.89
7,8-dimethyl-pterin	-2.82±0.42	-3.87±0.88	-1.05±1.30
6,8-dimethyl-pterin	-2.11±0.29	-4.14±0.68	-2.03±0.97
8-methyl-pterin	-4.08±0.26	-4.01±0.39	0.07±0.65

[a]Mean value and standard error from the FEP calculations (ref. 46).

on the strength of binding. This saturation effect is most likely due to the fact that the methyl substituents are in close proximity to one another. Thus, the first hydrophobic substituent effectively produces a solvent cage in its vicinity[25] which allows adjacent methyl groups to be accommodated with a much smaller free energy change. Also, with regard to the stability of ligand binding, the FEP calculations indicate only minor differences between similarly substituted pterins and N5-deazapterins, e.g. the binding constants of 6,8-dimethylpterin and 6,8-dimethyl-N5-deazapterin are predicted to be of the same order of magnitude.

8. ROLE OF SOLVENT IN LIGAND BINDING IN THE ACTIVE SITE OF DHFR

In MD simulation studies of NADPH and $NADP^+$ binding, solvent interactions in the active site of DHFR were found to have a significant effect on the relative binding thermodynamics.[47] However, most studies on DHFR-ligand binding have used limited numbers of water molecules.[3-10, 34, 46] We have examined how different ways of including solvent influence the structural details of the active-site region and also the computed free-energy change for the mutation between the substrate 8-methyl-pterin and the inhibitor 8-methyl-N5-deazapterin.[48]

In the method proposed by Solmajer and Mehler[49] for modeling the effects of bulk solvent in protein simulations, the functional form for the configurational energy of the system is obtained by adding an external potential V_{ext} for restraining the dynamics of solvent molecules to the standard potential energy function in Equation 1 to give

$$V = V_{bad} + V_{ele} + V_{vdW} + V_{ext} \quad (4)$$

The introduction of the external potential V_{ext} in Equation 4 is designed to mimic the effect of the surrounding (implicit) bulk solvent on the system by restricting the movement of any explicit water molecules.[49] Thus, V_{ext} is interpreted as arising from the force exerted on the explicit atoms by the implicit surrounding bulk solvent. This restraining potential has the simple harmonic form,[49]

$$V_{ext} = \sum_i \frac{k_i}{2}(R_{ix})^2 \quad (5)$$

where R_{ix} is the distance from atom i of a water molecule to a fixed reference point x, and k_i is the empirically-determined force constant. We consider two models for the inclusion of bulk-solvent effects in the MD simulations, each based on the use of the external potential V_{ext} given by Equation 5 with the same force constant $k = k_i$ assumed for all atoms. In the capped water (CW) model, R_{ix} is the distance of a water molecule's atom i from the ligand center minus the radius of the spherical solvation shell (R_w). Thus, the CW model simply prevents water from escaping the dynamics region, but otherwise the water molecule dynamics is not directly affected by an external potential. This model has been widely used in active-site simulations.[28] We use a value of 0.6 kcal/mol/Å.^{2} for this force constant.[48] In the "tethered" water (TW) model, R_{ix} is the distance of the water oxygen from its initial bulk solvent or crystallographic positions. Taking account of theoretical (MD simulation) studies[50] that suggest water dipoles align tangentially to protein surfaces, we apply V_{ext} only to the oxygen atoms, allowing water molecules to attain their preferred orientations without artificially hindering their librational degrees of freedom.

We found that smaller shells of dynamically-restrained water molecules yielded active sites that are in broad structural terms similar to those obtained using much larger solvent shells of unrestrained molecules, without the artifacts and distortions that were apparent when restraining potentials were not used.[48] This trend suggests that shells of restrained water would be an efficient model for bulk solvent in both active-site MD simulations and unconstrained-protein dynamics.[49] Moreover, we found that the value of

Table 6. Free energies (kcal/mol) calculated using solvated-protein models.

			ΔG_{com}[d]	
Model[a]	R_w[b]	N_w[c]	9 Å[e]	No cutoff[e]
CW	16	46	-4.88 ± 0.02	-3.74 ± 0.13
CW	22	392	-3.83 ± 0.04	-2.98 ± 0.25
CW	26	1033	-4.19 ± 0.03	-3.04 ± 0.07
TW (k=2.5)	16	46	-4.22 ± 0.19	-4.01 ± 0.08
TW (k=0.005)	16	46	-4.78 ± 0.01	-3.85 ± 0.18
TW (k=0.005)	22	392	-4.23 ± 0.30	-2.56 ± 0.04
TW (k=0.005)	26	1033	-3.43 ± 0.12	-2.67 ± 0.23

[a]CW = capped water, TW = tethered water (see text). k = force constant for restraining potential (kcal/mol/Å^2). [b]Radius (Å) of solvation sphere. [c]Numbers of dynamical water molecules within solvation sphere. [d]Mean and standard error for the forward (i.e. 8-methyl-N5-deazapterin → 8-methylpterin) and reverse mutation of the electrostatic force field
[e]Cutoff for protein-ligand and solvent-ligand interaction; all other interactions are subject to a 9 Å cutoff.

0.005 kcal/mol/Å^2 for the restraint used by Solmajer and Mehler[49] in BPTI simulations works quite well for the active site of DHFR. This value gave satisfactory results over a range of solvent shell sizes. In accord with the BPTI calculations,[49] larger values of the force constant resulted in protein structures that are too similar to the initial X-ray structure.

It is known from our other studies on 8-substituted pterins and N5-deazapterins that the contributions to the free energies due to mutation of vdW terms is quite small and these contributions cancel when differences are taken.[30] Consequently, in Table 6 we report the free-energy change for mutation of electrostatic terms only, with a 9 Å cutoff, and no cutoff for ligand-protein and ligand-solvent interactions. As may be seen, a range of binding free energies spanning ~2 kcal/mol is obtained. We have also performed other calculations with a cutoff of 8 Å for the nonbonded interactions (see Table 5).[46] However, it is clear from the results in Table 6 that substantial errors (ca. 1.5 kcal/mol) may arise from truncation of the long-range electrostatic forces.

With no cutoff for the nonbonded interactions between ligand and the rest of the system, the TW (R_w = 22 Å) model gives the smallest free energy change of -2.56 ± 0.04 kcal/mol, while the largest change (-4.01 ± 0.08 kcal/mol) is obtained for the unsolvated CW (R_w = 16 Å) model. The solvated models give absolute values of free energies about 1 kcal/mol below those obtained from the unsolvated (R_w = 16 Å) models. However, the variation of the free energy with the type of solvent model is much less than 1 kcal/mol. The difference between CW and TW models is ca. 0.4 kcal/mol, while the difference between R_w = 22 Å and R_w = 26 Å for both models is only ca. 0.1 kcal/mol. These results suggest that while the free energies are clearly affected by the presence of explicit solvent, the fine details of the solvent distribution and protein structural changes are of lesser importance. This conclusion also tends to be supported by comparing the results obtained for the tethering potentials k = 2.5 and k = 0.005. Although the results for k = 0.005 (R_w = 16 Å) deviate more from the initial X-ray structure as measured by overall rms deviations,[48] the difference in the computed free energy is less than 0.2 kcal/mol. Note, however, that significantly larger free energy differences between solvent/protein structures are obtained with the 9 Å cutoff.

The relative thermodynamic stability of the binding is given by the difference for ligand bound to the DHFR complex (ΔG_{com}) and free ligand in solution (ΔG_{aq}). This latter solvation free energy has been calculated previously to be in the range 3.7 to 4.2 kcal/mol, using parameter sets derived from the AM1 model for the ligands.[30, 46] Assuming a value of -4.0 kcal/mol for ΔG_{aq} (AM1 in Table 1) a value of -2.7 kcal/mol (TW model, R_w = 26 Å, k=0.005 in Table 6) for ΔG_{com} gives a difference (ΔG_{com} - ΔG_{aq}) in binding free energy of 1.3 kcal/mol, which may be compared with the experimental value of ± 0.2 kcal/mol.[15, 19, 20]

9. CATALYTIC MECHANISM OF DHFR

DHFR catalyses the hydride-ion transfer between the nicotinamide adenine dinucleotide phosphate (NADPH) cofactor and a substrate molecule (S) according to

$$NADPH + S + H^+ \rightarrow NADP^+ + SH_2 \qquad (6)$$

where the normal substrates are folate and dihydrofolate. Note that the hydride-ion transfer is pH dependent requiring the transfer of a proton to the substrate. As the natural substrates, folate and dihydrofolate (DHF), are unprotonated when bound to the enzyme at physiological pH, how the proton finds its way to the substrate in the active site appears to be critical to an understanding of the mechanism of activation towards hydride-ion transfer. The novel 8-substituted pterins, which, unlike folate and DHF, have a high pK_a (ca. 5.5) in solution and, on the basis of both experimental and theoretical studies to date, bind strongly to the active site in the protonated form. This provides strong evidence for a pre-protonation mechanism, i.e. the enzyme-bound substrate exists in the N3 protonated form (SH^+). We have used combined QM and molecular mechanics (QM/MM) methods to compute the free energy change for the hydride-ion transfer step between NADPH and the 8-methylpterin substrate.[51] Of critical importance in understanding the reaction mechanism and ligand binding process is the protonation behaviour of the functional groups within the active site of DHFR,[3, 12, 52, 53] particularly the relative protonation energies of various sites in the binding of the natural substrates folate and dihydrofolate.[54]

10. FUTURE PROSPECTS FOR BINDING FREE ENERGY STUDIES ON DHFR

As we have found for DHFR-binding ligands, the results of free energy calculations often depend on variations in simulation conditions. The free energy changes due to mutation of the electrostatic terms converge relatively quickly, i.e. over short (50 ps to 100 ps) simulation times. However, where larger structural perturbations are involved, such as may occur in the mutation of the vdW terms, sampling becomes more problematic. Structural changes (vdW terms) that involve creation and annihilation of H bonds, e.g. 6,8-dimethyl-N5-deazapterin → 6-methyl-N5-deazapterin, in the enzyme require substantially longer simulation times than those used to date (200 ps). Although free energy contributions due to mutation of the electrostatic potential energy terms converge relatively quickly they are sensitive to the atomic charges and cutoffs used for the neglect of non-bonded interactions. This applies not only to the ligand-solvent interactions but also solvent-solvent interactions. Even for relatively small enzymes such as DHFR, large numbers of solvent molecules may be required in order to predict relative free energies accurately. With ever increasing computer power, these size-related restrictions should no longer be a problem for a 150 to 200 residue enzyme such as DHFR.

While increases in the speed of computer hardware is promising for the future of free energy calculations on ligand binding to DHFR there are other more methodologically-based issues that need addressing. The binding geometries obtained from the modeling of large side-chains in the active site by simulated annealing require verification by X-ray crystallography. In general, our studies on DHFR indicate that moderate variations in force field parameters may also have a significant effect on the computed free energies. The use of MM potentials in itself limits the application of the FEP and TI methodology. The standard MM energy term V_{bad} (Equation 1) cannot accurately predict the difference between the highly conjugated pterin and N5-deazapterin ring systems. Our strategy has been to simply not include these terms in the calculation of free energy differences. This amounts to assuming that the change in the internal energy of the molecular fragment being mutated is about the same in both solution and enzyme-bound states and can, therefore, be safely neglected. Consequently, studies of DHFR-binding ligands have not yet proceeded beyond small polar to nonpolar transformations. The mutation (FEP and TI) approach is less general for QM methods but the LRA method can be adapted to a larger number of problems. As we are currently building on experience in using QM/MM methods in the study of the catalytic mechanism, the use of QM/MM-based methods combined with faster computers may also offer some future prospects for rational design of DHFR inhibitors.

Acknowledgments

Most of this work was supported by The National Health and Medical Research Council (NHMRC) of Australia.

11. REFERENCES

1. R. L. Blakley, Dihydrofolate reductase, in: Folates and Pterins, Chemistry and Biochemistry of Folates, vol. 1, R. L. Blakley and S. J. Benkovic, eds., Wiley, New York (1984), pp. 191-253.
2. J. Kraut and D. A. Matthews, Dihydrofolate reductase, in: Biological Macromolecules and Assemblies, vol. 3, F. A. Jurnak and A. McPherson, eds., Wiley, New York (1987), pp. 1-72.
3. U. C. Singh, Probing the salt bridge in the dihydrofolate reductase-methotrexate complex by using the coordinate-coupled free-energy perturbation method, Proc. Natl. Acad. Sci. USA **85**:4280 (1988).
4. U. C. Singh, and S. J. Benkovic, A free energy perturbation study of the binding of methotrexate to dihydrofolate reductase, Proc. Natl. Acad. Sci. USA **85**:9519 (1988).
5. C. L. Brooks, Thermodynamic calculations on biological molecules, Int. J. Quantum Chem. Quantum Biol. Symp. **15**:221 (1988).
6. C. L. Brooks and S. H. Fleischman, A theoretical approach to drug design. 1. Relative solvation thermodynamics for the antibacterial compound trimethoprim and ethyl derivatives substituted at the 3',4' and 5' positions, J. Am. Chem. Soc. **112**:3307 (1990).
7. S. H. Fleischman and C. L. Brooks, Protein-drug interactions: Characterization of inhibitor binding in complexes of DHFR with trimethoprim and related derivatives, Proteins: Struct. Funct. Genet. **7**:52 (1990).
8. J. J. McDonald and C. L. Brooks, Theoretical approach to drug design. 2. Relative thermodynamics of inhibitor binding by chicken dihydrofolate reductase to ethyl derivatives of trimethoprim substituted at 3',4'- and 5'-positions, J. Am. Chem. Soc. **113**:2295 (1991).
9. P. R. Gerber, A. E. Mark, and W. F. van Gunsteren, An approximate but efficient method to calculate free energy trends by computer simulation - application to dihydrofolate reductase inhibitor complexes, J. Comput.-Aided Mol. Design **7**:305 (1993).
10. J. Marelius, M. Graffnernordberg, T. Hansson, A. Hallberg, and J. Åqvist, Computation of affinity and selectivity - binding of 2,4-diaminopteridine and 2,4-diaminoquinazoline inhibitors to dihydrofolate reductases, J. Comput.-Aided Mol. Design **12**:119 (1998).
11. H. -J. Bohm, The development of a simple empirical scoring function to estimate the binding constant of a protein-ligand complex of known three-dimensional structure, J. Comput.-Aided Mol. Design **8**:243 (1994).
12. W. R. Cannon, B. J. Garrison, and S. J. Benkovic, Consideration of the pH-dependent inhibition of dihydrofolate reductase by methotraxate, J. Mol. Biol. **271**:656 (1997).
13. J. E. Gready, Theoretical studies on pteridines. 2. Geometries, tautomer, ionisation, and reduction energies of substrates and inhibitors of dihydrofolate reductase, J. Comput. Chem. **6**:377 (1985).
14. J. E. Gready, Theoretical studies on the activation of the pterin cofactor in the catalytic mechanism of dihydrofolate reductase, Biochemistry **24**:4761 (1985).

15. V. Thibault, M. J. Koen, and J. E. Gready, Enzymic properties of a new mechanism-based substrate for dihydrofolate reductase, Biochemistry **28**:6042 (1989).
16. J. E. Gready, Design of new mechanism-based substrates for dihydrofolate reductase, in: Chemistry and Biology of Pteridines 1989, H.-Ch. Curtius, S. Ghisla, and N. Blau, eds, de Gruyter, Berlin (1990), pp. 23-30.
17. J. E. Gready, P. L. Cummins, and P. Wormell, Computer-aided design of mechanism-based pterin analogues and MD/FEP simulations of their binding to dihydrofolate reductase, in: Advances in Experimental Medicine and Biology, (Chemistry and Biology of Pteridines and Folates), vol. 338, J. E. Ayling, M. G. Nair, and C. M. Baugh, eds, Plenum, New York (1993), pp. 487-492.
18. P. L. Cummins and J. E. Gready, Novel mechanism-based substrates of dihydrofolate reductase and the thermodynamics of ligand binding: A comparison of theory and experiment for 8-methylpterin and 6,8-dimethylpterin, Proteins: Structure, Function, and Genetics **15**:426 (1993).
19. M. T. G. Ivery and J. E. Gready, Structure-activity relationships for the 8-alkylpterins: A new class of mechanism-based substrates for dihydrofolate reductase (DHFR), Biochemistry **34**:3724 (1995).
20. M. T. G. Ivery and J. E. Gready, Structure-activity relationships and pH dependence of binding of 8-alkyl-N5-deazapterins to dihydrofolate reductase, J. Med. Chem. **37**:4211 (1994).
21. S. R. Stone, J. A. Montgomery, and J. F. Morrison, Inhibition of dihydrofolate reductase from bacterial and vertebrate sources by folate, aminopterin, methotrexate and their 5-deaza analogues, Biochem. Pharmacol. **33**:175 (1984).
22. P. Wormell and J. E. Gready, Electronic spectra of some pterins and deazapterins, Chem. Phys. **179**:55 (1994).
23. M. T. G. Ivery and J. E. Gready, An improved procedure for the preparation of 2-amino-8-alkylpyrido-[2,3-*d*]pyrimidin-4(3*H*)-ones (8-alkyl-N5-deazapterins), J. Heterocyc. Chem. **31**:1385 (1994).
24. S. -S. Jeong and J. E. Gready, Ionization state and pK_a of pterin-analogue ligands bound to dihydrofolate reductase, Eur. J. Biochem. **221**:1055 (1994).
25. B. G. Rao and U. C. Singh, Hydrophobic hydration: A free energy perturbation study, J. Am. Chem. Soc. **111**:3125 (1989).
26. W. F. van Gunsteren, Methods for calculations of free energies and binding constants: Successes and problems, in: Computer Simulation of Biomolecular Systems, W. F. van Gunsteren and P. K. Weiner, eds., ESCOM, Leiden (1989), pp. 27-59.
27. D. A. Pearlman and P. A. Kollman, Free energy perturbation calculations: Problems and pitfalls along the gilded road, in: Computer Simulation of Biomolecular Systems, W. F. van Gunsteren and P. K. Weiner, eds., ESCOM, Leiden (1989), pp. 101-119.
28. C. A. Reynolds, P. M. King, and W. G. Richards, Free energy calculations in molecular biophysics, Mol. Phys. **76**:251 (1992).
29. W. F. van Gunsteren, Molecular dynamics studies of proteins, Curr. Opin. Struct. Biol. **3**:277 (1993).
30. P. L. Cummins and J. E. Gready, Thermodynamic integration calculations on the relative free energies of complex ions in aqueous solution: Application to ligands of dihydrofolate reductase, J. Comput. Chem. **15**:704 (1994).
31. T. P. Straatsma and J. C. Berendsen, Free energy of ionic hydration: Analysis of a thermodynamic integration technique to evaluate free energy differences by molecular dynamics simulations, J. Chem. Phys. **89**:5876 (1988).
32. C. A. Reynolds, J. W. Essex, and W. G. Richards, Errors in free-energy perturbation calculations due to neglecting the conformational variation of atomic charges, Chem. Phys. Lett. **199**:257 (1992).
33. S. G. Lister, C. A. Reynolds, and W. G. Richards, Theoretical calculation of electrode potentials: Electron-withdrawing compounds, Int. J. Quant. Chem. **41**:293 (1992).

34. P. L. Cummins and J. E. Gready, Influence of starting coordinates in free energy simulations of ligand binding to dihydrofolate reductase, Mol. Sim. **15**:155 (1995).
35. T. J. Bolin, D. J. Filman, D. A. Matthews, R. C. Hamlin, and J. Kraut, Crystal structures of *Escherichia coli* and *Lactobacillus casei* refined at 1.7 angstroms resolution. I. General features and binding of methotrexate, J. Biol. Chem. **257**:13650 (1982).
36. E. E. Howell, J. E. Villafranca, M. S. Warren, S. J. Oatley, and J. Kraut, Functional role of aspartic acid-27 in dihydrofolate reductase revealed by mutagenesis, Science **231**:1123 (1986).
37. C. Bystroff, S. J. Oatley and J. Kraut, Crystal structure of *Escherichia coli* dihydrofolate reductase: The $NADP^+$ holoenzyme and the folate.$NADP^+$ ternary complex. Substrate binding and a model for the transition state, Biochemistry **29**:3263 (1990).
38. J. F. Davies, T. J. Delcamp, N. J. Prendergast, V. A. Ashford, J. H. Freisheim, and J. Kraut, Crystal structures of recombinant dihydrofolate reductase complexed with folate and 5-deazafolate, Biochemistry **29**:9467 (1990).
39. D. A. Matthews, T. J. Bolin, J. M. Burridge, D. J. Filman, K. N. Volz, B. T. Kaufman, C. R. Beddell, J. N. Champness, D. K. Stammers, and J. Kraut, Refined crystal structure of *Escherichia coli* and chicken liver dihydrofolate reductase containing bound trimethoprim, J. Biol. Chem. **260**:381 (1985).
40. M. A. McTigue, J. F. Davies, B. T. Kaufman, and J. Kraut, Crystal structure of chicken liver dihydrofolate reductase complexed with $NADP^+$ and biopterin, Biochemistry **31**:7264 (1992).
41. C. Oefner, A. D'Arcy, and F. K. Winkler, Crystal structure of human dihydrofolate reductase complexed with folate, Eur. J. Biochem. **174**:377 (1988).
42. V. M. Reyes, M. R. Sawaya, and J. Kraut, Isomorphous crystal structures of *Escherichia coli* dihydrofolate reductase complexed with folate, 5-deazafolate and 5,10-dideazatetrahydrofolate: Mechanistic implications, Biochemistry **34**:2710 (1995).
43. A. T. Hagler, J. R. Maple, T. S. Thacher, G. B. Fitzgerald, and U. Dinur, Potential energy functions for organic and biomolecular systems, in: Computer Simulation of Biomolecular Systems, W. F. van Gunsteren and P. K. Weiner, eds., ESCOM, Leiden (1989), pp. 149-167.
44. J. Aqvist, C. Medina, and J.-E. Samuelsson, New method for predicting binding affinity in computer-aided drug design, Protein Eng. **7**:385 (1994).
45. A. -D. Gorse and J. E. Gready, Molecular dynamics simulations of the docking of substituted N5-deazapterins to dihydrofolate reductase, Protein Eng. **10**:23 (1997).
46. P. L. Cummins and J. E. Gready, Computer-aided drug design: A free energy perturbation study on the binding of methyl-substituted pterins and N5-deazapterins to dihydrofolate reductase, J. Comput.-Aided Mol. Design **7**:535 (1993).
47. P. L. Cummins, K. Ramnarayan, U. C. Singh, and J. E. Gready, Molecular dynamics / free energy perturbation study on the relative affinities for the binding of reduced and oxidized NADP to dihydrofolate reductase, J. Am. Chem. Soc. **113**:8247 (1991).
48. P. L. Cummins and J. E. Gready, Solvent effects in active-site molecular dynamics simulations on the binding of 8-methyl-N5-deazapterin and 8-methyl-pterin to dihydrofolate reductase, J. Comput. Chem. **17**:1598 (1996).
49. T. Solmajer and E. L. Mehler, Modeling solvent effects in molecular dynamics simulations of proteins, Int. J. Quant. Chem. **44**:291 (1992).
50. P. Ahlström, O. Teleman, and B. Jönsson, Molecular dynamics simulation of interfacial water structure and dynamics in a parvalbumin solution, J. Am. Chem. Soc. **110**:4198 (1988).
51. P. L. Cummins and J. E. Gready, Molecular dynamics and free energy perturbation study of hydride-ion transfer step in dihydrofolate reductase using combined quantum and molecular mechanical model, J. Comput. Chem. **19**:977 (1998).

52. P. L. Cummins and J. E. Gready, Combined quantum and molecular mechanics (QM/MM) study of the ionization state of 8-methylpterin substrate bound to dihydrofolate reductase, J. Phys. Chem. B **104**:4503 (2000).
53. P. L. Cummins and J. E. Gready, QM/MM and SCRF studies of the ionization state of 8-methylpterin substrate bound to dihydrofolate reductase: Existence of a low-barrier hydrogen bond, J. Mol. Graph. Mod. **18**:42 (2000).
54. P. L. Cummins and J. E. Gready, Energetically most likely substrate and active-site protonation sites and pathways in the catalytic mechanism of dihydrofolate reductase, J. Am. Chem. Soc. **123**:3418 (2001).

Chapter 19

Adenosine Deaminase: Calculation of Relative Hydration Free Energy Differences

Mark D. Erion and M. Rami Reddy
Metabasis Therapeutics, Inc., San Diego, CA 92121

1. INTRODUCTION

Adenosine deaminase (ADA) catalyzes the deamination of adenosine to inosine and ammonia (Figure 1). Humans genetically deficient in ADA exhibit severe B- and T-cell deficiencies.[1] Accordingly, ADA inhibitors have long been sought as potential immunosuppressant agents for preventing organ transplant rejection or treating rheumatoid arthritis.[2] ADA inhibitors have also been evaluated as potential antiischemic agents based on the premise that inhibition of ADA elevates local adenosine levels which in turn limits the tissue damage resulting from a heart attack, stroke or other conditions associated with a brief period of decreased blood flow and local oxygen deficiency.[3]

The natural product 2'-deoxycoformycin (**1**, Figure 2) is an extraordinarily potent ADA inhibitor (K_i = 10 pM).[4] The high binding

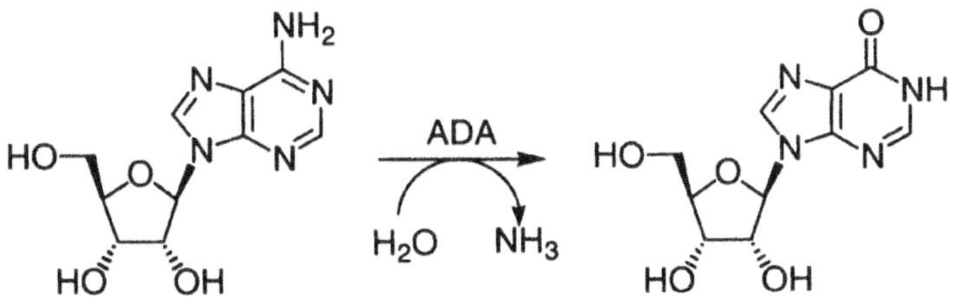

Figure 1. Adenosine deaminase reaction

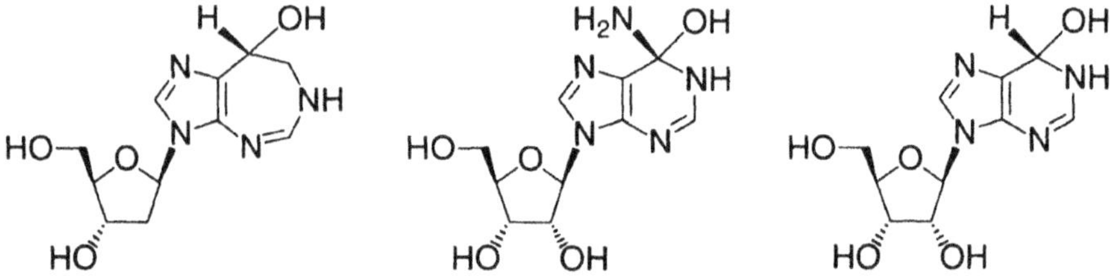

1 2 3

Figure 2. Adenosine deaminase inhibitors and tetrahedral intermediate **2**.

affinity of **1** is postulated to arise from its mimicry of the ADA transition state (TS) structure, which resembles the tetrahedral intermediate **2** produced during the rate-limiting, zinc-assisted hydration of the 1,6-double bond of adenine. The covalent hydrate of purine riboside, i.e. **3**, bears an even closer structural resemblance to the TS structure than **1** and correspondingly exhibits an even greater estimated inhibitor potency ($K_i = 10^{-13}$ M).[5] The molecular basis for the high affinity is apparent from the X-ray structure of the ADA complex, which shows favorable interactions between the 6-hydroxyl group and both the active-site zinc ion and nearby residues.[6] The hydrated form of purine riboside is, however, unstable in solution with an equilibrium constant that strongly favors the unhydrated form ($K_{eq} = 10^{-7}$).[7] Accordingly, since the hydrated species is responsible for inhibition of ADA, but exists only at low concentrations in solution, the apparent inhibition constant of purine riboside is a rather modest 10^{-5} M.

2. ADA INHIBITOR DESIGN STRATEGY

Our inhibitor design strategy was based on the premise that structural modifications in the base of purine riboside that enhance purine base hydration without impairing the binding of the hydrated species to the ADA binding site would result in purine riboside (PR) analogues with high ADA inhibitory potency. Since the apparent inhibition constant (K_i (app)) is related to the hydration equilibrium constant (K_{eq}) and the inhibitory constant for the hydrated molecule (K_i^*) by

$$K_i\,(\text{app}) = K_i^*(1 + 1/K_{eq}) \approx K_i^* Keq^{-1} \qquad (1)$$

groups that shifted the equilibrium constant from 10^{-7} to 10^{-4} without diminishing K_i^* would result in an increase in the apparent inhibition by

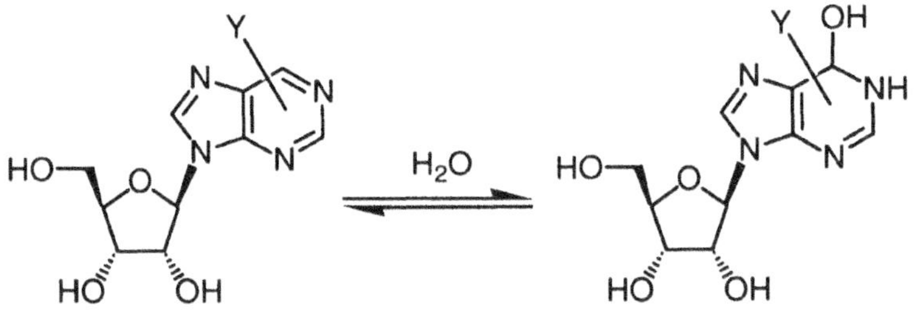

		PR	PR Analogue
K_{eq}	=	10^{-7}	10^{-4}
K_i* (hydrate)	=	10^{-13} M	10^{-13} M
K_i (app)	=	10^{-6} M	10^{-9} M

Figure 3. Purine riboside (PR) hydration and the effect of base modification Y.

greater than 1000-fold, i.e. shifting the K_i (app) for purine riboside from 1 μM to 1 nM (Figure 3).

Integral to our drug design strategy was the development of computational methodology that could be used to predict the effect of purine base modifications on both the hydration equilibrium and the binding affinity of the hydrated molecule for the deaminase binding site.[8] Relative hydration free energies were calculated using the methodology previously employed to calculate hydration free energies of carbonyl-containing compounds.[9] Calculation of relative binding affinities used the free energy perturbation method and a computer model derived from the X-ray coordinates of the ADA:6-hydroxy-1,6-dihydropurine riboside complex. The combination of the two calculations was shown to accurately predict the relative change in the apparent inhibition constant for the purine riboside analogue, 8-azapurine riboside.[8, 10]

Previous efforts to use free energy calculations in the design and analysis of ADA inhibitors have been reported. Prior to the ADA X-ray structure, Hansen and Kollman concluded that the hydroxyl in the hydrated form of purine riboside was unusually hydrophilic with a calculated solvation free energy difference relative to 6-dihydropurine (i.e. H instead of the C6 OH in **3**) of -11.8 kcal/mol.[11] Using an ADA binding site model designed only to mimic potential inhibitor-protein interactions, they further concluded that the potent ADA inhibition of **3** was a result of highly favorable binding site interactions. These interactions contributed at least 20 kcal/mol to binding free energy and therefore were sufficient to overcome the high desolvation costs. In another study, 2'-deoxy coformycin (**1**) was compared to coformycin using molecular dynamics and free energy simulations.[12] These studies suggested that the TS mimics shown in Figure 2 gain a large proportion of their binding affinity through the hydroxyl group at the position on the base corresponding to C6 of adenine as well as some binding affinity from the ribose moiety.

3. RELATIVE HYDRATION FREE ENERGIES

3.1 Methodology

Efforts to calculate absolute hydration free energies for the reversible addition of water to carbonyl-containing compounds[9] (Equation 2) entailed adding the gas phase free energy (ΔG_{gas}) for the reaction, which was calculated using ab initio quantum mechanics, and the free energy of solvation ($\Delta\Delta G_{sol}$) for the reaction (Equation 3), which was calculated using molecular dynamics (MD) simulations in conjunction with the free energy perturbation (FEP) approach.[13]

$$A + H_2O \rightleftharpoons A{\bullet}H_2O \tag{2}$$

$$\Delta G_{hyd} = -RT\ln K_{eq} = \Delta G_{gas} + \Delta\Delta G_{sol} \tag{3}$$

Despite inclusion of electron correlation energy contributions using third-order (MP3) and fourth-order (MP4) Moller-Plesset perturbation theory and QCISD(T) correlation methods at the 6-31G** basis set level, the calculated free energies deviated significantly from experimental results. Inaccuracies in the calculations were attributed in part to the slow convergence of the calculation of $\Delta\Delta G_{sol}$ using the FEP method and explicit solvent. These errors were minimized by calculating relative hydration free energies using Equation 4

$$\Delta\Delta G_{hyd} = \Delta\Delta G_{gas} + \Delta\Delta\Delta G_{sol} \tag{4}$$

where $\Delta\Delta G_{gas}$ and $\Delta\Delta\Delta G_{sol}$ are the relative differences between two structurally similar molecules in gas-phase quantum mechanical free energy and solvation free energy, respectively. Greater accuracy was presumably due to the cancellation of systematic errors and to increased convergence in the solvation free energy calculation. Importantly, the results were in good agreement with experimental results even at lower levels of quantum mechanical theory.

The computational details for both the quantum mechanical and FEP studies using molecular dynamics simulations are described elsewhere.[9] In brief, the gas-phase free energies (ΔG_{gas}) were calculated using energies

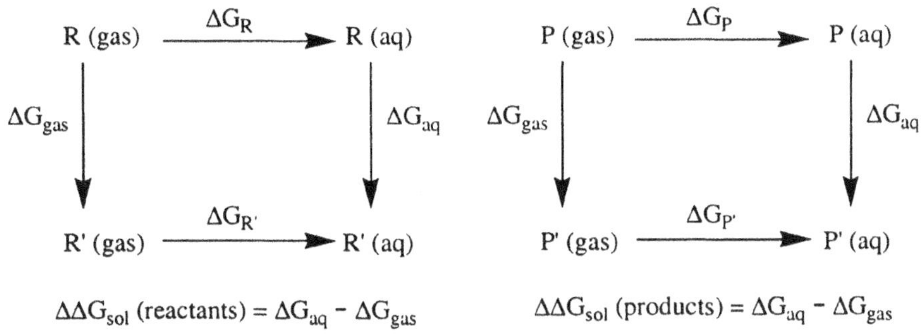

Figure 4. Thermodynamic cycles used for calculating $\Delta\Delta\Delta G_{sol}$

obtained from ab initio quantum mechanical calculations at the 6-31G** basis set level on fully-geometry optimized anhydrous and hydrated compounds. The solvation free energy differences ($\Delta\Delta\Delta G_{sol}$) were calculated using Equation 5, where

$$\Delta\Delta\Delta G_{sol} = \Delta\Delta G_{sol}(\text{products}) - \Delta\Delta G_{sol}(\text{reactants}) \quad (5)$$

$\Delta\Delta G_{sol}$(products) and $\Delta\Delta G_{sol}$(reactants) are the relative solvation free energy differences between the two products and two reactants of the two hydration reactions, respectively. The calculations were carried out using MD simulations and the window method implemented in the AMBER program. The thermodynamic cycles used for calculating $\Delta\Delta G_{sol}$(reactants) therefore entailed a computational transformation of the unhydrated molecule (R) to the unhydrated molecule (R') in the gas-phase and in the presence of SPC/E water[14] (Figure 4). The structural similarity between R and R' greatly aided calculation convergence and presumably accuracy relative to earlier attempts at calculating absolute hydration free energies, which required a transformation of the unhydrated molecule (R) to the corresponding hydrated molecule (P). The large geometrical change associated with hydration and the inclusion of one water molecule in the calculation were avoided by calculating relative solvation free energies.

3.2 Heteroaromatic Hydration Free Energies

Our next step was to assess whether the methodology used to calculate hydration free energy differences for simple carbonyl-containing compounds[9] was suitable for heteroaromatic bases. Since our drug design strategy entailed analysis of purine riboside hydration, a series of azanaphthalenes was initially selected for analysis in part because of their structural similarity to purines and in part because of the extensive experimental database reported for these compounds.[15] Hydration equilibrium constants for the series ranged from 10^{-7} to greater than 10^{3}. For example, approximately 20% of pteridine (1,3,5,8-tetraazanaphthalene) exists as the 3,4-double bond hydrated species (**4**) in aqueous solution. Hydration of the pyrimidine ring is highly dependent on several factors, including the number and location of nitrogens in the aromatic ring fused to the pyrimidine ring. Calculation of relative hydration free energies for these analogues relative to pteridine therefore represented a relatively small structural perturbation (CH to N) distal to the site of hydration. As shown in Table 1, the calculated results were consistent with the experimental data. Accurate results were also obtained for a triazanaphthalene compound that fails to hydrate in the neutral form, but exists almost exclusively in the hydrated form (**5**) as the cationic species.[16]

In another study, 4-methylpteridine (**6**) and 4-trifluoromethyl pteridine (**7**) showed that substituents can, in some instances, not only change the extent of hydration but also the preferred site of hydration (Figure 5).[17] Experimental results indicate that the 4-trifluoromethyl analogue hydrates initially across the 5,6- and 7,8-double bonds to form the dihydrate. Over

Table 1: Hydration free energies of heteroaromatic compounds (kcal/mol)

P	P'	Calc (P-P')	Expt
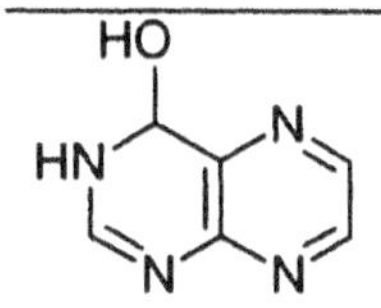 (**4**)	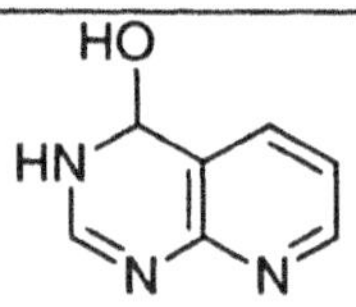	-2.0	-2.5
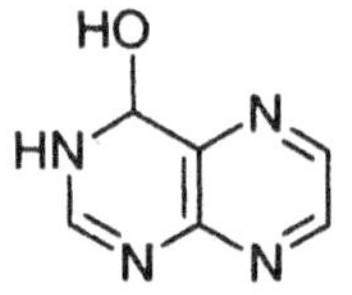	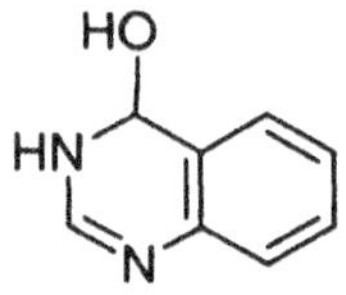	-4.0	-5.1
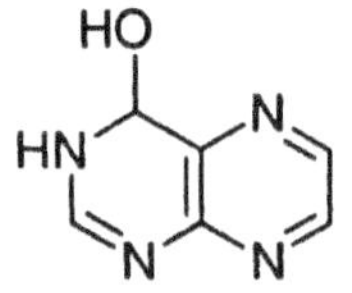	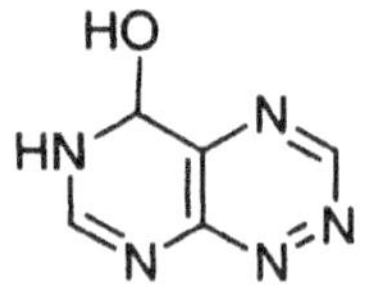	6.1	6.2
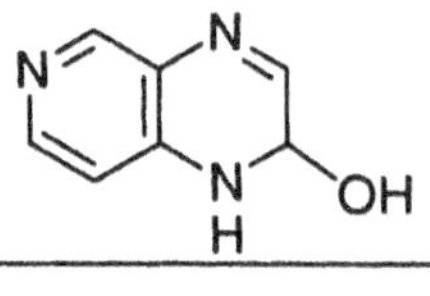	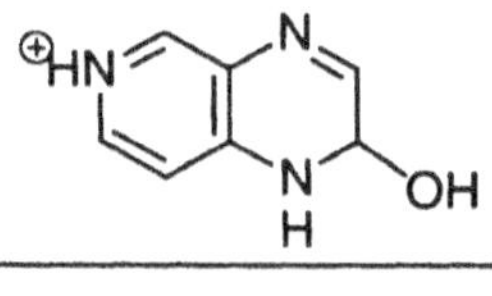 (**5**)	11.0	8.2

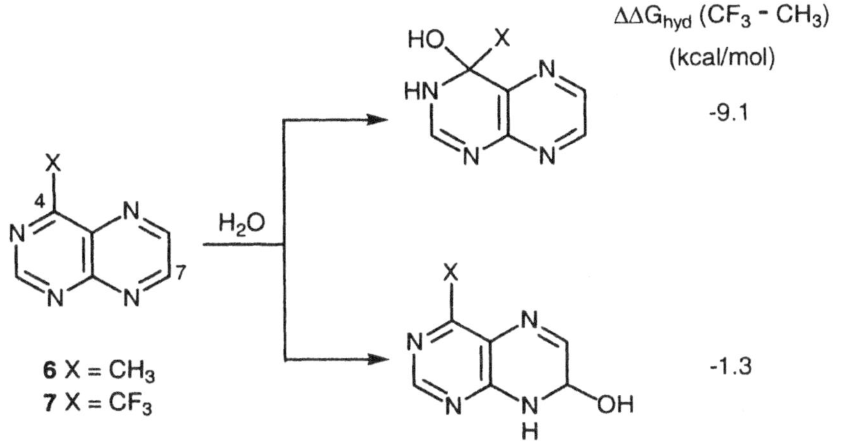

6 X = CH_3
7 X = CF_3

Figure 5: Hydration of 4-substituted pteridine analogues

time the dihydrate is converted to the 3,4-double bond hydrate. The anhydrous form is not detected. In contrast, the 4-methyl analogue is poorly hydrated and the only hydrated species detected in solution is the dihydrate and not the 3,4-double bond monohydrate. The molecular factors attributed to the difference is reported to be a mixture of the methyl group sterically destabilizing the 3,4-hydrate and the trifluoromethyl group stabilizing the hydrate through inductive effects. The calculated results are consistent with these findings, showing a 9.1 kcal/mol difference between the methyl and trifluoromethyl analogues for hydration across the 3,4-double bond and a 5.9 and -1.9 kcal/mol difference between the 3,4-double bond hydrate and the 7,8-double bond hydrate for the methyl and trifluoromethyl substituted compounds, respectively. Thus, the method can be used to accurately predict the site of hydration as well as the extent of hydration.

3.3. Hydration Free Energy Difference for Purine and Pteridine

Unlike pteridine, the hydrated form of purine riboside is an exceedingly rare species in aqueous solution.[5] Experimental data suggests that the hydration equilibrium constant for purine riboside is about 10^{-7} or nearly 7 orders of magnitude less favorable than the K_{eq} for pteridine (Figure 6). This finding is remarkable considering that both compounds hydrate across the 1,6-double bond of the pyrimidine ring and both contain a heteroaromatic ring fused at the 4- and 5-positions of the pyrimidine ring. Moreover, the

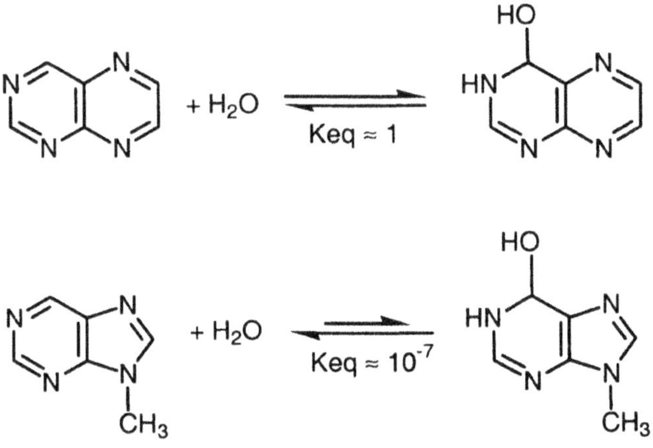

Purine - Pteridine (kcal/mol)		
$\Delta\Delta G_{hyd}$ (calc)	=	9.3
$\Delta\Delta G_{hyd}$ (expt)	≈	8.8

Figure 6: Hydration of pteridine vs. 9-methylpurine

fused heteroaromatic ring for both compounds contains nitrogens attached to the 4- and 5-positions of the pyrimidine ring. Assuming that the ribose group does not influence the hydration equilibrium, the major difference between pteridine and purine is an extra ring carbon in pteridine, which makes the fused ring a six-membered ring.

In contrast to previous calculations, calculation of the hydration free energy difference between pteridine and 9-methylpurine required use of the thread methodology.[18] Accordingly, these calculations entailed "threading together" the two molecules such that a single topology was used to describe atoms defined as common to both molecules, i.e. atoms having approximately the same force constants and equilibrium geometries (but not necessarily the same partial charges) whereas the remaining portions of the molecules were treated separately. In the calculation, the starting (reactant) and ending (product) topologies were defined with the appropriate geometries with one beginning and the other ending the simulation entirely as dummy atoms. Dummy atoms are identical to real atoms except that the Lennard-Jones parameters and atomic charges are set to zero. At intermediate points during the transformation, all atoms, in both topologies, have fractional Lennard-Jones parameters and charges. Molecules with both topologies interact with the environment, but not with each other.

The results of the calculations showed that the methodology could accurately predict the difference between purine and pteridine hydration equilibrium constants. This suggests that the methodology was not only capable of predicting the effects of ring atoms and various ring substituents, but also differences in the propensity of various heteroaromatic rings to hydrate.

4. RELATIVE ADA INHIBITORY POTENCY OF 8-AZAPURINE RIBOSIDE

The apparent inhibitory potency (K_i (app)) as measured experimentally is related to both the hydration equilibrium constant and the intrinsic binding affinity of the hydrated molecule (K_i^*) by Equation 1. Accordingly, the relative inhibitory potency between two analogues, in this case purine riboside (PR) and a purine riboside analogue (PR') (Figure 7), is related to the relative free energy difference (ΔG_{rel}) by Equation 6. Computationally, ΔG_{rel} is determined simply by summing the relative free energies for hydration and hydrate binding affinity as shown in Equation 7.

$$\Delta G_{rel} = -RT\ln[K_i(app)/K_i'(app)] \quad (6)$$

$$\Delta G_{rel} = \Delta\Delta G_{hyd} + \Delta\Delta G_{bind} \quad (7)$$

To test whether we could accurately calculate the fold-difference of ADA inhibitory potency between purine riboside (**8**) and analogues of purine riboside, we selected 8-azapurine riboside (**9**) for our studies.[8] Compound **9** was reported to be a 400-fold more potent ADA inhibitor relative to **8** despite differing from **8** only by the replacement of C8 with a nitrogen (Figure 8).[19] The molecular reason for this enhancement in potency was not determined, but could be due either to enhanced hydration or enhanced ADA binding affinity of the hydrated species or both. To determine the reason,

$$\mathrm{PR} \underset{K_{eq}}{\overset{H_2O}{\rightleftharpoons}} \mathrm{PR{\bullet}H_2O} \underset{K_i^*}{\overset{ADA}{\rightleftharpoons}} \mathrm{ADA{:}\ [PR{\bullet}H_2O]}$$

$$\mathrm{PR'} \underset{K_{eq'}}{\overset{H_2O}{\rightleftharpoons}} \mathrm{PR'{\bullet}H_2O} \underset{K_i^{*'}}{\overset{ADA}{\rightleftharpoons}} \mathrm{ADA{:}\ [PR'{\bullet}H_2O]}$$

Figure 7: Relative inhibitor potency

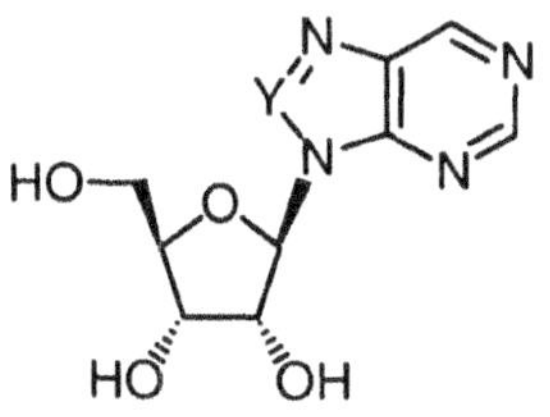

8 Y = CH
9 Y = N

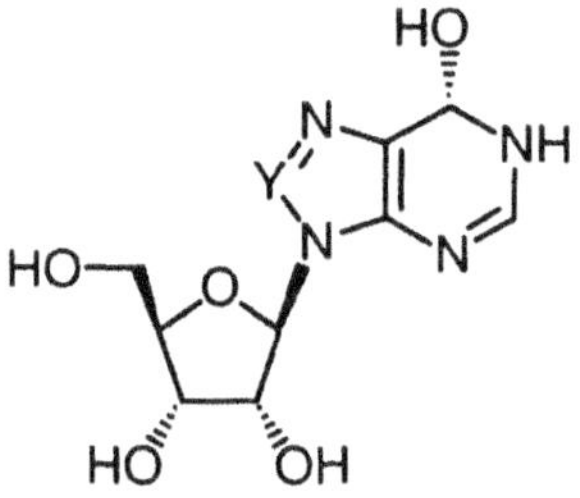

8H Y = CH
9H Y = N

Figure 8. Purine riboside/8-azapurine riboside and corresponding hydrates

we calculated the relative difference in free energies for both hydration and binding.

4.1 Relative Binding Free Energy

The FEP approach was used to calculate the relative binding free energy difference between the 1,6-double bond hydrates of purine riboside (**8H**) and 8-azapurine riboside (**9H**) with adenosine deaminase (ADA).[8] The X-ray structure of murine adenosine deaminase complexed with **8H**[6] provided the initial atomic coordinates used to generate the computer model and conduct the energy calculations. As detailed elsewhere, all molecular dynamics, molecular mechanics and FEP calculations were carried out with the AMBER program using an all atom force field and SPC/E potentials to describe water interactions.[8] The charge on the active-site zinc ion was determined using the 3-21G* wave function calculated for zinc coordinated with residues present in the ADA-**8H** complex. The calculated charge on zinc was +1.2 e, whereas the net charge on histidines 15, 17 and 214 was +0.2 e each and -0.8 e on the aspartic acid, Asp295. Thus, the net charge on the zinc complex was +1 e.

The average structure for the ADA-**8H** complex generated after energy minimization and equilibration with 20 ps of MD simulation showed considerable movement of atoms in the vicinity of the zinc ion when these residues were not constrained. Virtual bonds between zinc and the atoms that coordinate to the zinc ion were added to provide necessary atomic constraints prior to energy minimization and MD simulations. After constraints were applied, the average structure was similar to the X-ray

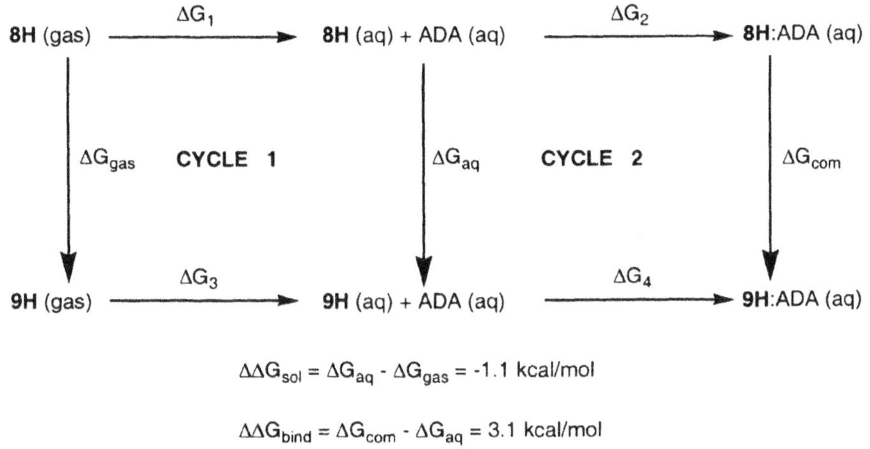

Figure 9: Thermodynamic cycles used to calculate solvation and binding free energies.

structure with root mean square deviations of less than 1.0 Å for backbone atoms and 1.45 Å for side-chain atoms

Using this structure and the FEP method, the relative binding free energy for the hydrated purine riboside (**8H**) and its 8-aza analogue (**9H**) was calculated (Figure 9, Cycle 2). The results indicate that **9H** has a 190-fold lower intrinsic binding affinity relative to **8H** ($\Delta\Delta G_{bind}$ = 3.1 ± 0.7 kcal/mol). The decreased binding affinity is attributed in part to a 1.1 ± 0.5 kcal/mol increase in desolvation energy for the 8-aza analogue as calculated from the gas phase and solvent phase free energies. The remainder of the lost binding energy associated with 8-azapurine riboside hydrate binding (2 kcal/mol) is attributed to a loss in intrinsic binding affinity. This decrease may arise from an unfavorable electrostatic interaction between the 8-nitrogen and Asp296 as is observed in the energy minimized, MD-equilibrated, ADA complex. While the relative binding free energy is not experimentally measurable due to the instability of the hydrated species, it is clear that the calculated result is in the opposite direction to the free energy difference derived from the apparent inhibition constants and therefore that differences in ADA binding affinity are not likely to be the reason for the improved ADA inhibitory potency exhibited by **9**.

4.2. Relative Inhibitory Potency of 8-Azapurine Riboside

An alternative explanation for the 400-fold improvement in inhibitory potency exhibited by 8-azapurine riboside (**9**) is that the 8-aza analogue hydrates to a much larger extent than purine riboside (**8**). Calculation of the relative hydration free energy difference between 9-methylpurine and 8-aza-

Table 2: Inhibition Potency of Purine Riboside and its 8-aza Analogue

Purine Riboside (**8**)		**9** - **8** (calc)[a]		8-Azapurine Riboside (**9**)	
K_{eq}	$= 1.1 \times 10^{-7}$	$\Delta\Delta G_{hyd}$	$= -7.1$	K_{eq}	$= 1.8 \times 10^{-2}$
K_i*	$= 1.8 \times 10^{-12}$ M	$\Delta\Delta G_{bind}$	$= 3.1$	K_i*	$= 3.4 \times 10^{-10}$ M
K_i (app)	$= 1.6 \times 10^{-5}$ M	ΔG_{rel}	$= -4.0$	K_i (app)	$= 1.9 \times 10^{-8}$ M
				K_i (app)	$= 4.0 \times 10^{-8}$ M (expt)

[a] units are kcal/mol

9-methylpurine strongly supported this possibility, since the difference was 7.1 kcal/mol or approximately a 5-order of magnitude rightward shift in the equilibrium constant for the 8-aza analogue.[8] Calculation of bond separation energies for the hydrated and unhydrated molecules suggested that the large difference in hydration arises from a large relative loss in resonance energy incurred by purine compared to 8-azapurine during the hydration reaction.[8]

The calculated results provide a clear explanation for the difference in inhibitory potency between purine riboside and its 8-aza analogue (Table 2). The relative hydration free energy difference indicates that 8-azapurines hydrate about 160,000-fold greater than the corresponding purine analogue ($\Delta\Delta G_{hyd} = -7.1$ kcal/mol) whereas the hydrate of 8-azapurine riboside suffers approximately a 200-fold decrease in binding affinity relative to the corresponding purine analogue ($\Delta\Delta G_{bind} = 3.1$ kcal/mol). The net effect is a 4.0 kcal/mol enhancement in inhibitory potency in favor of the 8-aza analogue, which translates to a predicted K_i(app) for 8-azapurine riboside of 2×10^{-8} M; a value very close to the experimental result of 4×10^{-8} M.

5. CONCLUSIONS

Rational drug design using computational methods is useful in prioritizing potential target compounds and thereby shortening the drug discovery timeline, if results accurately predict the experimental findings. Our study analyzing the difference in ADA inhibitory potency between purine riboside and 8-azapurine riboside illustrates the importance of calculating both the relative hydration free energy and the relative binding free energy for molecules that act as enzyme inhibitors only after undergoing covalent hydration. Last, our studies of heteroaromatic hydration suggest that various ring substituents and ring modifications can significantly enhance hydration and that strategies that increase hydration could be useful in the design of potent ADA inhibitors.

6. REFERENCES

1. N. M. Kredich and M. S. Hershfield, The Metabolic Basis of Inherited Disease, vol. 6, R. Scrivner, et al., eds., McGraw-Hill, New York (1989), pp. 1045-1075.
2. R. P. Agarwal, Inhibitors of adenosine deaminase, Pharmac. Ther. **17**:399 (1982).
3. J. Barankiewicz, A. M. Danks, E. Abushanab, L. Makings, T. Wiemann, R. A. Wallis, P. V. Pragnacharyulu, A. Fox, and P. J. Marangos, Regulation of adenosine concentration and cytoprotective effects of novel reversible adenosine deaminase inhibitors, J. Pharmacol. Exp. Ther. **283**:1230 (1997).
4. R. P. Agarwal, T. Spector, and R. E. Parks, Jr., Tight-binding inhibitors. IV. Inhibition of adenosine deaminase inhibitors, Biochem. Pharmacol. **26**:359 (1977).
5. W. M. Kati and R. Wolfenden, Major enhancement of the affinity of an enzyme for a transition-state analog by a single hydroxyl group, Science **243**:1591 (1989).
6. D. K. Wilson, F. B. Rudolph, and F. A. Quiocho, Atomic structure of adenosine deaminase complexed with a transition-state analog: Understanding catalysis and immunodeficiency mutations, Science **252:**1278 (1991).
7. W. Jones, L. C. Kurz, and R. Wolfenden, Transition-state stabilization by adenosine deaminase: 1,6-addition of water to purine ribonucleoside, the enzyme's affinity for 6-hydroxy-1,6-dihydropurine ribonucleoside, and the effective concentration of substrate water at the active site, Biochemistry **28**:1242 (1989).
8. M. D. Erion and M. R. Reddy, Calculation of relative hydration free energy differences for heteroaromatic compounds: Use in the design of adenosine deaminase and cytidine deaminase inhibitors, J. Am. Chem. Soc. **120**:3295 (1998).
9. M. D. Erion and M. R. Reddy, Calculation of relative free energy differences for the covalent hydration of organic compounds: A combined quantum mechanical and free energy perturbation study, J. Comp. Chem. **16**:1513 (1995).
10. M. D. Erion and M. R. Reddy, Calculation of relative hydration free energy differences for heteroaromatic compounds: Use in the design of AMP deaminase inhibitors, in: Rational Drug Design: Novel Methodology and Practical Applications, A. L. Parrill and M. R. Reddy, eds., ACS Symposium Series, Washington DC (1999), pp. 107-120.
11. L. M. Hansen and P.A. Kollman, Free energy perturbation calculations on models of active sites: Applications to adenosine deaminase inhibitors, J. Comp. Chem. **11**:994 (1990).
12. T. J. Marrone, T. P. Straatsma, J. M. Briggs, D. K. Wilson, F. A. Quiocho, and J. A. McCammon, Theoretical study of inhibition of adenosine deaminase by (8*R*)-coformycin and (8*R*)-deoxycoformycin, J. Med. Chem. **39**:277 (1996).
13. B. L. Tembe and J. A. McCammon, Ligand-receptor interactions, Comput. Chem. **8**:281 (1984).
14. H. J. C. Berendsen, J. R. Grigera, and T. P. Straatsma, The missing term in effective pair potentials, J. Phys. Chem. **91**:6269 (1987).
15. A. Albert, Covalent hydration in nitrogen heterocycles, Adv. Heterocyc. Chem. **20**:117 (1976).
16. D. D. Perrin and Y. Inoue, Kinetics of the reversible hydration of 2-hydroxypteridine, J. Phys. Chem. **66**:1689 (1962).
17. J. Clark and W. Pendergast, Heterocyclic studies. Part IX. Synthesis and covalent hydration of 4-trifluoromethylpteridine and some methyl derivatives, J. Chem. Soc. C. **6**:1751 (1969).
18. U. C. Singh and S. J. Benkovic, A free energy perturbation study of the binding of methotrexate to mutants of dihydrofolate reductase, Proc. Natl. Acad. Sci. USA **85**:9519 (1988).
19. D. S. Shewach, S. H. Krawczyk, O. L. Acevedo, and L. B. Townsend, Inhibition of adenosine deaminase by azapurine ribonucleosides, Biochem. Pharmacol. **44**:1697 (1992).

INDEX

GPSR Compliance

The European Union's (EU) General Product Safety Regulation (GPSR) is a set of rules that requires consumer products to be safe and our obligations to ensure this.

If you have any concerns about our products, you can contact us on ProductSafety@springernature.com

In case Publisher is established outside the EU, the EU authorized representative is:

Springer Nature Customer Service Center GmbH
Europaplatz 3
69115 Heidelberg, Germany

Batch number: 10755461

Printed by Printforce, the Netherlands